MICHIGAN SHRUBS AND VINES

T0313510

MICHIGAN
Shrubs & Vines

A GUIDE TO SPECIES OF THE GREAT LAKES REGION

Burton V. Barnes,
Christopher W. Dick,
and Melanie E. Gunn

University of Michigan Press
Ann Arbor

Published in the United States of America by the
University of Michigan Press
Printed and bound by CPI Group (UK) Ltd, Croydon, CR0 4YY

2019   2018   2017   2016   4   3   2   1

A CIP catalog record for this book is available from the British Library.

ISBN 978-0-472-11777-2 (hardcover : alk. paper)

ISBN 978-0-472-03625-7 (paperback : alk. paper)

ISBN 978-0-472-12107-6 (e-book)

PREFACE

Michigan Shrubs and Vines is a guide to shrub and vine identification and natural history for species of Michigan and the Great Lakes region. It is the long-awaited companion to *Michigan Trees*, and the distillation of more than 50 years of teaching about woody plants and forest ecology at the University of Michigan. Similar to *Michigan Trees*, both native and introduced species are placed into ecological context within Michigan's rich diversity of regional landscapes. We hope that in the process of discovering more about their natural world—and having a lot of fun in the process—users of *Michigan Shrubs and Vines* will become better botanists, ecologists, and stewards of our natural heritage.

Michigan Shrubs and Vines departs from many recent field guides in its use of line drawings. Despite the trend toward photography, we find that drawings are better suited for highlighting key characters and discouraging a "gestalt" approach that can lead to poor identification skills. Additionally, one of the most effective identification details is geographic range, i.e. knowing if a species is found in a local area. In an improvement from *Michigan Trees*, this volume includes county level maps for each species based on University of Michigan Herbarium (UMH) specimen records. We encourage the reader to visit the UMH's Michigan Flora website (michiganflora.org) for updates on geographic distributions and taxonomy, along with photographs vetted by UMH staff.

Burton V. Barnes (1930-2014) was the architect and driving force behind this book and its companion volume, *Michigan Trees*. Burt, also known as BVB, played a significant mentorship role for both coauthors and had a lasting impact on the way we both think about plants, ecology, teaching and scholarship. In 1965, Burt and Warren H. Wagner initiated the Woody Plants class, which Burt taught with enthusiasm, devotion, and fanfare for 40 years. Burt worked on this book during his last decade of teaching and subsequently as Professor Emeritus. *Michigan Shrubs and Vines* is BVB's final gift to the past, present, and future students of Woody Plants.

ACKNOWLEDGMENTS

Burt worked closely with artists Ed Trager and John Megahan to produce original artwork, often personally collecting subject specimens. He also took care in selecting line drawings from published sources. We gratefully acknowledge the Hanes Foundation, University of Michigan Herbarium, and University of Michigan Press for funding original artwork and permissions to reproduce drawings. Peter Widen and Na Wei helped to organize and digitize drawings. Hillary Butterworth used considerable artistic talent to render scanned drawings into publishable form.

The following people and institutions provided permissions for artwork reproductions: Royal Ontario Museum for use of drawings from Soper and Heimberger (1981); University of Minnesota Press for Rosendahl (1963); Estate of R. G. Brown for Brown and Brown (1972); New York Botanical Garden for Holmgren (1998); Ohio State University Press for Braun (1989); West Virginia University Press for Core and Ammons (1958); The Flora of North America Association and John Myers, illustrator, for *Salix* drawings; Stackpole Company for Grimm (1966); and Cranbrook Institute of Science for Billington (1977). We obtained several original drawings from artist Vera Wong. Jason Tallant generated the distribution maps for this volume. UMH curator Anton A. Reznicek provided expert botanical advice and critically reviewed the entire book. Virginia Laetz assisted BVB and coauthors for many years. The authors would like to acknowledge the following individuals for their support and feedback: Dennis Albert, Lenora Barnes, Daniel Buonaiuto, Julia Caroff, Howard Crum, Charles Davis, Rex Frobenius, Sondra Gunn, Mike Penskar, Melida Dick-Ruiz, Sylvia Taylor, Edward Voss, and Florence Wagner. Additionally, we are grateful for the support of editor Scott Ham and the University of Michigan Press.

This book is dedicated to students of Woody Plants.

Christopher W. Dick

Melanie E. Gunn

January 2016

Michigan county map adapted from Wilbert B. Hinsdale, *Archaeological Atlas of Michigan*, University of Michigan Press, 1931.

CONTENTS

WOODY PLANT IDENTIFICATION

Two forms of woody plants, shrub and vine, are the focus of this field guide for Michigan and the Great Lakes Region, including Ontario. Successful identification of shrubs and vines depends on knowing where they occur geographically and ecologically and what characteristics to examine. Most species can be determined by first noting where you are in the landscape and the physical features surrounding and available to the plant (i.e., site or habitat) and then observing key distinguishing characteristics of the plant itself. Important in reliable identification is an understanding of the geography and physical environment of any given region—geology, physiography and major landforms, macro- and microclimate, soils, hydrology, and microsite features. These are primary components of ecological systems to which plants are adapted genetically or are physiologically acclimated.

Therefore, when in the field, a keen sense of place is very important. First, where am I in the regional landscape (Fig. 19)—in lower or upper Michigan, in coastal or interior districts? In what distinctive landform are the plants located—wetland or upland, river floodplain, sand dune, moraine, or outwash plain? Site factors of topography (dry upper vs. moist lower slope) and soil are useful, as well as your examination of the whole plant habit before its parts. These physical factors provide a solid basis for applying the vegetative and morphological characteristics to best advantage. Resources and information on regional landscape ecosystems and their physical factors are given under the headings "Distribution" and "Site-Habitat" in this introduction and in the section "Ecology of Shrubs and Vines."

An initial goal of the guidebook is identification and recognition of shrubs and vines. More important and rewarding is building on recognition to understand the natural history of these organisms. For example, why do they grow where they do? What site factors limit their occurrence and regeneration? What features of their sexual and asexual reproduction have enabled their persistence to the present? What morphological and physiological features characterize their mutualistic and competitive relationships with associated species? What features characterize their response to diverse natural and human disturbances?

The relatively simple task of identification or recognition proceeds from the top down (from the large to the small)—with the plant's characteristics of size and habit, bark, leaves, stems-twigs, buds, flowers, fruits, and so on. Accordingly, our descriptions of species are organized this way. With the challenging and rewarding understanding of natural history, we also proceed from the top down but first with an overarching focus on the geography and physical site factors of the habitat or "home place"—thinking from a landscape ecosystem perspective. Within the regional and local landscape perspectives, we study in detail the plant's characteristics but in the new light of "place" and their relation to the natural history questions we seek to answer.

Many species, especially obligate wetland ones, have distinctive habitats and rarely colonize adjacent, disturbed uplands. Others, especially shade-intolerant, pioneer species are adapted to colonize open (competition-free) sites following natural and human disturbances.

All shrubs and vines have a characteristic home place of regional and local geographic occurrence, physical site conditions, and associated natural disturbances. They evolved as integral parts of landscape ecosystems thousands to millions of years before humans arrived in North America and the Great Lakes Region. Despite disturbances by Native Americans and European settlement, understanding regional and local geography and physical site conditions is exceedingly important in understanding the natural history of shrubs and vines, as well as their identification. Therefore, understanding the landscape, that is, "knowing the territory," is a major priority in learning plant natural history and ecology, especially in this time of climate-mediated change. Shrubs and vines are at the forefront of woody species responding to change. Their reliable identification is an important baseline and provides a heightened sense of accomplishment in understanding nature.

In the main body of the book, following the general keys, we present detailed descriptions of each species and a concise list of its key characters. In addition, we specifically contrast pairs of species that might be difficult to distinguish. Line drawings accompany the species description and illustrate characteristics of leaves, twigs, buds, fruits, and often flowers.

The book is organized in several parts. Following general introductory topics, we explain each of the useful identifying features, such as size and form, bark, leaves, stems-twigs, and so on, in the exact order in which they appear in the description of each species. Following this section, we consider fall coloration and hybrids. Then the general keys and species descriptions are presented in the main body of the text. Shrubs and vines are markedly different in life form from trees, not only in habit and size but in reproductive biology, dispersal, and ecology. Therefore, following the species descriptions, in the section "Ecology of Shrubs and Vines," we first present an overview of the historical biogeography of Michigan shrubs and vines. Then we examine the natural history features of shrubs and vines, their present occurrence in regional landscape ecosystems of Michigan, their relationship with Native Americans prior to European settlement, and their ecological role in landscape ecosystems.

Woody plants are grouped as *gymnosperms* and *angiosperms*. The major shrub gymnosperms in Michigan are three conifers: ground juniper (*Juniperus communis* var. *depressa*), creeping juniper (*J. horizontalis*), and Canada yew (*Taxus canadensis*). The conifers evolved long before the angiosperms. Their lineage can be traced back to the late Carboniferous period some 300 million years ago (Ma) (Judd et al. 2008). In the Mesozoic era of 250 to 170 Ma, conifers evolved into many current families and genera and achieved wide distribution and great abundance. Conifers declined markedly, and today they comprise only seven families, sixty to sixty-five genera, and just over six hundred species worldwide. Nevertheless, their great abundance, dominance, and ecological significance in fire-prone and nutrient-poor ecosystems of the northern hemisphere contrast with their low taxonomic diversity.

Angiosperms, flowering plants, arose later. Estimates place the origin of angiosperms in early Cretaceous times (ca. 135 Ma), and by the late Cretaceous (70–65 Ma), they had become widely distributed and achieved ecological dominance (Judd et al. 2008). Angiosperms now account for approximately 350,000 species (The Plant List 2013) and comprise most of the Earth's terrestrial biomass and plant diversity. Flowering plants are a remarkably flexible group. The transition from herbaceous to woody habit involves relatively minor genetic changes (Petit and Hampe 2006), and the evolution of shrubs and vines from either trees or herbs, as well as the reverse process, has occurred many times (Stebbins 1972; Judd et al. 2008). Biogeographic history, which helps to explain the origins and geographic distributions of our flora, is treated in greater depth in the section "Ecology of Shrubs and Vines."

Angiosperms that shed their leaves during winter in temperate zones or due to severe drought stress are termed *deciduous*. The term *broadleaf deciduous* is often used because some conifers exhibit deciduous foliage. Many angiospermous shrubs in subtropical and tropical climates retain their leaves for several years and are referred to as *evergreen* or *broadleaf evergreen*. Several broadleaf evergreen shrubs occur in our region, such as bearberry, trailing arbutus, leatherleaf, and Labrador-tea. The evergreen condition in angiosperms in northern temperate regions is generally associated with plants in places with cold, short growing seasons or very nutrient-poor sites. This condition is an adaptation for several things, including conservation of water and nutrients.

Besides fundamental differences in reproductive biology and time of evolution, conifers and angiosperms exhibit distinctive geographical and ecological patterns of occurrence and abundance. Conifers dominate the Boreal Forest Region (Rowe 1972; Elliott-Fisk 2000), the temperate rain forest of the Pacific Northwest (Franklin and Halpern 2000), the Rocky Mountains (Peet 2000), and the fire-prone plains of eastern North America south of the boreal forest (Braun 1950; Christensen 2000). In contrast, deciduous broadleaf forests characterize much of temperate eastern North America (Braun 1950; Barnes 1991; Delcourt and Delcourt 2000) and certain regional and local eco-systems of western North America (Barnes 1991).

Conifers tend to outcompete deciduous broadleaf plants under harsh con-ditions that limit photosynthesis, such as cold temperature, short and cool growing season, aridity, fire, high wind velocity, midsummer drought, and low nutrient availability. Deciduous angiosperms are generally favored by long, warm, humid growing seasons and favorable nutrient conditions. They are also favored by soil-site conditions of river floodplains and seasonally inun-dated bottomlands. However, in the Great Lakes Region, some angiosperms also are highly competitive under cold conditions (alders, willows, blueber-ries, cranberries, and wintergreen). In river valleys that are seasonally flood-ed, conifers are absent. Here the characteristic tree angiosperms are eastern cottonwood, silver maple, American elm, red ash, and sycamore. Shrubs in-clude willows, alders, dogwoods, and many vine species. The extensive Great Lakes coastlines provide remarkably diverse ecosystems of limestone bedrock (alvar shrublands and savannas), alternating dunes and swales, coastal wet-lands, and open and wooded dunes that favor many shrub and vine species (Albert 2003; Weatherbee 2006). Warm southern Michigan is especially rich in woody angiosperms, whereas conifers are rare (Denton and Barnes 1987). In contrast, conifers dominate fire-prone, cold, and diverse wetland ecosys-tems of northern lower Michigan and the Upper Peninsula. The shrub excep-tion is ground juniper, which is now widely distributed in warmer parts of the Midwest where fire is absent. In the species descriptions that follow, the geographic and ecological distribution of each is briefly considered under the headings of "Distribution" and "Site-Habitat," respectively.

ARCHITECTURAL FORM

Woody plants are arbitrarily and usefully characterized by three kinds of *form* or *habit*: tree, shrub, and vine. A tree is a woody perennial plant with a single trunk (typically unbranched near the base, usually exceeding 5 m (16.4 ft) in height. A *shrub* is traditionally defined as a woody perennial plant usually branched at or near the base and without a single trunk, giving it a bushy appearance of many ascending or erect stems. Height often is specified as not exceeding 5 m. A *vine* is a plant with a long thin stem, which is not self-supporting and grows along the ground or climbs vertically, either twining or using an attachment of some kind (e.g., tendril, adhesive disk, or root). Miss Alice Coats (1992), English gardener, provides an insightful definition.

> A shrub is defined as a woody plant with a number of growths from the base, as opposed to the tree's single stem: height is no criterion, for many tall shrubs, such as lilacs, may overtop small trees. For our purpose, any multiple-stemmed

woody plant that begins to produce its flowers when they are still at or below eye level, may fairly be regarded as a shrub, even though it may eventually attain a great height. In other respects, knowing it would be impossible to please everybody, I have pleased myself in the boundaries I have set.

Miss Coats emphasizes shrub form and the importance of flowers and fruit because of the overwhelming importance of shrubs in gardens and parks.

Shrubs are remarkably diverse and widespread on the Earth's surface, and they have been classified according to different characters whose significance is most important for a given region (i.e., arid, tropical, or temperate). These features include shrub height, degree of stem lignification, habit or form, location of renewal buds, parts periodically shed (leaves, shoots, or branches), life duration and seasonality of assimilating organs, leaf size, and leaf variation (Orshan 1989). These features indicate the greater range of conditions that shrubs occupy compared to trees and their different roles in the respective ecosystems that support them.

Woody vines (also termed *lianas*), usually of trailing or climbing form, are much less common than shrubs in northern temperate regions. Occasionally they adopt the shrub life form. For example, the well-known organism poison-ivy is given species status either as a vine (*Toxicodendron radicans*) or a shrub (*T. rydbergii*). Either is potent enough to cause acute skin irritation or an itchy rash. Woody vines are especially characteristic of tropical rain forests and make up a significant proportion of the vegetation. Vines in our region root in the soil, and with rapid growth and adaptations of aerial roots, tendrils, and adhesive discs they climb trees and other structures to obtain light, where they flower and fruit abundantly. They often smother shrubs and young trees by bridging thickets and forming a mass of foliage in open understories. Vines require not only high light irradiance but nutrient-rich soils as well. In the Great Lakes Region they are most abundant and diverse in river floodplains.

NAMES

Common names for each kind of shrub or vine have been passed along from one generation to another—a distinctive plant language at first used by the native peoples of a given region. Common names are often useful in identification to emphasize some prominent characteristic (velvetleaf blueberry, prickly gooseberry), form (dwarf blackberry, running strawberry-bush), habitat (sandbar willow), or color (red-osier). A limitation is their usage for one species while the same name is applied to another species in a different region. For example, the name huckleberry is applied to species of the genus *Gaylussacia* in our region and in the East generally but to species of *Vaccinium* in western North America. When the common name refers to an entirely unrelated taxonomic group, usually because of a superficial similarity, hyphens or compound words are used to indicate this distinction. Examples include sweetfern (not a fern), blue-beech (not a beech), and running strawberry-bush (neither strawberry nor bush). Dirr (1990) observes, "Common names are a constant source of confusion and embarrassment." Nevertheless, we use the most widely applied and accepted common names for species of the Great Lakes Region (Gleason and Cronquist 1991; Voss and Reznicek 2012). A second common name may be given to recognize long-standing regional or cultural significance.

The currently accepted *scientific name* in Latin is the one used worldwide to ensure reliable communication. The scientific name of a species (never a *specie,* the use of which will reveal your wretched ignorance) always consists of two parts (i.e., a binomial name), as in *Sambucus canadensis,* the common elder. The genus name is a noun. It comes first and is always written with a capital letter. The second name, the *species epithet,* is usually an adjective, and is always written in lowercase. It is customary to add the author name, that is, the name of the person or persons who first gave the name to the plant, for example, *Sambucus canadensis* Linnaeus (abbreviated L.). Explanations of names of botanical authors are given by Soper and Heimburger (1982), Gleason and Cronquist (1991), Farrar (1995), and The International Plant Names Index (2014). Two intraspecific ranks below that of species are *subspecies,* usually applied to geographic races within a variable species, and *variety,* a rank below that of subspecies. When certain variants of a shrub species are propagated for ornamental use, they may be given a cultivated variety or cultivar name, which, for example, may be written *Sambucus canadensis* 'Rubra' to indicate a red-fruited individual. In current convention, the cultivar name is always enclosed in single quotation marks; it is not italicized. A detailed treatment of the taxonomy, nomenclature, and classification of plants is presented by Judd et al. (2008).

Hybrids between two species are designated by means of the multiplication sign × (meaning "hybrid"). If a binomial name has been designated for a hybrid, the × is placed before the species epithet. For example, *Betula* × *purpusii* Schneider is the hybrid between the shrub *Betula pumila* L. and the tree *Betula alleghaniensis* Britton (i.e., *Betula alleghaniensis* × *B. pumila*). When in written form, *Betula* × *purpusii* is pronounced *Betula* HYBRID *purpusii* (Wagner 1975). However, in conversation it is usually abbreviated to *Betula purpusii,* thereby omitting the distinction of its hybrid origin.

DESCRIPTIONS OF INDIVIDUAL SPECIES

Before describing the specific identification characteristics and natural history features of the shrubs and vines (i.e., the major headings on each page devoted to a given species), an overview comparing the approach to shrubs and vines versus trees is useful.

In identifying shrubs and vines, once you know the regional landscape (see Fig. 19) and local site conditions, we urge you to view the whole plant at once—something rarely possible with forest trees. We can see and often touch and examine the whole crown and its parts. At the right time of year, we can examine individual flowers, striking floral displays, and fruits, as well as foliage and stem characteristics. Shrub and vine habit—prostrate, erect, arching, straggling, climbing, clump-forming—provides a rich diversity of characteristic forms. With snow covering parts or all of some shrubs in winter, trees and their characteristic bark and crown forms have the advantage for identification. Therefore, the shrub and vine illustrations emphasize growing season characteristics and winter twigs and buds as well. In the following sections, each identification characteristic or other feature is explained exactly in the order that it occurs in the description of each species.

Size and Form

Size and *form* (i.e., plant habit) refer to the general appearance of a plant. In this section, we identify the species as a shrub or vine and give characteristics of size (height of *adult* individuals, i.e., those in the adult phase of physiological development), crown form or branching habit, and the means of clone formation (i.e., sprouting from rhizome, root, or root collar; rooting of stems or branch tips; layering; or fragmentation). Physical and biotic environments play major roles in determining the size and form of a species or individuals of a species. All our shrubs evolved to grow and persist under the relatively extreme environmental conditions of shaded forest understories or, much more often, in open sites characterized by extreme conditions of cold temperature and short growing season (arctic and subarctic lands) or intense heat and drought (arid lands). In addition, lack of oxygen (wetlands and inundated floodplains) and repeated or catastrophic fires elicit distinctive adaptations. Shrubs are strongly affected in size and form by variations in any of these site factors, as well as age. Nevertheless, the field naturalist soon discovers what to look for, and these are the characteristics of shrub form we describe for each species. Because of marked variation in height, we arbitrarily identify dwarf (<0.5 m), low (0.5–1.5 m), medium (1.5–3 m), and tall (>3 m) shrubs, as well as indicating the habit of ones that climb, creep, trail, or are prostrate on the forest floor. Shrub size is not only influenced by age but especially by the soil-water and nutrient availability. Shrubs on moist, fertile soils grow faster and taller than those of the same species on sites that are dry and infertile or wet and poorly aerated. Despite these environmental influences, the relative size of a given species is an important identification characteristic, especially in winter. For example, tall shrubs such as the highbush blueberry are easily distinguished from low shrubs such as low sweet blueberry and bearberry.

The branching pattern of woody plants is controlled by the interacting combination of genetic factors, physiological phase (i.e., juvenile or adult), and the plant's physical environment of atmosphere, landform, soil, hydrology, disturbance, herbivory, and other biotic factors such as mutualism and competition with associates. Each species has an inherently determined growth plan or architectural model that is adapted to display the trunk or stem, woody crown, leaves, and roots for optimum capture of light for photosynthesis and soil-water (i.e., moisture), soil-oxygen, and nutrients. The patterns for shrubs are not as well understood as those for trees with their distinctive excurrent or deliquescent form (Zimmerman and Brown 1971; Wilson 1984; Barnes et al. 1998, 122–26; Barnes and Wagner 2004). Therefore, we use terms such as *erect* or *upright*; *spreading* or *arching*; and *trailing, creeping,* or *prostrate* to assist in the identification of the habit of each species. Proceeding from the "top down" (i.e., from large to small), after habitat, habit or form is the next most important characteristic for reliable identification and understanding of the natural history and ecological role of a species.

Virtually all shrubs and vines employ at least one method of asexual reproduction to maintain themselves and often spread spatially. Shrubs often form clonal clumps by sprouting from the root collar at the base of the plant. Other methods include sprouting from roots or rhizomes, tip rooting, layering (i.e., a lower branch comes in contact with soil and takes root), and seed apomixis.

Large size in diameter and height is highly prized by humans as special, remarkable, or outright amazing. Finding Big Trees has become a national pastime, and shrubs also are given state and national recognition. We have included the dimensions of the biggest individuals that have been published for Michigan (Ehrle 2006). These and other large individuals of a given species can be accessed using the website of the Michigan Botanical Club, http://www.michbotclub.org.

Bark

Bark is the outside covering of the stems and roots of woody plants, which protects the cambium and phloem tissues. As for trees, characteristics of the bark of the main stem and large branches are well suited for year-round identification, unlike deciduous leaves and flowers. Compared to trees, however, the bark of shrubs is often less useful. Nevertheless, because of the relatively small stature of shrubs, we can observe and examine the outer covering of the main stems and most of the branches and twigs. Bark is a superb distinguishing feature of trees because with increasing age and trunk size they produce multiple layers of inner and outer bark whose pattern, color, shape, and taste are distinctive. However, the life-span of most shrub stems is short, circa two to fifteen years, and stems are rapidly recycled as old ones die and new ones arise. Following severe disturbance or stress, shrubs have the ability to regenerate new main stems by sprouting from the root collar or by other means. Thus, compared to trees, there is little time in which to form thick bark layers. Even shrubs regarded as small trees have relatively thin bark (e.g., the sumacs, choke cherry, serviceberry, nannyberry, and witch-hazel). In almost all low or tall shrubs, the epidermal layer remains more or less intact, in contrast to long-lived trees. This outer layer, together with the cortex of stems and twigs, takes on many distinctive properties at the fine scale. Lacking a large trunk, in many cases, "bark" can hardly be distinguished from the color or outside peeling-shredding characters of main stems and branches. Therefore, when the heading "Bark" is absent, we describe the useful features of the outside covering of main stems under the heading "Stems-Twigs." The fine-scale features of the outside covering provide excellent clues to the genus and species: texture (smooth, rough, warty with lenticels, pubescent, prickly, spiny, or ridged), epidermis break-up (peeling, shredding, or scaly), and color of the cortex (green, bright red, gray, reddish-brown, or black). As with tree crowns, the thin-barked branches and twigs have photosynthesizing chlorophyll in the cortex layer that provides these small plants with additional energy. The warty-corky lenticels of the main stem of elderberries and the bright red epidermis of red-osier stems are key characters. The ridged versus stringy bark of the closely related exotic honeysuckle species, Maack's and Tartarian honeysuckles, is highly diagnostic in summer and winter. The strongly ridged outer bark of a twenty-four-year-old Maack's honeysuckle is illustrated in Figure 1.

Leaves

Leaves are important means of discriminating between species because of their many distinctive features and the convenience of using them—especially for shrubs. However, leaves are variable and in our climate are present on deciduous shrubs for only about seven months of the year. Nevertheless, they can be used to distinguish all but the most closely related spe-

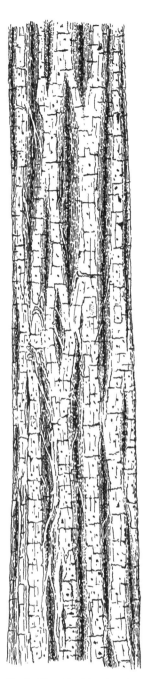

Fig. 1. Strongly ridged and anastomosing outer bark of an individual of Maack's honeysuckle, *Lonicera maackii*, approximately twenty years old. Stem diameter 16.0 cm (6.3 in).

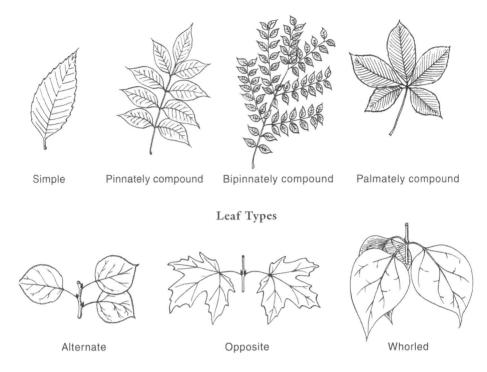

Simple   Pinnately compound   Bipinnately compound   Palmately compound

Leaf Types

Alternate   Opposite   Whorled

Fig. 2. Leaf types and leaf arrangement. Leaf types are simple or compound (pinnately or palmately). The types illustrated are simple, pinnately compound with many leaflets, bipinnately compound, and palmately compound. The three recognized leaf arrangements are alternate, opposite, and whorled.

cies. Most important are the *leaf type* and *leaf arrangement* (Fig. 2). Next, leaf size is easily observed.

Leaf size is influenced by external conditions: the availability of light, soil-water, oxygen, and nutrients. Within an individual, size and shape are affected by age; position in the crown; vigor and size of the branch or twig; size of the vegetative buds; whether leaves are borne on vegetative or flowering shoots; the kind, position, and length of the shoot on which the leaves are borne; and when they are formed—preformed in the winter buds or during the growing season (neoformed). When examining leaves of an individual or comparing leaves of different species, it is important to use leaves from the same position in the crown and from the same kind and size of shoot. Leaves of evergreen coniferous shrubs are markedly less variable than those of deciduous ones and are present throughout the year.

Leaves of shrubs and vines exhibit many characters that are useful in identification. In the species descriptions that follow, not every possible character is given for each species. Nevertheless, the pertinent characters are described in a systematic way in the following sequence.

Leaf arrangement on the stem (alterative, opposite, or whorled)

Leaf type (simple or compound)

Blade size and shape for simple leaves, leaf length and leaflet size and shape for compound leaves

Blade apex and blade base (the shape of these characters is separated by a semicolon)

Blade margin (entire, lobed, serrate, etc.)

Blade texture, as appropriate

Blade color of upper and lower sides in the growing season

Blade pubescence, as appropriate

Venation (number, pattern branching)

Petiole (shape, color, presence of pubescence or glands, length)

Stipules (presence, prominence, morphology, persistence)

Diagrams illustrating leaf type and leaf arrangement are presented in Figure 2, and the most common types of blade shape, tip, and margin are shown in Figures 3, 4, and 5, respectively. Leaf size and shape are important characters for many species. However, leaf dimensions are highly variable, even within a single shrub or from base to tip of an individual shoot. Thus the dimensions presented serve only as a relative guide to leaf and blade size.

In conifers, leaf shape is most important, and four shapes are recognized—needle (acicular), linear, awl-shaped, and scalelike (Fig. 3). One of the three conifer species we describe has linear leaves, Canada yew, *Taxus canadensis*.

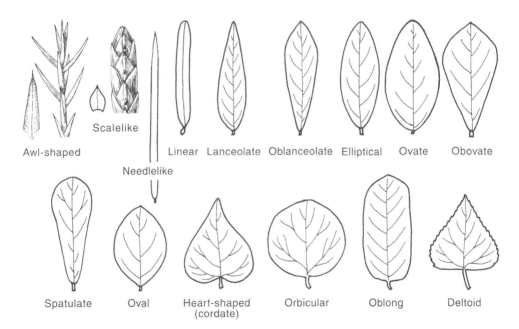

Fig. 3. Leaf shapes

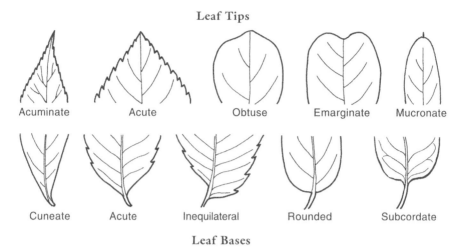

Leaf Tips

Acuminate Acute Obtuse Emarginate Mucronate

Cuneate Acute Inequilateral Rounded Subcordate

Leaf Bases

Fig. 4. Selected leaf blade tips and bases

Of the two juniper species, ground juniper, *Juniperus communis* var. *depressa*, has only awl-shaped needles—short and tapered to a sharp point, flat, and stiff. The creeping juniper, *J. horizontalis*, has both juvenile needles that are awl-shaped and scalelike leaves in the adult physiological phase.

In angiosperms, flowering plants, it is best to examine the most striking characters first, that is, whether leaf type is simple or compound and whether leaf arrangement is alternate, opposite, or whorled (Fig. 2). Most shrub and vine species have alternate leaf arrangement. Therefore, when one finds that leaf arrangement is opposite, identification is greatly simplified. Genera with opposite leaf arrangement include *Euonymus, Lonicera, Sambucus, Symphoricarpos,* and *Viburnum.* Because buds arise in the axils of leaves, species with opposite leaf arrangement also have opposite twig and branch arrangement (if both branches at a node develop and remain on the plant). Thus, if leaves are absent or too high to reach, the branching pattern will indicate the leaf arrangement. Most shrubs have simple leaves, and the presence of compound leaves narrows the task of identification considerably, as in *Dasiphora, Ptelea, Rhus, Rosa, Sambucus, Sorbus, Toxicodendron,* and *Zanthoxylum.*

Leaves of deciduous shrubs are only borne directly on shoots of the current year—never directly on twigs or branches of past years. In winter, shoots and leaves of dormant broadleaf deciduous shrubs are already preformed and tightly packed and telescoped inside the terminal and lateral buds, which are located on twigs of the previous growing season. Thus the terminal bud, or vegetative bud that developed in the axil of a leaf on the current shoot, contains the shoot and leaf primordia that will form next year's shoot and leaves. These preformed leaves develop early in the next growing season and are termed *early leaves.* These leaves, located at nodes along the shoot, are the best ones to use in identification when they are mature. The most useful early

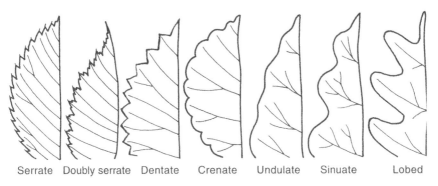

Serrate   Doubly serrate   Dentate   Crenate   Undulate   Sinuate   Lobed

Fig. 5. Selected leaf blade margins

leaves are those borne on *determinate shoots*. Such shoots develop, extend, form leaves, and then set buds relatively early in the growing season—never to extend again in length. Many species develop *indeterminate shoots,* which continue to grow during the entire season. They bear early leaves along their basal (i.e., proximal) part and then give rise to new, neoformed leaves as the shoot continues to develop during the middle to latter part of the growing season. These are termed *late leaves*. Late leaves may be markedly different in shape, size, and other characteristics from early leaves that were preformed in the dormant winter buds. Such differences are often overlooked and thus are a main source of frustration, indecision, and confusion when grabbing the nearest, or "any-old," leaves for identification. The marked difference between early and late leaves along an indeterminate shoot of downy arrow-wood is il-lustrated in Figure 6A. For comparison, a determinate shoot with early leaves is shown in Figure 6B. The early leaves are typical of downy arrow-wood, *Viburnum rafinesquianum*, whereas the late leaves at the tip could be readily mistaken for leaves of the gray dogwood, *Cornus foemina*! The leaves in the line drawings for each species are virtually all of determinate shoots with early leaves to facilitate identification. Nevertheless, in the field expect to find some surprising differences in leaf size and shape within a given individual. For example, the leaves on flowering shoots (e.g., in *Crataegus Ribes, Rosa, Rubus,* and *Salix*) are typically different in size and shape from those of typical early leaves of a given species.

Indeterminate shoots continue to grow in late summer and into autumn, given favorable conditions of soil-water, light, and temperature, until their succulent shoots and new leaves are killed by freezing temperatures. They are markedly longer than determinate shoots and are located in the upper part of the crown and on lateral branches that spread into the open. Late leaves are initially relatively small in comparison to early leaves. They progressively increase in size with favorable conditions and then decrease in size toward the tip of the shoot as growth slows in autumn. Most shrub species exhibit the indeterminate pattern of shoot growth. This strategy is evidently highly suc-cessful for increasing plant size and vigor, providing platforms for flowering

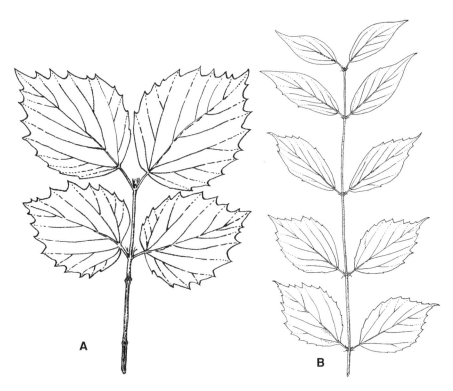

Fig. 6. Shoots of downy arrow-wood, *Viburnum rafinesquianum*. A, a typical determinate shoot with early leaves, which were preformed in the terminal bud of the past year; B, part of an indeterminate shoot illustrating marked differences in leaf blade morphology between preformed early leaves (leaf pairs at two lower nodes), a transitional pair, and the neoformed late leaves (leaf pairs at upper two nodes), which were formed during the later part of the present growing season.

and sexual reproduction, occupying space otherwise available to competitors, and arching to the ground where the tips may take root.

Stems-Twigs

The three main anatomical and structural parts of woody plants are stem, root, and leaf. *Stem* is defined at several scales, and several terms besides *stem* are applied to different contexts of woody plant organization—*trunk, branch, twig,* and *shoot*—all of which are anatomically stems. The main above ground, upright-growing stem serves as support and to transport and store water and food materials. In trees, by convention, the single stem is termed a *trunk* or *bole*. Shrubs typically have many upright stems, and these terms are not usually applied, except in the case of a small tree with a single stem. The multiple stems of a clone of silky dogwood (*Cornus amomum*) are illustrated in Figure 7. Notice that all stems, crowded at the base, arise from the root collar or enlarged basal plate. The main stem of a woody plant at increasingly finer scales bears branches, twigs, and shoots—all alternate names for *stem*. All stems

Fig. 7. Multiple stems of a clonal clump of silky dogwood, *Cornus amomum*. The stems arise from the root collar, which after many years of sprouting form an enlarged basal plate. (After Smith 2008.)

have *pith,* the soft, spongy tissue in the center, except in the few species whose stems are hollow between the nodes (e.g., *Lonicera, Solanum dulcamara,* and species of the genus *Smilax).* Belowground stems are termed *rhizomes.* They are distinguished from roots by the presence of pith.

Located at the tip or distal end of a shrub branch is the newly developed and herblike stem—the current shoot or simply shoot. The complete shoot consists of three parts: stem, leaves, and buds. As the current shoot matures in summer and fall it becomes woody, and we recognize it as a *twig.* It bears a terminal or end bud, which in winter contains the tiny, telescoped shoot and early leaves that emerge in the next growing season. Generally a twig refers to a small segment of a branch that is made up of several to many modular stem units, all woody. Winter twigs bear lateral buds which give rise to new shoots with their complement of leaves and buds. As twigs age and become larger they become what are termed branches. Therefore, from the main stem, or *trunk,* outward, and with decreasing age, we encounter a series of stems: branches, twigs, and finally at the ends of last year's twigs the newly developed current shoots, which support leaves and buds in their axils.

Shoots and twigs provide good characteristics throughout the year. In the spring, the stems of new shoots are green, herbaceous, and often covered with protective pubescence, which may be shed as the growing season progresses and the succulent stem hardens into a woody twig. They change color rapidly during the growing season, and the same stem, late in the first year or the following year, takes on a different color as it gradually becomes woody. The principal characteristics for identification are diameter (slender or stout), color, taste, odor, form (straight, drooping, zigzag, angled), smooth, or with lenticels or corky warts, armature (prickles, spines, or thorns), presence or absence of bristles or pubescence, leaf scar shape and size, number of bundle scars, and pith.

LONG AND SHORT SHOOTS

The kinds of shoots are important not only in determining growth habit of the plant but also for identification. Woody plants are characterized by an open system of growth whereby the shoot is the modular unit of construction. Shoots develop following the shedding of bud scales, absorption of water, and expansion to their normal size for a given species. Similar to the situation in trees, shrubs have vegetative shoots, which are technically termed *long* and *short*. These terms do not refer to absolute length but to the special property of internode length. In long shoots, the distance between internodes is relatively long and leaves are clearly separated. All trees and shrubs bear long shoots, and some species bear both long and short shoots. Short shoots exhibit little or apparently no internode elongation (i.e., the portion of the stem between successive leaves or leaf scars); they are termed *spur shoots* in broadleaf deciduous shrubs and trees—or simply *short shoots*. Short shoots are characteristic of several species, including prickly-ash, *Zanthoxylum americanum*; sand cherry, *Prunus pumila*; and common buckthorn, *Rhamnus cathartica*. Short shoots of prickly-ash are illustrated in Figure 8. The boundary between consecutive annual shoots is marked by groups of ringlike scars left by outer and inner bud scales. Annual shoot growth is only circa 1 mm, so internodes are extremely short, and the annual ring scars lie virtually on top of one another. All other shoots are long shoots regardless of whether they are 2.5, 25, or 50 cm long.

In contrast to short shoots, long shoots typically have 1 cm or more elongation between leaf nodes. Long shoots may be either seasonally determinate or indeterminate. In summary:

Seasonally determinate long shoots.
They are completely preformed within the end buds and vegetative lateral buds, make substantial growth in length, bear only early leaves, and occur in all species.

Seasonally indeterminate long shoots.
The basal part is predetermined in end or lateral buds. After this proximal part (bearing early leaves) elongates in the spring, the shoot continues to grow in the

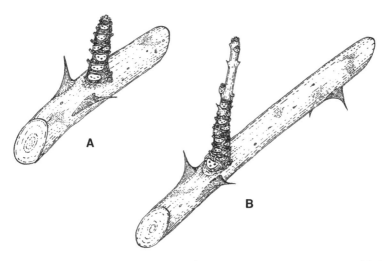

Fig. 8. Short shoots of prickly-ash, *Zanthoxylum americanum*. A, a seven-year-old short shoot; B, after nine years of short shoot growth a long shoot with relatively long internodes has developed in the tenth growing season.

summer (often 20 cm or more) and gives rise to late leaves (i.e., leaves borne in the latter part of the growing season). The terminal (i.e., distal) portion of this shoot continues to grow throughout the summer under favorable conditions and only ceases growth as the shortening days of late summer and autumn signal the approach of winter or when killed by freezing temperatures. These shoots "forage" for light and significantly extend the plant into new space. Not surprisingly, such indeterminate shoots are especially well developed in well-lighted portions of the crowns of fast-growing pioneer species such as buttonbush, *Cephalanthus occidentalis*; raspberries and blackberries, *Rubus* spp.; and willows, *Salix* spp.

RHYTHMIC AND CONTINUOUS GROWTH: SYLLEPTIC SHOOTS

In climates with clearly defined seasons and alternating periods favorable or unfavorable for growth, shoot extension is usually rhythmic. Growth ceases during a cold or dry season, at which time the apical meristem is protected within an end or axillary bud. In contrast, in a uniformly favorable environment, a plant may grow continuously with constant production of shoots and leaves, and axillary meristems may have no dormancy or resting phase. The axillary shoots that develop and grow continuously are termed *sylleptic shoots* and their extension *sylleptic growth*. Therefore, each axillary shoot primordium will usually have one of two immediate fates other than abortion. It may become a temporarily dormant bud in the axil of a leaf or it may develop and grow without delay, forming a new shoot, which diverges from the current

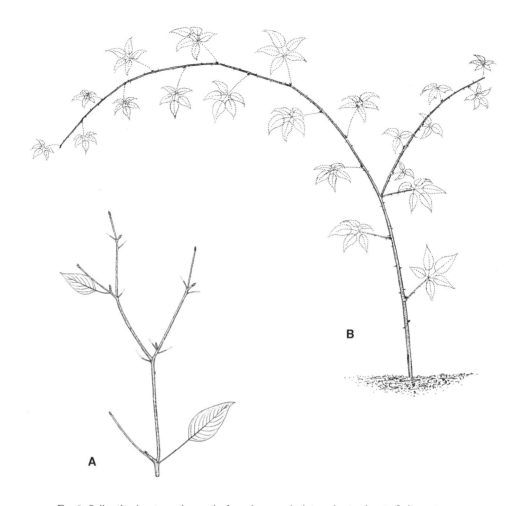

Fig. 9. Sylleptic shoots and growth. A, a vigorous indeterminate shoot of alternate-leaved dogwood, *Cornus alternifolia,* extending upward and to the left, has given rise to three sylleptic shoots--lower left, middle right, and upper left. Instead of the standard condition of setting a bud in the axil of the leaf, occasionally a new shoot (termed a sylleptic shoot) with its own leaves that extends the plant's foraging capacity for light. B, sylleptic shoots of black raspberry, *Rubus alleghaniensis,* may form along the rapid growing *primocane.* Just one sylleptic shoot is illustrated here, but several others may typically form in leaf axils of the arching primocane.

shoot and expands into new space. This pattern of growth is common in the moist tropics, rare in temperate trees, but more common in temperate shrubs. Sylleptic shoots extend the crown into well-lighted areas and therefore increase the light-gathering and photosynthetic ability of the plant. Look for sylleptic shoots in well-lighted outer and upper crowns of alternate-leaved dogwood, willows, alders, buttonbush, and some *Rubus* and *Ribes* species. Sylleptic shoots and growth are illustrated for *Cornus alternifolia* and *Rubus alleghaniensis* in Figure 9.

Shrubs, unlike trees, have a low stature as adult plants. With few exceptions they are primary targets for herbivory by animals as long as they live. Trees soon "escape" predation by developing thick bark or growing out of the reach of deer and certain other herbivores, except those insects, mammals, and birds that live in or visit tree crowns. Although growth in height appears to be the most striking difference in avoiding herbivory for trees compared to shrubs, the rapid and continuing ability to replace stems and other plant parts is a growth trait common to both. Thus all woody plants face a dilemma: grow or defend? Plants must outgrow herbivory or defend against it (Herms and Mattson 1992). In favorable sites, plants allocate photosynthate to growth and only relatively small amounts to defense. The allocation to defense is high in harsh sites where resource availability (soil-water, nutrients, oxygen) is low; where plant habit is relatively low, <1–3 m; and where the effects of herbivores, pathogens, and physical conditions on growth may be severely limiting. Avenues of defense include physical armature, production of secondary chemical compounds, and reliance on arthropods that serve as "bodyguards."

Sharp-pointed armatures such as prickles, spines, and thorns are important identification features (Fig. 10). They serve to protect stems from herbivory by mammals, especially in the juvenile physiological phase (Figure 11). *Prickles* are an outgrowth of the epidermis or cortex and are relatively easy

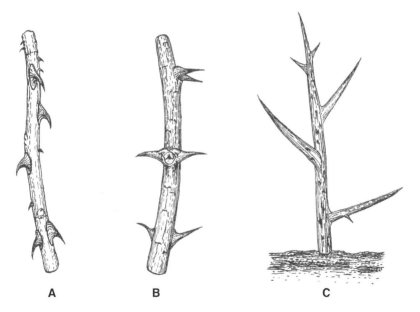

A   **B**   **C**

Fig. 10. Prickles, spines, and thorns. A, prickles (*Rosa* spp.) are outgrowths of the epidermis and separate cleanly from the stem (and into your flesh!); B, spines of prickly-ash, *Zanthoxylum americanum,* are modified leaves; C, thorns of the hawthorns, *Crataegus* spp., are modified shoots and have vascular connections into the supporting stem.

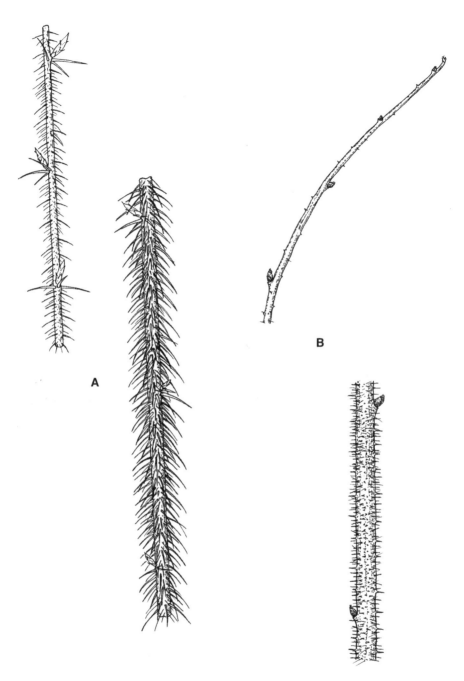

Fig. 11. A, dense prickles cover the basal 30 cm (total length 60 cm) of the young stem of prickly gooseberry, *Ribes cynosbati*, with spines at the internodes. Even in the upper part of this new stem in the juvenile phase, the abundance of prickles and spines is characteristic; and both are effective deterrents to herbivory by small mammals (rabbits, mice, voles). B, the basal portion of a young stem of red raspberry, *Rubus strigosus,* is covered with short bristles, which gradually decrease in density toward the shoot tip. At the tip of the same shoot, and above the reach of small mammals at 90 cm, and to the shoot apex at 121 cm, there are very few bristles. Often, bristles may be lacking at the apex.

to dislodge from the twig (as in roses). They lack vascular strands. *Spines* are specialized leaves or parts of leaves (e.g., *Berberis thunbergii and Ribes lacustre*). *Thorns* are less common than prickles and are specialized stems (e.g., *Crataegus* and *Elaeagnus*). Both spines and thorns have vascular bundles just as in typical leaf and stem tissues. Hair, including bristles and various kinds of pubescence, is also a form of protection, especially against insects likely to attack the soft, herblike tissue of new shoots and flowers (e.g., willows). In the juvenile phase, armature is densely expressed at the base of newly formed stems in *Ribes, Rosa,* and *Rubus.*

Chemical defenses include the presence of hormonelike compounds that affect the development of insects. In all higher plants, a great number and diversity of secondary chemicals are known—alkaloids, terpenes, phenolics, cyanogenic glycosides—and thousands (four thousand alkaloids alone) have been described (Levin 1976; Herms and Mattson 1992). Similar to physical defenses, some secondary chemicals have their highest concentration during early stages of leaf expansion and seedling growth, as well as in shoots produced following severe browsing or other injury. Secondary chemicals produce taste and aroma, which are excellent characteristics for woody plant genera, including *Comptonia, Gaultheria, Juniperus, Lindera, Myrica, Prunus, Rhus, Taxus,* and *Zanthoxylum.*

Some shrub and vine species defend themselves through coevolutionary relationships with so-called bodyguards. By providing nectar to carnivorous insects such as wasps and ants, the plants obtain access to predators, which consume some sugars and otherwise consume or fend off herbivorous insects. The nectar is provided through glands called extrafloral nectaries located in leaf blades, petioles, and/or shoots and ranging in size from microscopic trichomes to vascularized cuplike structures (as in choke cherry and trumpet creeper). Extrafloral nectaries can provide diagnostic characters and are found in *Lonicera, Prunus, Rosa, Salix, Sambucus, Smilax, Solanum, Staphylea, Vaccinium, Vitis,* and *Viburnum.*

A lesser-known form of indirect defense occurs in the form of housing for predatory mites (O'Dowd and Willson 1991). In the midrib leaf vein in species of *Prunus, Quercus,* and *Vitis,* one may find small tufted structures called acarodomatia, within which, under magnification, communities of mites are visible. These typically are carnivores that consume herbivorous mites and fungi to the benefit of the host plant. For example, acarodomatia and their mite associates in riverbank grapes have been shown to slow the spread of grapevine powdery mildew (English-Loeb et al. 2005). Domesticated grapes do not contain mite domatia and are more strongly impacted by this fungal pest.

LEAF SCARS AND BUNDLE SCARS

The arrangement, size, and shape of leaf scars and the number and arrangement of *bundle scars* within the leaf scar are often useful in shrub identifica-

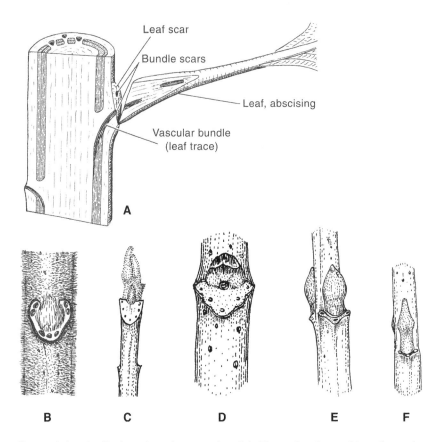

Fig. 12. A, longitudinal section of stem and petiole illustrating the position of vascular tissues; B–F, leaf scars illustrating the number and arrangement of bundle scars. B, staghorn sumac (many bundle scars); C, poison-ivy (6-7 bundle scars); D, common elder (five bundle scars), E, highbush-cranberry (three bundle scars); F, red-osier dogwood (three bundle scars). (A, after Core and Ammons.)

tion. Deciduous shrubs shed leaves in autumn through the action of a corky abscission layer. This living layer actively causes the leaf to snap off and then seals the living tissues of the stem. The scar of the leaf attachment varies in size and shape (e.g., small or large and circular-, shield-, or crescent-shaped) depending on leaf size, requirements of water and nutrients, and shape of the petiole attachment to the stem. Within a leaf scar in winter condition, there are one or more small raised or depressed dots, which, though sometimes inconspicuous, are often prominent and distinctive. The bundle scars are left by the vascular strands or bundles that pass through the stem into the petiole and leaf blade (Figure 12A). They transport water and nutrients into and from the veins of the leaves. Bundle scars have a characteristic size, number, and arrangement (Figure 12B, C, D, E, F). There may be one (blueberries), three (red-osier and American highbush-cranberry), five (common elder, poison-ivy), seven (staghorn sumac), or many (poison-sumac). Bundle scars are typically arranged in U- or V-shaped patterns.

Recall that in the center of the stem a mass of soft-walled parenchyma cells, termed pith, provides important distinguishing features. Usually it is white, but in may be brown, greenish, or other colors. In cross section, pith is usually circular, but it may be triangular, as in the alders. When cut longitudinally, pith is typically continuous and homogeneous. Rarely, pith may be lacking between nodes, as in *Solanum dulcamara*, *Smilax* spp., and some species of *Lonicera*.

Winter Buds

Buds are like tiny cocoons. They protect the dormant apical and axillary meristems and the undeveloped young shoot by means of special hard flaps, called *bud scales*. The majority of buds, especially lateral or axillary buds, remain quiescent and never form shoots unless induced to do so by hormones or injury (e.g., young leaves killed by freezing temperature). Buds are formed of protective outer bud scales, which cover and protect additional but softer inner bud scales, embryonic leaves, and in the very center the delicate embryonic meristem and growing point. The formation of a new shoot in the spring of the year is accomplished by shedding the bud scales, enlarging the leaves or flowers, and (except in short or spur shoots) elongating the stem axis.

Buds have important identification characteristics in autumn and winter. Their arrangement, size, shape, color, presence or absence of pubescence, and the number, kind, and arrangement of bud scales are major features (Figure 13). All flowering plant species have lateral buds, which are situated on the sides of the shoot in the axils of the leaves (or the leaf scars in winter). The buds may be closely appressed to the stem or diverge from it at various angles. Barring injury, all species have buds at the end of the current shoot, *true terminal buds* or *end buds* (i.e., false terminal buds). However, many shrubs lack true terminal buds because the end bud in many species is actually a lateral bud. Recall that in many shrub species the tip of the current shoot grows until it is killed by freezing temperature. It dies back to the next living lateral bud, leaving a dead remnant of the indeterminate shoot or a small bump or scar at the end of the twig immediately adjacent to this lateral bud. The closest lateral bud takes over the terminal position and is known as a false terminal bud, pseudoterminal bud, or simply end bud. Figure 14 illustrates examples of dieback of the indeterminate shoot caused by winterkill for shining sumac and common elder. The lateral bud nearest the end of the living shoot, now the false terminal or end bud, assumes the dominant position, and the leading shoot of the next year comes from this bud. In some species, the flower buds are borne in structures at the ends of the shoots. Compared to vegetative buds, flower buds are often markedly larger or swollen (e.g., in nannyberry, Figure 15) because packed inside are the diverse organs of the flowers.

Lateral or axillary buds are typically similar to true terminal buds in col-

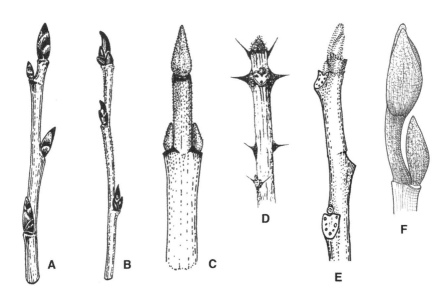

Fig. 13. Characteristic winter terminal buds of shrub species, A, choke cherry; B, black chokeberry; C, silky dogwood; D, prickly-ash; E, poison-ivy; F, witch-hazel. Buds of poison-ivy and witch-hazel are naked; buds of prickly-ash are woolly and indistinct.

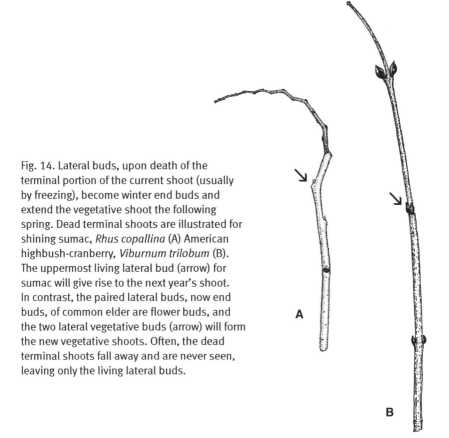

Fig. 14. Lateral buds, upon death of the terminal portion of the current shoot (usually by freezing), become winter end buds and extend the vegetative shoot the following spring. Dead terminal shoots are illustrated for shining sumac, *Rhus copallina* (A) American highbush-cranberry, *Viburnum trilobum* (B). The uppermost living lateral bud (arrow) for sumac will give rise to the next year's shoot. In contrast, the paired lateral buds, now end buds, of common elder are flower buds, and the two lateral vegetative buds (arrow) will form the new vegetative shoots. Often, the dead terminal shoots fall away and are never seen, leaving only the living lateral buds.

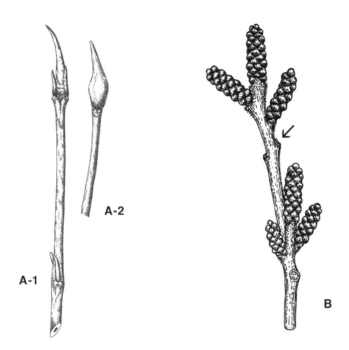

Fig. 15. Distinctive winter flower buds or catkins may assume the terminal or end-bud position for many species. For example, in A, nannyberry, *Viburnum lentago*, the terminal vegetative (A-1) and flower buds (A-2) are compared. The vegetative bud (A-1) encloses next year's terminal shoot, whereas the swollen flower bud (A-2) encloses the preformed bisexual flower. In B, a catkin of fragrant sumac, *Rhus aromatica*, enclosing male flowers occupies terminal and lateral positions. In spring a vegetative shoot will develop from the uppermost live lateral vegetative bud (arrow). Catkins also are excellent winter identification characters for birches and alders.

or, form, and pubescence, but they are usually smaller and have fewer outer bud scales. In the species descriptions, it is the terminal or end buds that are described unless otherwise indicated. Lateral buds are either vegetative (containing embryonic lateral shoots and their leaves) or reproductive (containing embryonic flowers or cones). Terminal or end buds also may be flower buds, as in dogwoods (*Cornus* spp.) and arrow-woods (*Viburnum* spp.). In some species, the lateral vegetative and flower buds are positioned one above the other, a condition termed *superposed*. The flower bud is typically located above the vegetative bud on the stem.

Flowers (Strobili in Conifers)

Flowers are almost always excellent features for plant classification and identification because they tend to vary less than twig, leaf, and other characters. They are under strong genetic control and less influenced by the physical environment. They are excellent for recognizing genera and distinguishing species within genera. Unlike trees, you don't have to be 16 m (53 ft) tall to observe them. However, the main drawback for identification is that their availability for observation lasts a relatively short time, usually in the spring. Therefore, we have described floral features, especially those closely related to fruit type and position, but rarely use them in keys for field identification. Type of inflorescence, position as terminal or axillary, and structure are certainly important. Most flowering shrubs are bisexual and insect pollinated. The flower cluster is termed an *inflorescence*. Fruits develop from the fertilized female flowers and therefore are borne in the same way; the structure is termed an *infructescence*. The parts of a bisexual flower and the different ways flowers are borne (i.e., inflorescences) on broadleaf deciduous shrubs are illustrated in Figure 16.

Strobili, the young reproductive structures of conifers, develop into fleshy cones in the three conifer species described. They are not flowers because floral parts are lacking and the seed does not develop within an ovary. The seed or ovule enclosing the embryo is borne naked on a scale inside the fleshy covering of *Juniperus* and *Taxus*.

The transfer of pollen from male organs to the appropriate receptors in the female organs is called pollination and occurs predominantly in the spring. Gymnosperms are wind pollinated. Flowering plants may be pollinated by insects or wind, and most shrubs are insect pollinated. Of the 138 species we describe, 91 percent are insect pollinated. In the northern temperate zone, this feature is likely due in part to favorable site conditions near the ground and in the understory, where delicate, diverse, and visually attractive flowers thrive compared to those in the exposed and harsh wind climate of treetops. The understory is also the favorable home of diverse insect pollinators of shrubs. A few species, willows in particular, may be pollinated either by insects or wind. Insects tend to visit flowers of a particular kind and thus usually transfer pollen with considerable precision. Therefore, insect-pollinated plants produce relatively small amounts of pollen compared to those of wind-pollinated species. Insect-pollinated flowers can be recognized because they are more or less showy, the petals well developed and usually white or brightly colored. They tend to be fragrant, and they usually produce nectar, which insects use as food. The nectaries often appear as shiny swellings in the flower, and at the time of pollination they secrete a sugary liquid. Insect-pollinated flowers tend to have both sex organs present, the *bisexual* condition. Shrubs in our region are pollinated by a variety of insects, including flies, bees and wasps, butterflies, moths, and beetles. The ruby-throated hummingbird, *Archilochus colubris,* pollinates the trumpet creeper, *Campsis radicans*, and species of *Lonicera.*

Wind-pollinated strobili and flowers are generally very simple. They lack or have only slightly developed petals or attractive structures (e.g., nectaries), and they are usually unisexual, having either only male or only female organs. Only 9 percent of the species we describe are pollinated only by wind; the genera are *Juniperus, Alnus, Celtis, Comptonia, Corylus, Myrica,* and *Taxus.*

Shrubs and trees in the juvenile physiological phase do not flower or pro-

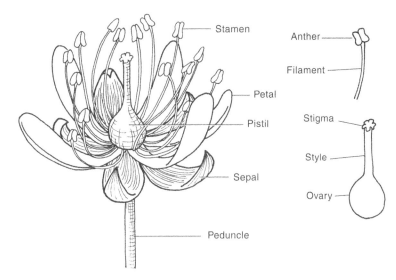

Fig. 16A. Parts of a bisexual flower.

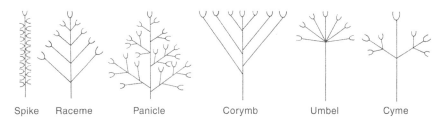

Spike   Raceme   Panicle   Corymb   Umbel   Cyme

Fig. 16B. Types of inflorescences of woody plants.

duce strobili and hence bear no fruit or cones. Shrubs have a significantly shorter juvenile phase than most trees, but to flower and produce fruit the woody plant crown typically requires high light irradiance. Full sunlight is usually the best for abundant flower and fruit (or cone) production. Unlike trees, most shrubs and vines are adapted to live and regenerate in full or abundant sunlight, and many are regarded as shade-intolerant or understory-intolerant species: 72 percent of our 138 species. For example, shrubs native in bogs, fens, and open sand dunes, or well-lighted edges of such open areas, are typically shade-intolerant. These species typically produce flowers and fruits annually in contrast to many tree species, which have a definite periodicity of flowering: high production followed by one or more years of low to moderate production. Although shade- or understory-intolerant species may persist for years in lightly shaded forests, they are unlikely to produce abundant flowers and fruit (e.g., *Vaccinium* spp.). Shrub species adapted to live and reproduce in moderate or deeply shaded forest understories are notably "patient" or "laid-back" plants. Such species include leatherwood, running strawberry-bush, striped and mountain maples, and creeping-snowberry. These shade-tolerant species flower and extend their shoots vigorously in early spring when deciduous trees of the overstory are leafless and light irradiance is high. They are also favored by small- to large-sized gaps in a dense canopy, which are caused by disturbances. Then they quickly take advantage of open conditions and flower and fruit abundantly until light levels are again too low for such vigorous reproduction.

Fruits and Cones

The seed-bearing products of plants, termed *fruits* in angiosperms and *cones* in conifers, are very important identification features. One or more fruit or cone characters are almost always the key characters of each species. Also they are markedly less variable than leaves, twigs, and habit. Fruits and cones are more accessible than flowers because they stay on the shrub for a longer period of time. Because so many fruits are fleshy and attractive to birds and animals, they are less often found on the ground as are cones or tree fruits such as nuts and samaras.

In the coniferous shrubs, the seeds are borne in a fleshy cone of *Juniperus* or surrounded by a fleshy covering termed an *aril* (in *Taxus*). In all conifers the seeds are borne naked on the scales of a woody cone. In angiosperm shrubs and trees, seeds are borne in fruits, which differ markedly in morphology due to the many ways in which the ovary walls develop. The major fruit types of angiosperms are illustrated in Figure 17.

For most species described, an illustration of the fruit accompanies the line drawings of foliage and stems. Fleshy fruit types—berry, drupe, and pome—predominate in shrubs and are primarily dispersed by birds and mammals. Drupes have a stony inner wall (stone or pit) with an outer fleshy covering. Berries and pomes are composed mostly of fleshy tissue that surrounds the seed(s). Dry fruits that split open along suture lines (dehiscent) include the capsule, legume, and follicle. Dry fruits that do not split (indehiscent) are the nut, samara, and achene. Some species with fruits that split open release seeds that have tufts of cottony hairs attached to them such as the willows. These seeds are primarily wind dispersed. Simple fruits are sometimes permanently clustered to form compound fruits. They are termed either *aggregate* or *multiple* depending on whether the fruits are derived from a single flower with many pistils (aggregate) or many separate flowers (multiple). In the genus *Rubus*, raspberries and blackberries exhibit an aggregate of drupes.

Distribution

A general statement of the present distribution of each species in Michigan is based on the range map of Voss and Reznicek 2012, specimens at the Univer-

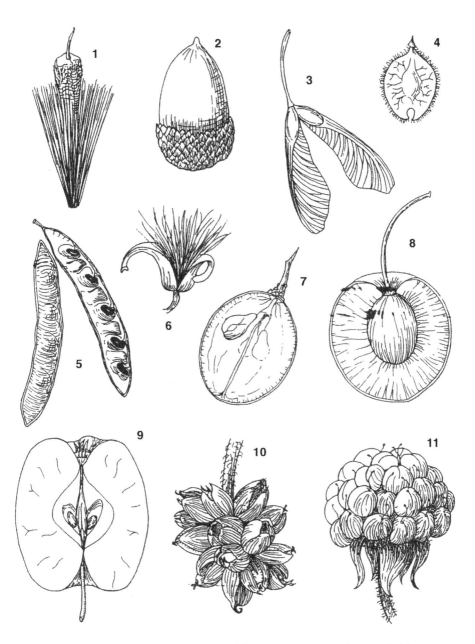

Fig. 17. Selected fruit types of woody plants. (1) achene of sycamore (*Platanus*); (2) acorn of oak (*Quercus*); (3) double samara of maple (*Acer*); (4) single samara of elm (*Ulmus*); (5) legume of black locust (*Robinia*), open; (6) capsule of willow (*Salix*); (7) berry of grape (*Vitis*), cut open; (8) drupe of cherry (*Prunus*), cut open; (9) pome of apple (*Malus*), cut open; (10) multiple drupes of mulberry (*Morus*); (11) aggregate of drupes of black raspberry (*Rubus*).

sity of Michigan Herbarium, and observations of the authors. A qualitative estimate of the abundance of each species is given based on a continuously grading five-part scale (Barnes and Wagner 2004).

Rare   Very seldom encountered, occurring singly or in small numbers wherever found, often absent in the expected habitats
Occasional Infrequent in the habitats and areas where it grows
Frequent   Usually found if one looks for it, but not commonly occurring
Common   Usually and easily found, widespread and reasonably plentiful throughout its range and expected habitats
Abundant Very common to always present in its expected range and habitats

For each species described, we provide a map (Voss and Reznicek 2012) showing its county of occurrence in each of the four regional landscape ecosystems of Michigan (Figure 19). The percentage of county occurrence for each region is given, as well as the total percentage for all counties of the state. The general North American range is briefly described. A discussion of the ecosystem classification and map, together with examples of several patterns of shrub distribution, are presented in the section "Ecology of Shrubs and Vines."

Site-Habitat

Site has a specific meaning, characterizing the geographic place where a plant or population of plants grows and the sum total of *physical site factors* available to and surrounding the plant(s) as modified by the biota. These ecosystem components thus control or strongly affect plant establishment, growth, reproduction, and death. The biota of associated plants and animals also plays a significant role, especially their mutualistic relationships and competition with a given plant or population. *Habitat* is a conveniently general and fuzzy term used to describe the location of plants or animals and various conditions, typically a "dealer's choice," associated with the plant or animal of interest. Habitat may include selected physical factors, biota, or both—for example, the environment of a plant or group of plants.

Of the physical ecosystem components, the availability of light, soil-water, nutrients, and oxygen is of particular importance together with disturbances such as fire, flooding, wind and ice storms, and insect or disease epidemics. After citing the major physical site factors that are important in the establishment, growth, and reproduction of each species, we identify the ecosystems, sites, or communities where we would likely find it in Michigan today. Shrubs and vines have spread widely into many favorable areas of human creation, such as old fields, roadsides, power lines, ditches, and fencerows, among others. We include them where appropriate. In addition, characteristic tree and shrub associates of each species are usually given.

Site relationships of shrubs and vines are very complex, and our descriptions are necessarily oversimplified, depending on factors such as macro- and microclimate, physiography (surface topography and parent material), soil, and water-table conditions—and in addition the biotic community of plants and animals with which a given shrub is associated. Furthermore, favorable conditions of one factor may compensate for less favorable conditions of one or more physical or biotic factors. Nevertheless for the keen observer, the specific geographic site and its physical ecosystem components are powerful and useful means of distinguishing species in the field. A consideration of specific site factors is presented in the section "Ecology of Shrubs and Vines."

Notes

Pertinent information about the natural history that was not presented in earlier sections is included in the notes. First, the section includes an estimate of the ability of individuals of the species to establish and maintain themselves in the shade of competing individuals such as in the understory of a forest. We use the arbitrary continuum of classes of shade tolerance (Barnes et al. 1998): very tolerant, tolerant, midtolerant, intolerant, and very intolerant. A discussion of the importance of light is presented in the section "Ecology of Shrubs and Vines." Often the origin of the generic name or specific epithet is given. Because of the multitude of gardening and landscaping books about shrubs, we do not include a section on ornamental use. However, the great importance of shrubs in horticulture and landscape design cannot be overemphasized. Compared to most trees, their superiority in many traits was quickly recognized: easily manipulated habit (e.g., for barriers, mazes, hedges, and ornamental design as in topiary art and bonsai); ease of propagation vegetatively or by seed; enormous genetic variation in habit, flowers, leaf morphology, and color; efficient use of space for fruit production; low maintenance for trained gardeners; ease of replacement; and lack of liability due to windstorm breakage or uprooting. Because of the extreme interest in the use of exotic shrubs for gardens of the British Isles and Central Europe, seventeenth-century (and earlier) plant explorers visited Asia, Europe, North America, and other parts of the world seeking new plants for horticulture (Fairchild 1947; Coats 1970, 1992). Shrubs were enormously popular, and genetic variants were discovered in nature by the thousands. Such cultivated shrub varieties, or "cultivars," are especially important in *Euonymus, Ilex, Rhododendron, Rosa, Spiraea, Syringa,* and *Viburnum,* among others.

For selected species medicinal uses are described, but a systematic consideration is beyond our scope. General and specific traditional medicinal uses by the Great Lakes Ojibwa peoples for counties of western upper Michigan, northern Wisconsin, and eastern Minnesota near Lakes Superior and Michigan are summarized by Meeker et al. (1993).

Chromosome Number

The chromosome numbers of species provide a window into their evolutionary history, their relationships with one another, and their biology and ecology. Over millions of years chromosome numbers of plants evolved from relatively low to higher numbers through processes of natural selection, adaptation, interspecific hybridization, and chromosome doubling. Many of our angiospermous species are of ancient or modern polyploid origin—as indicated by their chromosome numbers. For each species, where information is available, we give three chromosome numbers. Individuals of a species have two characteristic chromosome numbers. The key number in sexual reproduction is the number of chromosomes in each egg cell and pollen grain, a *set* that is termed the *gametic* or *haploid* number—n. At fertilization each individual inherits a set of matched or homologous chromosomes from the female and male parents. This union of two sets of chromosomes gives rise to individuals with the $2n$ or *diploid* number. Each plant carries in its *body* or *somatic cells* (in stems, roots, and leaves) two sets of homologous chromosomes, one inherited from each parent.

Also usually given is a third number, the baseline or *base chromosome number* for the genus, which is known as the x number. The base number has been established for each genus, usually the lowest gametic number for the genus. The base number provides the basis for understanding, in part, evolutionary relationships among species of a genus and family because not all species within a genus (especially angiosperms) have the same body cell ($2n$) chromosome number due to their evolutionary history. For example, the x number for the genus *Pinus* (pine) is 11, and all pines have the same gametic (n) number of 11 chromosomes and body cell number of 22 (the $2n$ number and *also* the $2x$ number). However, for the genus *Betula* ($x = 14$), the $2n$ number ranges from 28 (sweet birch, *B. lenta*), 56 (bog birch, *B. pumila*), and 84 (yellow birch, *B. alleghaniensis*) to 112, or eight times the base number in Murray birch (*B. murrayana*). Thus, to characterize and study the difference in the number of chromosome sets among species within a genus, we use the base or x number as the standard for comparison. Species whose individuals have more than two sets of the base number are termed *polyploids* (i.e., multiple ploids [sets] of the x number), and species whose individuals have only two sets of the base number are termed *diploids* in an evolutionary sense.

Therefore, using the established base or x number for a genus, we can determine the number of homologous chromosome sets, or the *ploidy* level, of a given species. Individuals of triploid species have three sets ($3x$); a tetraploid, four sets ($4x$), a hexaploid, six sets ($6x$), and an octoploid, eight sets ($8x$). For example, in the genus *Vaccinium* ($x = 12$) most species are diploid ($2n$ and $2x$ = 24), but the low sweet blueberry, *V. angustifolium,* and highbush blueberry, *V. corymbosum,* are tetraploids ($2n$ and $2x = 48$). For the genus *Salix,* the base number is 19, and of the species we describe individuals of five have body cells with more than two sets of 19 chromosomes. The five are tetraploids, $4x = 76$ (*Salix discolor, S. humilis, S. lucida, S. pedicellaris,* and *S. serissima*); for each the gametic (n) number is 38 and the body cell ($2n$) number is 76.

Polyploidy is of important evolutionary and ecological significance (Mosquin 1966; Soltis et al. 2009). Knowing the chromosome number of individuals of a species and the base number for the genus gives insights into its evolutionary development and ecological occurrence. Polyploidy is rare in conifers but relatively common in angiosperms, especially *Rubus, Salix,* and *Viburnum.* New species are often formed quickly, in an evolutionary sense, by hybridization of individuals of related species followed by doubling of chromosomes of the hybrid. For example, the shrub bog birch (*Betula pumila,* $2n$ and $4x = 56$) and the yellow birch tree (*B. alleghaniensis,* $2n$ and $6x = 84$) hybridized to form the Purpus' birch, (*B.* × *purpusii,* $2n$ and $5x = 70$)—a shrub or small tree. Gametic chromosome doubling produced individuals of *B. murrayana* ($2n$ and $8x = 112$ chromosomes), which was discovered in 1965 in Michigan (Barnes and Dancik 1985). Some polyploid species are known to grow faster or have larger leaves than others, but this is not true for all polyploids. Ecologically, polyploids are favored especially in relatively uniform environments of boreal, arctic, and disturbed or weedy habitats (Stebbins 1985; Mosquin 1966; Jackson 1976; Barnes et al. 1998).

Similar Species

Similar species are those that are nearly alike in habit, in many morphological characters, and often in the physical site conditions in which they occur. Species that are native in Michigan, the Great Lakes Region, the Northeast, and associated parts of Canada often resemble or are closely related to species that occur in the southern United States, western North America, eastern Asia, or Europe. Some of these species have worldwide distribution in northern and boreal regions (if boreal, they are termed circumboreal). Although they have disjunct ranges, many are nearly indistinguishable morphologically from our counterpart and have the same scientific name (e.g., *Vaccinium oxycoccos, V. ovalifolium, Kalmia polifolia, Symphoricarpos albus,* and *Rhododendron groenlandicum,* among others, of the Pacific Northwest and often Alaska; and *Sambucus racemosa of the* Pacific Northwest and Central Europe). Others have only minor morphological differences and have different scientific names (e.g., Eurasian *Rubus idaeus* and North American *Rubus strigosis*). A biogeographic perspective on the intercontinental distributions of similar species is presented in the section "Ecology of Shrubs and Vines."

Key Characters

Reliable field identification requires knowing what characters to use. In order of importance and depending on season of the year, these are site, shrub or vine habit, leaves, stems-twigs, fruit, and flowers. For each species we have listed key characters primarily for growing season identification. Characters for habit and leaves, which are typically present and distinctive, are almost always presented. Because key characters are never quite the same for each species, other characters that are most appropriate for each species, often stems-twigs, fruit, and site, are given. For some closely related species or those likely to be confused with one another, we have provided a brief paragraph with a set of specific distinguishing characters.

FALL COLORATION

All native woody plants undergo a distinctive color change of foliage in autumn. Northern temperate species closely monitor physiologically the changes in day length versus night length. Shorter days and longer nights are an early warning signal that "hard times" are ahead. Woody plants are very finely tuned to changes in day length, and photoperiodic races are well differentiated in northern latitudes where fall and winter freezing temperatures are lethal to living cells and plants. Color change is an outward sign that a plant is entering dormancy. The stages of dormancy, function of the abscission layer, and the coloration process were described in detail in *Michigan Trees* (Barnes and Wagner 2004). Here we describe briefly the pigments and their role in plant acclimation to seasonal climate change.

In summer the green shades of leaves are due to the presence and dominance of *chlorophyll*. Its two forms, termed A and B, occur in the plastids of leaf cells together with other pigments, *carotene* and *xanthophylls*, which provide the yellow and orange hues of autumn. As nights lengthen, woody plants in northern latitudes and high elevations (i.e., wherever short growing seasons occur) prepare for winter by entering early rest, the first stage of dormancy. With prolonged subfreezing temperatures, they enter true dormancy to survive winter. A shrub in true dormancy will not burst dormant buds (i.e., leaf out) if brought into the warm conditions of a greenhouse or home. The pigment producing the brilliant red and purple hues, *anthocyanin*, is actively produced where low and often freezing temperatures are common in autumn together with warm, sunny days. Produced in the vacuole of epidermal and mesophytic leaf cells, anthocyanin performs several useful functions that serve the continued life of the plant (Hoch et al. 2001; Milius 2002). The functions that prolong the life of leaves include:

- Protecting the leaf from high light intensity by acting as antioxidants to absorb free radicals, which can damage cell membranes, DNA, and the photosynthesis apparatus.
- Protecting the leaves from freeze damage, a biological "antifreeze," so that they are not shed prematurely.

When leaves are senescing in autumn, foliar nutrients, especially nitrogen, are reabsorbed (termed phloem loading) into the stems for use the next year. Anthocyanins thus prolong the life of the leaf to allow nutrients to be reabsorbed. Not surprisingly, most of the world's tree and shrub species with the reddest foliage occur in the northern freeze-prone regions of North America. Red color in the leaves of some woody plants is observed in the early spring, for anthocyanins are protecting the leaves from freezing and damaging their "expensive" photosynthetic apparatuses.

Striking fall color in shrubs, noted in the descriptions of leaves where appropriate, is seen in smooth, staghorn, and fragrant sumacs (brilliant orange-red); burning-bush and Virginia creeper (scarlet, red to reddish-purple), bearberry (bronze to reddish), poison-sumac (orange); species of arrow-woods (rose, pinkish, orange-red to maroon); and leatherwood, river-bank grape, spicebush, and witch-hazel (yellow). Once-popular shrub species from Asia and Europe, (e.g., honeysuckles, common and glossy buckthorn, and common privet, among others) retain green leaves long into autumn and even winter. These species are native in low latitudes that have long growing seasons with little danger of freezing temperatures. Thus they are not genetically adapted to monitor the increasing night length as a warning signal and typically retain live green leaves in fall until freezing temperatures kill the leaves. Many shrubs and trees planted in parks and gardens, having brought a part of their "home place" with them, are "oblivious" to the freezing temperatures to come. The added length of their growing seasons, compared to native species of the same genus and associated plants, likely gives them an added advantage to become naturalized and even detrimental to the regeneration of native plants.

HYBRIDS

Woody plants are fascinating because of their ability to produce hybrids, which are important not only in evolution but for occupying new habitats, as well as their importance and wide use in horticulture. Nevertheless, hybrids in many shrub groups complicate identification. Hybridization is the cross between individuals of populations having different adaptive gene complexes, that is, species, subspecies, and races. Hybridization is frequent in natural populations in zones of contact between species and where disturbance provides open, competition-free habitats. In these areas, parent species colonize vigorously and hybridize, and hybrid seedlings find suitable sites for establishment. Often such habitats are intermediate between those of the parent species and are termed hybrid habitats. In pre–European settlement times, hybrids were likely rare due to the rapid recolonization of sites following natural disturbances and the lack of hybrid habitats. Hybrids are usually weak individuals and soon would have been out-competed. The great number of reported hybrids during the twentieth century undoubtedly reflects the widespread human disturbance of ecosystems, thereby providing relatively competition-free sites for their establishment. Because of vigorous vegetative reproduction, shrubs increase greatly their spread, persistence, and potential for the creation of new species by means of polyploidy.

It has been popular to find and report natural hybrids, usually based on morphological characteristics. The excitement of a new discovery was perhaps based on its presumed rarity, or, as the botanist Herb Wagner (1968) related, seeking hybrids "was all part of the 'game,' and added to the thrill of the chase, like adding a new stamp to the collection."

Hybridization is especially common among angiosperms. Shrub genera notorious for hybridization in the Rosaceae are *Amelanchier, Crataegus, Rosa*, and *Rubus*. Other notable genera are *Aronia, Ribes, Salix, Spiraea*, and *Viburnum*. Some of the best-known hybrids among native shrubs is *Rhus × pulvinata* (smooth × staghorn sumac, *Rhus glabra × R. typhina*).

Hybrids may be sterile or fertile; they are usually weak, not vigorous. In most cases, they are intermediate in characteristics between the parent species—habit, twigs, buds, leaves, flowers, and fruit. Field identification may be difficult because of the great natural variation in characteristics of a "good" species—especially leaves—due to location on the plant or environmental modification. Identification can be further complicated because fertile or partially fertile hybrids may backcross to one of the parent species by means of introgressive hybridization, termed *introgression*. This process results in the infiltration of genetic material of one species into that of another. In such a case, the individual is not exactly intermediate in characteristics but may resemble closely—but not quite!—one of the parent species. Therefore, understanding the variation pattern of a given character, leaves for example, and the site conditions under which the individual is growing is of the utmost importance, especially in groups known to engage extensively in hybridization.

Hybridization is of major evolutionary significance, acting as an evolutionary catalyst. Approximately 70 percent of all flowering plants owe their existence to past natural hybridization between different species or genera (Arnold 1992). The evolutionary consequences of hybridization are discussed by Judd et al. (2008). Considering the quite different perspective of today, many shrub hybrids are important in horticulture and gardening due to their selection for resistance to disease, insects, or frost; unusual forms of foliage or color; and similar features of flowers.

INTRODUCTION TO THE
IDENTIFICATION KEYS

A series of keys for primary use in summer and fall is provided. In addition, for selected genera a key to species within each genus is provided in the introduction to these genera. A key is a useful method for identifying any unknown object under consideration. Our keys are intended to make it possible for any interested person to determine the commonly occurring shrubs and woody vines of Michigan. The keys will prove useful for identifying local and wide-ranging species throughout the Great Lakes Region, northeastern North America, and adjacent Canada. As this is a field guidebook, we have selected native genera and species that are most commonly encountered, together with important nonnative species. For comprehensive treatments of shrubs and vines, we recommend Voss and Reznicek 2012; Smith 2008; Gleason and Cronquist 1991; and Core and Ammons 1958. For cultivated plants, Dirr 1990; Flint 1997; and Cullina 2002 are excellent.

Keys are based on major differences and similarities of the distinctive parts of an organism. For shrubs and vines, these include size and form (i.e., habit), leaves, stems-twigs, flowers, and fruit. Our keys emphasize these plant features. The characteristic site (i.e., physical habitat) of a species is very often an excellent identification feature, obviously in distinguishing wetland versus dryland species but often in distinctive landform and soil-site conditions as well. With a little experience, and study of our section on the Site-Habitat of a given species, you will find that knowing where you are in the landscape is your best initial discriminator in narrowing identification choices.

The keys used here are modeled after the type used by Voss and Reznicek (2012). A key is characterized by paired statements called *couplets*. Each statement is called a *lead,* and the two leads of each couplet have the same number and the same indentation on the page. At each couplet, select the lead that best describes the plant and plant parts you are examining. In our keys, this statement may direct you to a genus name or a species name. If a name is not given, go to the next couplet and again carefully examine each lead. Follow this sequence until you reach a genus or species name and its corresponding page number(s). If you reach a species name (e.g., *Betula pumila*), the page number will bring you to the description of the species. If you reach a genus name, there are two possibilities.

1) If there is *one page number,* it will bring you to the genus description, and a similar key is provided for species of the genera. For example, if you reach *Cornus,* the page number directs you to the introduction to the genus *Cornus* and the key to the five species of *Cornus.* Using this key, determine the appropriate species of *Cornus* and the page number will direct you to its detailed description.

2) If you reach a genus name and there are two page numbers, turn directly to the species on these two pages. Then compare your plant's characteristics to those of both species using "Key Characters" and also to the characteristics listed in the paragraph just below "Key Characters," if there is one, which discusses *similar species in the genus.* If your plant does not key out, it may be a less common or introduced species, which is not described in this guidebook.

SPECIFIC KEYS

Keying requires keen observation of the whole plant in the field and the site where it grows; selection of typical stems, leaves, and other characters; accurate observation of multiple characteristics; and correct decisions at each step. Botanical homework, for those without appropriate background, is required to learn the basic morphological characteristics and terms. We suggest especially the sections of the text on *size and form, leaves, stems-twigs, winter buds, fruit,* and *habitat.* Leaves are notoriously variable, but they have a distinctive pattern of size and shape for each species. In selecting leaves avoid unusually small or large leaves and especially leaves from sprouts or those at the ends of very long shoots. Our keys emphasize knowing characteristics for field identification primarily in summer and fall when leaves are fully formed, stems and buds are mature, and fruits are often present. Flower and inflorescence characteristics are excellent ways to distinguish plant taxa at all levels, but we have only included a few. Their use requires being present in the narrow spring or summer window when they are available and that your target shrub or vine is in fact flowering. Shrubs and vines are typically sun-loving plants. In deep to moderate shade they may survive but often fail to flower regularly or entirely. We present the major flower characters for each species in its detailed description. For keys that are more strongly tied to floral characteristics, we recommend Voss and Reznicek 2012.

A good hand lens of circa 10 × is highly recommended, especially for studying a given species based on the details of the species description. A hand lens will enable you to see and distinguish among many small features of stems and buds, size and shape of leaf scars, stipules, number of bundle scars, and the size, shape, and color of buds.

We present five major keys based on characteristics of plant group (conifer vs. angiosperm—needle- or scalelike versus broad-blade leaves), leaf persistence (evergreen or deciduous), leaf type (simple or compound), and arrangement (opposite or whorled or alternate). All 138 species we describe are included in the five major keys. The first key includes three coniferous shrubs, whereas keys 2–5 include 135 species, which are all angiosperms (flowering plants). They are as follows.

Cone-Bearing Plants: Conifers
Key 1: Plants with Needlelike, Scalelike, or Awl-Shaped Leaves

Flower and Fruit-Bearing Plants: Angiosperms
Key 2: Broadleaf Evergreen Flowering Plants
Key 3: Broadleaf Deciduous Plants with Opposite or Whorled Leaves
Key 4: Broadleaf Deciduous Plants with Alternate, Compound Leaves
Key 5: Broadleaf Deciduous Plants with Alternate, Simple Leaves

Your choice among the five keys is the most important one.

- Choose Key 1 if your plant is a conifer with evergreen leaves (leaves persisting three years or more) of the narrow types (needlelike, linear, awl-shaped, or scalelike) and bears cones not flowers and fruit. There are few coniferous shrubs, so your choice is relatively straightforward once you recognize the shrub life form.
- Choose Key 2 if the leaves of your plant are retained on the stems for more than one year, are relatively broad (i.e., broader than needles or scales), are thick and leathery, and may have pubescence or scales on the lower surface.
- Choose Key 3 if the leaves are broad and leaf, twig, and branch arrangement is either opposite or whorled. The leaves may be either compound or simple.
- Choose Key 4 if the leaves are broad; leaf, twig, or arrangement is alternate; and leaf type is compound.
- Choose Key 5 if the leaves are broad; leaf, twig, and branch arrangement is alternate; and leaf type is simple.

QUICK KEYS FOR VINES AND SPECIES WITH ARMATURE

Two additional keys for flowering plants are presented. These include species that are embedded in keys 2–5 but have special features that are easily recognized: Key 6: Plants with Stems-Twigs Armed: Thorns, Spines, Prickles; and Key 7: Woody Vines: Plants Creeping, Trailing, or Climbing. If your target plant has one of these distinctive characteristics, keying will be much quicker than using keys 2–5.

KEY 1: PLANTS WITH NEEDLELIKE, SCALELIKE, OR AWL-SHAPED LEAVES

1. Low, straggly or bushy evergreen shrub, rarely to 1.5 m; leaves either all needlelike or awl-shaped.
    2. Leaves linear, spirally arranged but twisted at the base to appear opposite and 2-ranked; dark green above, yellow-green beneath with no white or yellowish stomatal lines . . . *Taxus canadensis*, p. 318
    2. Leaves awl-shaped in whorls of 3, jointed at base, loose or spreading; grayish or green, concave above with a broad white stomatal band . . . *Juniperus communis* var. *depressa*, p. 162
1. Decumbent shrub with long, trailing primary stems; leaves of two kinds, (1) mainly sessile, scalelike, closely appressed, overlapping, only on adult shoots, (2) awl-shaped, loosely ranged in whorls, only on juvenile shoots . . . *Juniperus horizontalis*, p. 164

KEY 2: BROADLEAF EVERGREEN FLOWERING PLANTS

1. Plants creeping, trailing, decumbent, or low (<50 cm).
    2. Leaves with wintergreen odor when crushed or chewed . . . *Gaultheria*, p. 146, p. 148
    2. Leaves without wintergreen odor.
        3. Leaves very small, 0.3–1.8 cm long; stems slender, threadlike; fruit borne on a pedicel with 2 small, paired bracts . . . *Vaccinium*, in part, p. 330, p. 336
        3. Leaves larger; stems not threadlike; fruit not borne as above.
            4. Leaves linear and revolute; very finely white-pubescent beneath.
                5. Leaves opposite; flowers rose-purple; fruit a subglobose capsule . . . *Kalmia polifolia*, p. 168
                5. Leaves alternate; flowers white to pinkish; fruit a capsule much depressed at apex . . . *Andromeda glaucophylla*, p. 70
            4. Leaves not linear or revolute; not white-pubescent beneath.
                6. Leaves small, 1–3 cm long, glabrous; petioles short, 2–4 mm long . . . *Arctostaphylos uva-ursi*, p. 74
                6. Leaves larger, 2–8 cm long, more or less bristly-hairy above and below; petioles to 3 cm long . . . *Epigaea repens*, p. 138
1. Plants erect (>50 cm).
    7. Leaves alternate, lower surface becoming brownish, coriaceous and scurfy or densely rusty-brown-wooly or brownish beneath.
        8. Leaves dull green, dotted with silvery scales above, brownish, coria-

ceous and scurfy beneath; not fragrant when crushed; twigs smooth and dark reddish-brown . . . *Chamaedaphne calyculata*, p. 94

    8. Leaves bright green and finely rugose, lacking silvery scales above, beneath white at first, later densely rusty-brown-wooly; fragrant when crushed; twigs brown-wooly . . . *Rhododendron groenlandicum*, p. 208

  7. Leaves in whorls of 3 or opposite; green and glabrous both sides . . . *Kalmia angustifolia*, p. 166

KEY 3: BROADLEAF DECIDUOUS PLANTS WITH OPPOSITE OR WHORLED LEAVES

1. Leaves compound.
   2. Erect or upright shrubs.
      3. Leaves pinnately compound, leaflets 5–11; flowers in erect clusters; fruit a small red or purple drupe . . . *Sambucus*, p. 288, p. 290
      3. Leaves trifoliate; flowers in drooping clusters; fruit an inflated, bladder-like, yellowish-brown capsule to 8 cm long . . . *Staphylea trifolia*, p. 312
   2. Vines with trailing, twining, or climbing stems.
      4. Slender vine climbing by leaf petioles which act as tendrils; leaves trifoliate; flowers small, white or bluish to pinkish purple, not trumpet-shaped; fruit a flattened achene, 3–5 mm long . . . *Clematis virginiana*, p. 96
      4. Large, vigorous vine climbing by aerial roots; leaves pinnately compound, leaflets 7–13; flowers large and showy, reddish-orange, trumpet-shaped; fruit a thick capsule, 10–20 cm long . . . *Campsis radicans*, p. 82
1. Leaves simple.
   5. Stems climbing, creeping, or low erect or semierect straggling shrub or vine to 1.5 m.
      6. Leaves evergreen, sessile, lower surface conspicuously white with dense pubescence . . . *Kalmia polifolia*, p. 168
      6. Leaves not evergreen or as above.
         7. Stems creeping, semierect, or trailing, climbing vine.
            8. Small creeping shrub, a few upright branches to 30–40 cm; stems green, 4-sided; petiole or leaf scars not meeting around twig; bundle scar 1; fruit a capsule, pink to crimson, each seed covered by a scarlet or orange aril . . . *Euonymus obovatus*, p. 142
            8. Semierect, usually a trailing or twining vine, 1–3 m long; stems gray or straw-colored; petiole or leaf scars meeting around the stem; bundle scars 3; fruit an orange-red berry . . . *Lonicera dioica*, p. 176
         7. Stems erect or semierect, not a climbing vine.
            9. Stems with exfoliating papery bark; leaves dotted with translu-

cent glands; fruit a dark brown, nonfleshy capsule . . . *Hypericum*, p. 154, p. 156

9. Stem bark not exfoliating; leaves without translucent glands; fruit a bright white, waxy, berrylike drupe or a red to purplish berry.

10. Leaves with petioles about 5 mm long; stems with pith continuous between nodes; fruit a red to purplish berry . . . *Lonicera canadensis*, p. 174

10. Leaves sessile or nearly so; stems hollow between nodes; fruit a bright white, waxy berrylike drupe . . . *Symphoricarpos albus*, p. 314

5. Stems erect, upright shrub or small tree >1.5 m.

11. Leaves with stipules, petioles with glands near the junction or lower blade surface covered with red to brown resin dots; flowers large, white, in broad terminal clusters . . . *Viburnum*, p. 340

11. Leaves without stipules, petioles lacking glands and leaves lacking brown resin dots; flowers not as above.

12. Leaves lobed.

13. Leaves longer than wide, margins coarsely and irregularly single-toothed; blade tips acute to short-acuminate; lower surface downy white pubescent; bark reddish to grayish-brown; flowers borne in erect clusters; samara wings forming an angle less than 90° (not strongly divergent) . . . *Acer spicatum*, p. 56

13. Leaves as wide as or wider than long, margins doubly serrate; blade tips long-acuminate; bark greenish-brown, marked by conspicuous white vertical stripes; flowers borne in drooping clusters; samara wings forming an angle of 90° (strongly divergent) . . . *Acer pensylvanicum*, p. 54

12. Leaves not lobed.

14. Leaves entire or nearly so, a few leaves irregularly toothed or wavy margined.

15. Twigs often tipped with a short, stout thorn as long as the end lateral bud; some leaves subopposite or occasionally alternate . . . *Rhamnus cathartica*, p. 206

15. Twigs without thorns; leaves opposite or whorled.

16. Leaves dotted with translucent glands; stems 2-angled with winglike ridges below the nodes . . . *Hypericum prolificum*, p. 156

16. Leaves without translucent glands; stems not 2-angled.

17. Leaves mostly or all in whorls of 3 or 4 at a node.

18. Low shrub, <2 m; leaves evergreen; apex blunt; stems without prominent lenticels; flowers borne in drooping clusters; fruit a capsule, 3–5 mm across . . . *Kalmia angustifolia*, p. 166

18. Tall shrub, 2–5 m; leaves deciduous; apex acute to acum-

inate; stems with prominent lenticels; flowers borne in dense, globose heads; fruit an achene borne in an aggregate of many achenes in a dense head . . . *Cephalanthus occidentalis*, p. 92

17. Leaves opposite, 2 at a node.
    19. Leaves mostly less than 4 cm long.
        20. Petiole bases of opposite leaves meeting around twig or connected by a definite line or ridge; pith hollow between nodes.
            21. Bundle scars 3; fruit scarlet or red . . . *Lonicera*, in part, p. 176, p. 178, p. 180
            21. Bundle scar 1; fruit white . . . *Symphoricarpos albus*, p. 314
        20. Petiole bases of opposite leaves not meeting and not connected; pith continuous . . . *Ligustrum vulgare*, p. 170
    19. Leaves mostly more than 4 cm long.
        22. Leaves with lateral veins strongly and evenly curving upward toward the tip as they near the margin (arcuate venation) . . . *Cornus*, p. 100
        22. Leaves with lateral veins not as above.
            23. Lateral buds concealed, very small, somewhat raised, dome-shaped, and appearing sunken in bark; stems moderately stout with prominent, vertical, raised lenticels . . . *Cephalanthus occidentalis*, p. 92
            23. Lateral buds clearly evident; stems without prominent raised lenticels.
      24. Leaves heart-shaped (cordate-ovate), especially at base, and tapering to a slender point . . . *Syringa vulgaris*, p. 316
      24. Leaves not heart-shaped.
                25. Bud scale remnants persistent at the base of the current year's shoot . . . *Lonicera*, in part, p. 174
                25. Bud scales or remnants not persisting.
                    26. Winter buds valvate with 2 scales, scurfy; lower surface of leaves dotted with dark reddish glands or often scurfy; petiole bases or leaf scars at each node meeting around the stem or connected by a transverse ridge . . . *Viburnum*, in part, p. 340
                    26. Bud scales imbricate; lower surface of leaves glabrous, not as above; petiole bases and leaf scars not meeting

around the stem . . . *Ligustrum vulgare,* p. 170

14. Leaves distinctly toothed to finely serrate or crenate-serrate.

27. Leaves coarsely dentate; lower leaf surface and current shoots densely pubescent . . . *Viburnum,* p. 340

27. Leaves not coarsely dentate; leaves and shoots glabrous or nearly so.

28. Twigs often tipped with a short, stout thorn as long as the end lateral bud; some leaves subopposite or occasionally alternate . . . *Rhamnus cathartica,* p. 206

28. Twigs not armed.

29. Stems and branchlets greenish, 4-angled; fruit a pinkish-red fleshy capsule . . . *Euonymus atropurpureus,* p. 140

29. Stems and branchlets grayish to yellowish-brown; fruit a greenish-brown capsule with a long, slender beak . . . *Diervilla lonicera,* p. 132

KEY 4: BROADLEAF DECIDUOUS PLANTS WITH ALTERNATE, COMPOUND LEAVES

1. Stems armed with prickles or spines.

2. Leaves bipinnately compound; low perennial shrub dying back in winter to the ground or a woody base; fruit a purplish-black drupe . . . *Aralia hispida,* p. 72

2. Leaves palmately or pinnately compound, leaflets 3–5.

3. Stems with paired stipular spines at nodes; tall shrub or small tree; leaves aromatic when crushed, margin of leaflets entire to crenulate . . . *Zanthoxylum americanum,* p. 358

3. Stems with prickles, not paired spines at nodes; low or medium shrub; leaves not aromatic; margin of leaflets toothed.

4. Stems (canes) biennial, flowering occurring only on floricanes produced the 2nd year from primocanes; flowers small, white; fruit a drupe, borne as an aggregate of drupelets on a receptacle . . . *Rubus,* p. 240

4. Stems not biennial; fruit a red hip enclosing an aggregate of achenes . . . *Rosa,* p. 230

1. Stems unarmed, lacking prickles or spines.

5. Stems climbing, trailing, creeping, or low erect or semierect straggling shrub to 1.5 m high.

6. Leaves trifoliate; tendrils lacking; fruit a white drupe . . . *Toxicodendron radicans,* p. 320

6. Leaves palmately compound with 5 leaflets; tendrils present; fruit a dark blue or bluish-blackberry . . . *Parthenocissus,* p. 186, p. 188, p. 190

5. Stems erect, upright shrub or small tree, >1.5 m; not vinelike.

7. Leaflets 3 or 5 –7, palmately or pinnately compound.

8. Medium shrub to 2 m, clone forming; fruit a drupe or head of achenes.
   9. All parts aromatic; leaflets broad, 3-lobed, coarsely toothed above the middle with rounded or abruptly pointed teeth, green and glabrate; fruit a reddish, hairy drupe . . . *Rhus aromatica*, p. 212
   9. Parts not aromatic; leaflets narrow, crowded, mostly 5, entire, pale and silky-pubescent below; fruit an achene borne in a compact head . . . *Dasiphora fruticosa*, p. 130
8. Tall shrub or small tree, 3–5 m; fruit a samara with a conspicuous membranous wing surrounding the seed cavity . . . *Ptelea trifoliata*, p. 200
7. Leaflets 6–31, pinnately compound.
   10. Margins of leaves not sharply toothed, entire, or blunt to wavy-toothed.
      11. Medium shrub to 2 m; leaf rachis pubescent, more or less winged between each pair of leaflets; stems downy-pubescent; fruit a dark red drupe . . . *Rhus copallina*, p. 214
      11. Tall shrub or small tree to 6 m; leaf rachis glabrous or nearly so, not winged; stems glabrous; fruit a white drupe . . . *Toxicodendron vernix*, p. 322
   10. Margins of leaves sharply toothed.
      12. Adult twigs glabrous; terminal bud present, large; leaf scars crescent-shaped; bundle scars 5; fruit a bright orange or red glabrous pome . . . *Sorbus*, p. 302, p. 304
      12. Adult twigs glabrous or densely pubescent; terminal bud absent; leaf scars large, horseshoe-shaped, nearly enclosing the bud; bundle scars many; fruit a drupe covered with dense hairs . . . *Rhus*, in part, p. 210, p. 216, p. 218

KEY 5: BROADLEAF DECIDUOUS PLANTS WITH ALTERNATE, SIMPLE LEAVES

1. Leaves deeply or shallowly cut or lobed, the segments or lobes blunt, rounded, or sharp-pointed.
   2. Trailing, scrambling, or climbing vines.
      3. Tendrils present; leaf margins toothed.
         4. Tendrils with each branch ending in a large adhesive disk, basal leaves trifoliate, a cultivated nonnative ornamental of urban areas and built environments . . . *Parthenocissus tricuspidata*, p. 190
         4. Tendrils lacking adhesive disks, all leaves simple, native species of diverse natural habitats . . . *Vitis*, p. 354, p. 356
      3. Tendrils lacking; leaf margins entire.
         5. Leaves orbicular or broadly ovate, about as wide as long; petioles

long, attached to lower surface of blade near the base; flowers creamy to greenish-white; fruit a bluish-black drupe . . . *Menispermum canadense*, p. 182

     5. Leaves elliptic, about half as wide as long; petiole short, attached at base of blade; flowers purple; fruit a translucent red berry . . . *Solanum dulcamara*, p. 298

2. Erect shrubs or small trees.

   6. Leaves palmately lobed.

     7. Stems unarmed.

       8. Leaves large, 10–20 cm across . . . *Rubus*, in part, p. 254

       8. Leaves smaller, 3–8 cm long and wide.

         9. Low shrub (<1.5 m); stems few branched, not conspicuously peeling or exfoliating; fruit a many-seeded berry . . . *Ribes americanum*, p. 222

         9. Medium to tall shrub, up to 3 m; stems much branched, conspicuously peeling or exfoliating in papery layers; fruit a light brown follicle borne in an aggregate . . . *Physocarpus opulifolius*, p. 192

     7. Stems armed with prickles, spines, or thorns.

      10. Low, creeping or erect shrubs, <1.5 m high; stems with bristles, prickles, or nodal spines; floral tube campanulate or saucer-shaped; fruit a many-seeded berry . . . *Ribes,* in part, p. 220

      10. Tall shrub or small tree to 10 m; stems with stout straight or curved thorns; flowers lacking a floral tube; fruit a dark or bright red or orange fleshy pome . . . *Crataegus,* in part, p. 116

   6. Leaves shallowly or deeply pinnately lobed.

     11. Leaves deeply pinnately lobed, fernlike; all parts fragrant; fruit an achene . . . *Comptonia peregrina*, p. 98

     11. Leaves shallowly lobed; not fragrant; leaf margins with 4–8 rounded lobes or teeth; fruit a large nut (acorn) . . . *Quercus prinoides*, p. 202

1. Leaves not lobed.

   12. Leaves silvery beneath with scattered rust-brown, peltate scales . . . *Elaeagnus umbellata*, p. 136

   12. Leaves lacking silvery scales beneath.

     13. Margin of leaves entire or slightly wavy.

       14. Plants with prickles, spines, or thorns.

         15. Vines with bristly or prickly stems; leaves green, without scales; native species . . . *Smilax*, p. 294, p. 296

         15. Low shrub, 0.5 to 1.5 m; spines borne singly at a node; leaves green, lacking scales; wood bright yellow; fruit a drooping, bright red berry, persistent into winter . . . *Berberis thunbergii*, p. 78

       14. Shrubs without prickles, spines, or thorns; leaves various.

         16. Leaves with arcuate venation, lateral veins strongly and

evenly curving upward toward the tip as they near the margin . . . *Cornus alternifolia*, p. 102

16. Leaves lacking arcuate venation.
    17. Leaves with golden-yellow resinous dots on both sides, especially lower surface; fruit a black, edible drupe . . . *Gaylussacia baccata*, p. 150
    17. Leaves lacking golden-yellow resinous dots; fruit various.
        18. Plants with spicy-aromatic parts; twigs, leaves, and bright red drupes with citronellalike odor when crushed or bruised . . . *Lindera benzoin*, p. 172
        18. Plants lacking aromatic odor.
            19. Stems with tough, flexible, leathery bark and soft wood; stem nodes conspicuously swollen; buds covered by base of petiole . . . *Dirca palustris*, p. 134
            19. Stems, nodes, and buds not as above.
                20. Buds covered with a single scale; plants unisexual, flowers borne in catkins, the male and female on separate plants . . . *Salix*, in part, p. 258
                20. Buds naked, valvate, or with two or more scales; flowers bisexual, flowers solitary or in few- to many-flowered, terminal or axillary clusters, borne on the same plant.
                    21. Stems green or green becoming reddish; predominantly low or creeping shrubs, (<1.5 m); fruit a blue to blue-black edible berry with many small seeds; flowers cylindric to bell-shaped . . . *Vaccinium*, in part, p. 324
                    21. Stems not as above; tall shrubs or small trees; fruit a drupe; flowers not as above.
                        22. Terminal bud not naked, 2 scales exposed; shoots and small twigs purplish; leaves elliptic-oblong, often with nearly parallel sides, apex acute or abruptly blunt with mucronate tip; drupe bright pink to purplish-red; native species . . . *Ilex mucronata*, p. 158
                        22. Terminal bud naked; shoots and small twigs not purplish; leaves oblong to obovate, apex acute not

mucronate; drupe red, turning dark purple or black; nonnative species invading wetlands and adjacent uplands . . . *Frangula alnus*, p. 144

13. Margins of leaves toothed or distinctly wavy.
    23. Vine with tough twining stems, low- or high-climbing; fruits orange, splitting and displaying scarlet arils covering the seeds . . . *Celastrus scandens*, p. 88
    23. Shrubs with creeping, trailing, or ascending stems; fruits not as above.
        24. Shrubs creeping or decumbent with ascending branches; buds covered with a single scale . . . *Salix*, in part, p. 258
        24. Erect shrubs or small trees; buds with more than one scale.
            25. Stems armed with spines or thorns.
                26. Leaves with wavy margins, silvery beneath with scattered light brown scales; fruit a drupelike achene; a naturalized Eurasian species that has spread into forests of lower Michigan . . . *Elaeagnus umbellata*, p. 136
                26. Leaves with finely to coarsely serrate, toothed, or lobed; leaves green, lacking brown scales; native species.
                    27. Plants armed with thorns borne singly along stems; margins of leaf blades finely to coarsely serrate, toothed, or lobed; stems lacking bitter almond odor or taste; fruit a pome . . . *Crataegus*, in part, p. 116
                    27. Plants with thorns not borne singly along stems; margins of leaf blades finely serrate; stems with bitter almond odor and taste when crushed; fruit a juicy drupe . . . *Prunus americana*, p. 194
            25. Stems unarmed.
                28. Buds naked; leaf base strongly asymmetrical; flowering in autumn with conspicuous bright yellow petals . . . *Hamamelis virginiana*, p. 152
                28. Buds not naked; leaf base not asymmetrical at base; not flowering in autumn.
                    29. Buds covered by a single scale; flowers borne in nonwoody catkins; fruit a capsule . . . *Salix*, in part, p. 258
                    29. Buds with 2 or more scales; flowers solitary or in various types of inflorescences if in catkins; fruit a nut or nutlet.

30. Foliage aromatic when crushed; leaves tapered toward base, slightly toothed toward apex, both surfaces dotted with lustrous, yellow resin glands; fruit a nutlet borne in dry resinous catkins . . . *Myrica gale*, p. 184

30. Foliage not aromatic; leaves and fruit not as above.

    31. Leaves with 3 prominent veins from the base.

        32. Two veins on either side of the midrib strongly and evenly curving upward toward the tip as they near the margin (arcuate); leaf margin serrate entire length; bundle scar 1; fruit a brown capsule persisting into winter . . . *Ceanothus*, p. 84, p. 86

        32. Two veins on either side of the midrib not as above (not arcuate); leaf margin entire or shallowly toothed above the middle; bundle scars 3; fruit a plump drupe . . . *Celtis tenuifolia*, p. 90

    31. Leaves with one main vein, midrib and multiple lateral branches.

33. Leaves elliptic; leaf margins with 4–8 rounded teeth; bundle scars numerous; fruit a large nut . . . *Quercus prinoides*, p. 202

33. Leaves various shapes; leaf, bundle scars, and fruit not as above.

    34. Leaf margins entire or shallowly toothed or wavy-undulate.

        35. Leaf margins entire or irregularly shallowly toothed above the middle; base rounded or moderately inequilateral; fruit a plump drupe with smooth surface when mature; pith chambered . . . *Celtis tenuifolia*, p. 90

        35. Leaf margins with 4–8 rounded teeth; base cuneate; fruit a large nut; pith continuous . . . *Quercus prinoides*, p. 202

    34. Leaf margins serrate.

        36. Leaf margins doubly serrate.

            37. Wide-spreading shrub or small tree to 8 m; lateral and end buds distinctly stalked; pith white, triangular in cross section; fruit a small nut, borne in drooping, woody, conelike catkins less than 2.5 cm long, persistent into the next growing season . . . *Alnus*, p. 58, p. 60

            37. Bushy shrub to 3 m with spreading branches; buds not stalked; pith tan to brown, not triangular in cross section; fruit a larger

nut to 4 cm long, enclosed by 2 leafy-bracted clusters or by a tubular involucre of united bracts . . . *Corylus,* p. 112, p. 114

36. Leaf margins not doubly serrate.

    38. Glands present on leaf blades or petioles.

        39. Leaves with small, dark, hairlike glands on the midrib of upper surface of leaf; stems lacking bitter almond odor and taste when crushed; lenticels not prominent; fruit a pome . . . *Aronia prunifolia,* p. 76

        39. Leaves with glands on petioles or at the blade base near junction with the petiole; stems with bitter almond odor and taste when crushed; lenticels prominent; fruit a drupe.

            40. Tall shrub or small tree to 10 m; leaves finely and sharply serrate from base to apex . . . *Prunus virginiana,* p. 198

            40. Low shrub, 0.5–1.5 m; erect or decumbent; leaves remotely and glandular-serrate above the middle . . . *Prunus pumila,* p. 196

    38. Glands lacking on leaf blades and petioles.

        41. Leaves sessile or borne on very short petioles, <0.5 cm.

            42. Stems green or green becoming reddish, glabrous; flowers bell-shaped, solitary or in few-flowered clusters; fruit a many-seeded berry . . . *Vaccinium,* in part, p. 324

            42. Stems yellowish-brown to reddish-brown, with or without pubescence; flowers minute, roselike, in many-flowered erect terminal panicles or umbellike racemes; fruit a persistent capsule when present . . . *Spiraea,* p. 306, p. 308, p. 310

        41. Leaves with longer petioles, 0.5–3cm.

            43. Low, few-branched shrub 0.3–0.6 m, occasionally spreading; leaves with blade margin crenate-serrate; stems slender, ridged below the nodes; flowers greenish-yellow, beginning in May; fruit a black subglobose drupe . . . *Rhamnus alnifolia,* p. 204

            43. Low to tall shrub or small tree, 1–10 m; leaf margins, stems, flowers, and fruit not as above.

                44. Low to tall shrub or small tree; leaf margins finely serrate to coarsely dentate; stems without ridges; flowers white beginning March–April; flowers bisexual; fruit a dark purple to blackish pome . . . *Amelanchier,* in part, p. 62

                44. Tall shrub or small tree to 5 m; leaf margins sharply serrate with short bristle tips that point upward perpendicular to blade surface; flowers usually unisexual; fruit a bright red drupe . . . *Ilex verticillata,* p. 160

KEY 6: PLANTS WITH STEMS-TWIGS ARMED: THORNS, SPINES, PRICKLES

1. Vine.
    2. Armed only on outer bud scales with long-pointed spinelike tips . . . *Celastrus scandens*, p. 88
    2. Buds unarmed; stems armed with bristlelike to stout prickles . . . *Smilax*, p. 294, p. 296
1. Shrub or small tree.
    3. Stems with one or more spines at a node.
        4. Stems with stout spines in pairs at nodes; leaves opposite; pinnately compound, aromatic when crushed . . . *Zanthoxylum americanum*, p. 358
        4. Stems with straight, needlelike spine(s) at nodes; leaves alternate, simple, not aromatic.
            5. Leaves small, blades 1–3 cm long, entire, not lobed . . . *Berberis thunbergii*, p. 78
            5. Leaves larger, blades 2–10 cm long, lobed, margins toothed . . . *Ribes*, p. 220
    3. Stems without spines at nodes.
        6. Leaves opposite or subopposite; stems with small, needlelike thorns only at ends of leaf-bearing twigs . . . *Rhamnus cathartica*, p. 206
        6. Leaves alternate.
            7. Leaves bipinnately compound, low perennial, dying back in winter to the ground or a woody base . . . *Aralia hispida*, p. 72
            7. Leaves not bipinnately compound, not dying back in winter.
                8. Plants with thorns at ends of leaf-bearing twigs and stout thorns (to 6 cm) along branches or twigs . . . *Prunus americana*, p. 194
                8. Plants with thorns or prickles occurring singly along branches and twigs.
                    9. Tall shrub or small tree; leaves simple; long, stout thorns on branches and twigs . . . *Crataegus*, p. 116
                    9. Low or medium shrub; leaves compound; lacking long, stout thorns.
                        10. Prickles stout; bristles and spines absent; stipules conspicuous, attached to petiole for half their length or more; fruit a red hip enclosing an aggregate of achenes . . . *Rosa*, p. 230
                        10. Prickles not stout (except in *Rubus alleghaniensis*); bristles and prickles present; stipules attached to petiole only at base; fruit an aggregate of drupes . . . *Rubus*, p. 240

KEY 7: WOODY VINES: PLANTS CREEPING, TRAILING, OR CLIMBING

1. Plants climbing by means of aerial roots.
    2. Leaves opposite, leaves pinnately compound . . . *Campsis radicans*, p. 82
    2. Leaves alternate, trifoliate or palmately compound.
        3. Leaves trifoliate . . . *Toxicodendron radicans*, p. 320
        3. Leaves palmately compound . . . *Parthenocissus quinquefolia*, p. 188
1. Plants climbing by other means.
    4. Plants climbing by twining petioles; leaves opposite, compound . . . *Clematis virginiana* p. 96
    4. Plants climbing by twining stems or tendrils; leaves alternate or opposite.
        5. Plants climbing by twining stems.
            6. Leaves opposite, leaves below inflorescences and infructescences united around the stem . . . *Lonicera dioica*, p. 176
            6. Leaves alternate, not united around the stem.
                7. Leaves finely serrate or crenulate . . . *Celastrus scandens*, p. 88
                7. Leaves mostly entire or toothed.
                    8. Plants woody only near the base; leaf blades of the upper stem with small basal lobes; fruit a red berry . . . *Solanum dulcamara*, p. 298
                    8. Plants woody; leaves without basal lobes; fruit a bluish-black drupe . . . *Menispermum canadense*, p. 182
        5. Plants climbing by tendrils.
            9. Tendrils attached to leaf petioles; stems green and usually prickly . . . *Smilax*, p. 294, p. 296
            9. Tendrils attached to stems opposite the leaves; stems not prickly.
                10. Leaves simple . . . *Vitis*, p. 354, p. 356
                10. Leaves compound.
                    11. Leaves glossy green above, not glaucous beneath; climbing by twining stems and tendrils, lacking aerial roots; tendrils not developing adhesive disks . . . *Parthenocissus inserta*, p. 186
                    11. Leaves dull green above, glaucous beneath; climbing by coarse aerial roots; tendrils developing adhesive disks . . . *Parthenocissus quinquefolia*, p. 188

SHRUBS AND VINES DESCRIPTIONS

Acer
Alnus
Amelanchier
Andromeda
Aralia
Arctostaphylos
Aronia
Berberis
Betula
Campsis
Ceanothus
Celastrus
Celtis
Cephalanthus
Chamaedaphne
Clematis
Comptonia
Cornus
Corylus
Crataegus
Dasiphora
Diervilla
Dirca
Elaeagnus
Epigaea
Euonymus
Frangula
Gaultheria
Gaylussacia
Hamamelis
Hypericum
Ilex
Juniperus

Kalmia
Ligustrum
Lindera
Lonicera
Menispermum
Myrica
Parthenocissus
Physocarpus
Prunus
Ptelea
Quercus
Rhamnus
Rhododendron
Rhus
Ribes
Rosa
Rubus
Salix
Sambucus
Shepherdia
Smilax
Solanum
Sorbus
Spiraea
Staphylea
Symphoricarpos
Syringa
Taxus
Toxicodendron
Vaccinium
Viburnum
Vitis
Zanthoxylum

SAPINDACEAE
Acer pensylvanicum Linnaeus
Striped Maple, Moosewood

Size and Form. Tall shrub or small tree, 6–12 m (20–40 ft) high and 12–24 cm (5–10 in) in diameter. Striped, upright branches form a broad, uneven, rounded crown. Roots shallow and wide spreading. Michigan Big Tree: girth 112 cm (44 in), diameter 36 cm (14 in), height 18.0 m (59 ft), Marquette Co.

Bark. Thin, smooth, greenish when young; marked vertically by conspicuous greenish-white stripes; becoming brownish-green with darkened stripes.

Leaves. Opposite, simple, large, blades 12–18 cm long and nearly as wide, occasionally wider than long, shallowly 3-lobed above the middle with long, fine tips pointing forward, the end lobe broadly triangular, sinuses rounded at base; early leaves often without lobes; base rounded or heart-shaped; uniformly doubly serrate, teeth forward; yellowish-green above, paler beneath, turning pale to golden yellow in autumn; palmately 3-nerved; glabrous; petioles stout, grooved, 2–8 cm long.

Stems-Twigs. Moderately stout, light green, with small to wide whitish stripes developing the second season, becoming reddish with whitish stripes; leaf scars crescent-shaped; bundle scars 3; pith white.

Winter Buds. Bright red, conspicuously stalked; terminal bud 0.8–1 cm long, about twice as long as wide; lateral buds much smaller, appressed; 2 scales visible, valvate, keeled.

Flowers. May–June, when the leaves are nearly full grown; functionally unisexual; plants monoecious or dioecious; large, bright yellow, bell-shaped, in slender, drooping terminal racemes 10–15 cm long; calyx 5-parted; petals 5; stamens 7–8; ovary downy. Sex expression is reported to change from year to year—bearing female flowers one year and male flowers the next (Hibbs and Fischer 1979). Insect pollinated.

Fruit. Samara; late July–August; dark reddish-brown; glabrous, wings 2.5–3 cm long, widely divergent (90°); seed cavity indented on one side; borne in long, terminal clusters; seeds 11,100/lb (24,420/kg).

Distribution. Occasional to common in eastern upper Michigan (and west in counties bordering Lake Superior to Keweenaw Co.) and the northern part of northern lower Michigan (south to Benzie and Alcona Co.), common and locally abundant in lake-moderated ecosystems of Districts 11 and 12 (Fig. 19) in northern lower Michigan. County occurrence (%) by ecosystem region (Fig. 19): I, 3; II, 33; III, 71; IV, 38. Entire state 23%. Widely distributed in e. USA and adjacent Canada, NS, s. QC, ON to n. MN, s. to WI, MI, NY, PA, and in the mts. to NC, TN to n. GA.

Site-Habitat. Characteristic of the understory of mesic deciduous forests; associates include sugar and red maples, beech, basswood, yellow birch, northern red oak, maple-leaved viburnum, Canadian fly honeysuckle, leatherwood. May also occur with white pine, pin cherry, paper birch, and aspens on mesic to dry-mesic disturbed sites. Requires cool, moist conditions for establishment; exhibits low nutrient requirements.

Notes. Shade-tolerant; slow-growing; short-lived, although individuals are known to occupy forest openings for more than 100 years (Hibbs 1979). Reproductive biology is described in detail by Sullivan (1983). Germination epigeal. Layering is common but mainly enables plants to survive in shaded understories. An attractive understory shrub or small tree, especially in autumn because of its large, clear-yellow leaves. A favorite food for moose and deer. Has potential as an ornamental in cool, moist, shady places.

Chromosome No. $2n = 26$, $n = 13$; $x = 13$

Similar Species. In mountainous areas of North Korea, northeastern China, and Russia, the counterpart is *Acer tegmentosum* Maxim. It exhibits similar yellow fall leaf color. The closely allied species in Japan are *A. capillipes* Maxim. and *A. rufinerve* Siebold & Zucc. Both are distinctively white-striped on reddish or green bark, but, in contrast to *A. pensylvanicum* and *A. tegmentosum*, leaves of both turn reddish to crimson, occasionally orange.

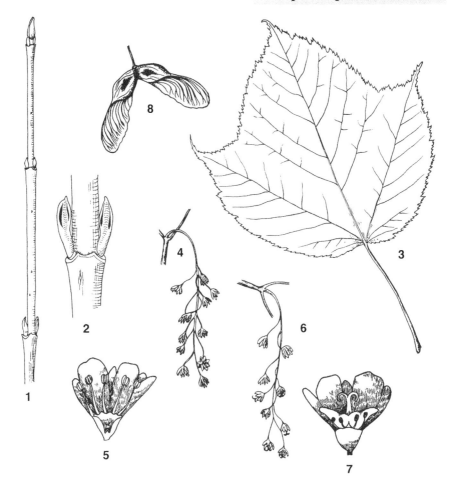

1. Winter twig, × 1/2
2. Portion of twig, × 1 1/2
3. Leaf, × 1/2
4. Male flowering raceme, × 1/2
5. Vertical section of male flower, × 1/2
6. Female flowering raceme
7. Vertical section of female flower
8. Fruit, samaras, × 3/4

KEY CHARACTERS

- large understory shrub or small tree
- leaves opposite; blades large, 3-lobed above the middle, lobes short with long tapering; tips doubly serrate, turning yellow in autumn
- bark green when young with conspicuous, vertical whitish stripes, becoming reddish
- winter buds bright red, conspicuously stalked; 2 visible scales, valvate
- samaras with widely diverging wings; seed cavities indented on one side

Distinguished from mountain maple, *Acer spicatum*, by leaves with finer, uniform, and doubly serrate teeth and glabrous lower surface; autumn color yellow; samara wings widely divergent; young stems with conspicuous, vertical, whitish stripes; less pubescent to glabrous in all its parts; tolerates a much wider range of site conditions and therefore is more common in northern forests than mountain maple.

55

SAPINDACEAE
Acer spicatum Lamarck

Mountain Maple

Size and Form. Tall bushy shrub or rarely a small tree, 6–9 m (20–30 ft) high and 8–18 cm (3–8 in) in diameter. Trunk short, crooked, giving rise to several small, upright branches, which form a small, irregularly rounded crown. Root system very shallow. Michigan Big Tree: girth 84 cm (33 in), diameter 27 cm (11 in), height 17.7 m (58 ft), Houghton Co.

Bark. Thin, dull, reddish to grayish-brown, smooth or slightly furrowed.

Leaves. Opposite, simple, blade 10–13 cm long and 2/3 as wide; palmately 3-lobed above the middle, often with 2 small lobes near the base, the central lobe triangular; the lobes singly, coarsely, and irregularly crenate-serrate with pointed teeth, the teeth 2–3 per cm, each tipped with a minute sharp gland, the sinuses usually wide-angled and acute at the base; tips acute to short-acuminate; base subcordate; early leaves often distinctly 5-lobed; thin; dark green and glabrous above, covered with a whitish down beneath; veins prominent; turning orange to scarlet in autumn; petioles long, slender, with enlarged base.

Stems-Twigs. Slender, yellowish-green to reddish, slightly hairy or velvety with short, gray hairs, especially near the tip; pith brown; leaf scars crescent-shaped; bundle scars 3; pith white.

Winter Buds. Bright red, more or less tomentose, the terminal bud slender, slightly stalked, 3–5 mm long, 2–3 times as long as wide, containing the flowers; lateral buds much smaller, appressed; 2 scales visible, valvate.

Flowers. June, after the leaves are full grown; functionally unisexual; plants monoecious; small, yellowish green; borne in fascicles of 2–4 in erect, slightly compound, many-flowered, long-stemmed, terminal panicles; calyx downy, 5-lobed; petals 5; stamens 7–8; ovary tomentose. Insect pollinated.

Fruit. Samara; July–August; bright red, turning brown in late autumn; often persistent on the tree into winter; samaras small, glabrous, paired, in pendulous, particulate clusters; samara wing 1–2 cm long; angle between wings <90°; seed cavity indented; seeds 22,130/lb (48,686/kg).

Distribution. Common in the Upper Peninsula; occasional in northern lower Michigan, south and east to Oakland Co. County occurrence (%) by ecosystem region (Fig. 19): I, 39; II, 77; III, 100; IV, 100. Entire state 57%. Widely distributed in New England and s. to VA, in Canada from NL to SK, s. to MN, WI, MI, NY, PA and in the mts. to NC, TN to n. GA.

Site-Habitat. Characteristic of cool, moist soils and sites of high humidity, along streams and adjacent ravines and conifer-hardwood swamps; protected, moist rocky hillsides. Associates include spruces, eastern hemlock, black ash, northern white-cedar, red maple, yew, speckled alder, Canadian fly-honeysuckle, red-osier, winterberry.

Notes. Shade-tolerant; slow-growing; short-lived. Reproductive biology is described in detail by Sullivan (1983). Germination epigeal. Sprouts vigorously following injury by fire and browsing. Reproduces vegetatively by layering. Not commonly used as an ornamental, but it has promise for use in moist, shaded sites.

Chromosome No. $2n = 26$, $n = 13$; $x = 13$

Similar Species. In China and Japan, the counterpart species is *Acer caudatum* subsp. *ukurunduense* (Trautv. & C. A. Mey.) A. E. Murr. It occurs in the understory of cool, moist forests together with many similar counterpart associates of *A. spicatum* that occur in northern forests and the Appalachian Mountains of eastern North America.

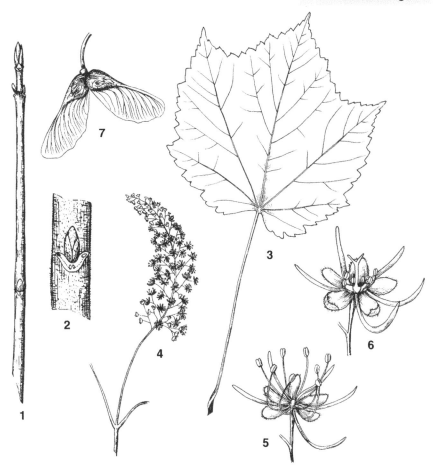

1. Winter twig, × 1
2. Portion of twig, enlarged
3. Leaf, × 1/2
4. Flowering panicle, × 1/2
5. Male flower, enlarged
6. Female flower, enlarged
7. Fruit, samaras, × 1

KEY CHARACTERS

- small understory shrub or tree
- leaves opposite; blades 3-lobed above the middle; singly, coarsely, and irregularly serrate; thin with prominent veins; soft, white pubescence on undersurface
- winter buds bright red, slightly stalked, 2 scales visible, valvate
- twigs velvety pubescent
- fruit a bright red samara turning brown in late autumn; angle between wings <90°

Distinguished from striped maple, *Acer pensylvanicum*, by leaves with single, coarse and more irregular teeth and conspicuous pubescence on the lower surface; turning orange and scarlet in autumn; angle between samara wings <90°; stem not striped; more pubescent in all its parts; much more restricted in site conditions and therefore less common.

BETULACEAE
Alnus incana (L.) Moench
Speckled Alder

Size and Form. Wide-spreading shrub or small tree, 2–8 m. Several crooked stems rising from the root collar form an open, irregular, sparsely branched crown. Clone-forming by sprouting from the root collar; thickets often dense with contorted stems. Roots very shallow. Michigan Big Tree: girth 97 cm (38 in), diameter 31 cm (12 in), height 20.1 m (66 ft), St. Clair Co.

Bark. Thin, rough, reddish-brown, with many conspicuous, horizontal whitish or orange lenticels up to 7 mm long.

Leaves. Alternate, simple, blades 5–10 cm long, half as wide; oval to obovate, acute; rounded; undulate or wavy, doubly and finely serrate, often appearing coarsely dentate; thick; glabrous, dull green above, pale to hoary beneath, pubescent or glabrous on veins beneath; veins straight, small veins forming a ladderlike pattern, 10–13 per side, impressed above, conspicuously projecting beneath, petioles hairy, 1–2 cm long.

Stems-Twigs. Moderately slender, reddish-brown, glabrous; leaf scars half-round, raised; bundle scars 3; pith white, triangular in cross section.

Winter Buds. Terminal bud absent; lateral and end buds distinctly stalked, blunt, 2–3 scales visible, reddish-brown, 5–8 mm long.

Flowers. March–May; unisexual, plants monoecious; small, borne in catkins; male catkins appear the preceding season at end of twigs, 3–4 in a short raceme, each catkin slender, conelike, scaly, about 1 cm long, purplish-brown in winter, elongating and shedding pollen the following spring in March–May before leaves develop, catkins pendent, 5–10 cm long, male flowers borne in the axils of bracts, each flower with a perianth of 4 parts, stamens 4; female catkins also formed prior to and exposed to winter, expanding in March–May before the leaves, borne on shoots below male catkins and at right angles to them, becoming woody; plants monoecious. Wind pollinated.

Fruit. Nutlet with thin, narrow wings; August–October; mature seed catkins ovoid to globular, conelike, woody, long-stalked, drooping, persistent into the next growing season, 1–1.5 cm long, 6–10 mm wide; catkin with woody, 5-toothed scales, about 4 mm long; nutlet orbicular or slightly ovoid without wings at maturity but with thin margins; seeds small, 300,000/lb (660,000/kg).

Distribution. Abundant in upper Michigan and northern lower Michigan; locally common to abundant in southern lower Michigan except for interior counties. County occurrence (%) by ecosystem region (Fig. 19): I, 71; II, 100; III, 100; IV, 100. Entire state 87%. Distributed from NL to SK, s. to New England and mts. of PA, WV, NC, w. to IL and IA.

Site-Habitat. Wet, open, seasonally inundated, poorly drained sites with moving water and slightly acidic to moderately basic soils; wet meadows, interdunal wetlands, prairies, southern and northern shrub thickets, diverse swamps, swales, along streams, limestone shores. Associates include an extreme range of species found in open, neutral to basic conditions and with characteristic species that root in acidic surface layers, especially ericaceous plants. Reported from 12 natural communities of Michigan (Kost et al., 2007).

Notes. Very shade-intolerant; moderately fast-growing; stems short-lived. Often forming dense thickets bending out over streams. Excellent colonizer following disturbance. Fine roots bear clusters of nodules containing nitrogen-fixing bacteria that transform nitrogen of the soil air into compounds, which are translocated into the plant. Named for the speckled appearance of lenticels on the bark. Our individuals are subsp. *rugosa* (Du Roi) Clausen; subsp. *incana* is restricted to Europe. The European black alder, *Alnus glutinosa* (L.) Gaertn., is the most common alder in southern Michigan. It is an erect tree to 12 m (40 ft) with leaves all broadly rounded to a notched or truncate apex.

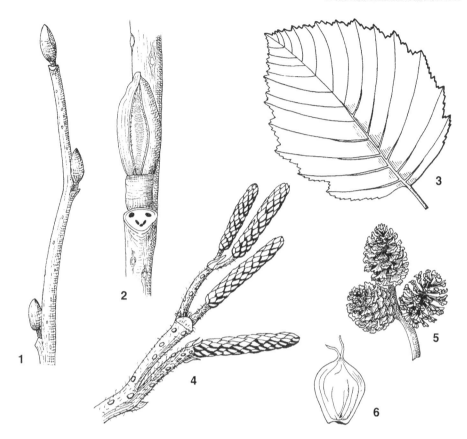

1. Winter twig, × 1
2. Portion of twig, × 4
3. Leaf, × 1/2

4. Male catkins, × 1
5. Female catkins, × 1
6. Fruit, nutlet, × 5

KEY CHARACTERS

- wide-spreading shrub or small tree, 2–8 m
- bark smooth, lenticels conspicuous
- leaf blades doubly and finely serrate, veins straight, impressed above, prominently projecting beneath, leaves not resinous-gummy and aromatic when young
- winter buds distinctly stalked, blunt at tip
- fruit a small nut lacking wings at maturity; borne in a conelike, woody catkin persistent into the next growing season
- habitat open, wet, circumneutral to basic, diverse sites with moving water

Distinguished from green alder, *Alnus viridis*, by presences of only long shoots; leaf blades doubly serrate with undulate, often coarsely dentate margins; leaves not glutinous; nutlets lacking a broad wing at maturity; occurs in lower and upper Michigan.

Chromosome No. $2n = 28$, $n = 14$; $x = 14$
Similar Species. Related to the European alder, *Alnus incana* subsp. *incana*. Related to and hybridizes with the mountain alder of western North America, *Alnus incana* subsp. *tenuifolia* (Nutt.) Breitung (Farrar 1995).

BETULACEAE
Alnus viridis (Chaix) DC

Green Alder, Mountain Alder

Size and Form. Tall bushy shrub to 3 m or more; often in clonal clumps due to basal sprouting and root suckering.

Bark. Glabrous, reddish-brown to gray with pale lenticels.

Leaves. Alternate, simple, deciduous; long shoots: blades 4–9 cm long, 2.5–5 cm wide; ovate or oval; obtuse or acute; rounded; finely, sharply, and occasionally irregularly serrulate, often "crispy-wavy" (i.e., curled), bright green glabrous above, shiny yellowish-green and usually pubescent beneath along the veins, more or less glutinous (resin dots) and aromatic beneath and when young; veins mostly 6–9 per side; leaves on short shoots similar but smaller, fewer veins per side; petioles 8–10 mm long.

Stems-Twigs. Moderate, reddish-brown to purplish, glabrous or nearly so; leaf scars half circular to triangular; bundle scars 3; pith white, triangular in cross section. Both long and short shoots present.

Winter Buds. Terminal bud absent; lateral buds virtually sessile; visible imbricate bud scales 3–5, overlapping, sharp-pointed, acuminate, curved at tip; dark red to purplish; resinous-gummy.

Flowers. May–June; unisexual, plants monoecious; small; borne in catkins; male catkins, scaly and about 1 cm long, appear at the ends of shoots the preceding fall, singly or 2–3 together on long stalks, and expand and shed pollen in May–June as leaves develop; male catkins slender, 6–10 cm long; male flowers, 3 in the axil of each bract subtended by 4 bractlets, each flower with a perianth of 4 parts, stamens 4, catkins shed after pollen is dispersed; female catkins develop within winter flower buds, appearing in spring in small clusters; female flowers, 2 in the axil of each bract, 2 small bractlets subtending each flower, perianth lacking, ovary 2-celled; plants monoecious. Wind pollinated.

Fruit. Nutlet; September–mid October; borne in woody, oval, conelike catkins, 1.5–1.8 cm long, 7–9 mm wide, with long pubescent stalks, drooping; scales firm and woody, irregularly 5-lobed, about 4 mm long; catkins persistent into the next growing season; the nutlet ovoid with a broad wing, 2–2.5 mm wide; seeds small, 1,280,000/lb (2,816,000/kg).

Distribution. Locally frequent in upper

Michigan. County occurrence (%) by ecosystem region (Fig. 19): I, 0; II, 0; III, 43; IV, 75. Entire state 11%. Transcontinental across n. USA and adjacent Canada to AK; Greenland, NL s. to NC in mts.

Site-Habitat. Open, typically extreme conditions; wet and wet-mesic to drier; sandy, gravelly, rocky; acidic to neutral sites; typical of central to western Lake Superior cobble and bedrock shores and cliffs; rooted in thin soil of cracks, joints, and depressions of bedrock. Usually growing on drier sites than *Alnus incana*. Tolerant of cold, windy sites; short growing season. Associates include white spruce, balsam fir, hemlock, northern white-cedar, white pine, paper birch, trembling aspen, willows, speckled alder, red-osier, thimbleberry, showy mountain-ash, ninebark, wild red raspberry.

Notes. Shade-intolerant, but more shade-tolerant than other alders. Moderately fast growing; stems short-lived. Our individuals are subsp. *crispa* (Aiton) Turrill; the subsp. *viridis* is restricted to Europe. The name *crispa*, "curled," refers to the margin of the leaf blades. A northern species of wet- to dry-mesic sites, it occurs on dry upland sites with jack pine in Canada.

Chromosome No. $2n = 28$, $n = 14$; $x = 14$

Similar Species. Closely related to the European green alder, *Alnus viridis* subsp. *viridis* DC. Also related to the Sitka alder and Siberian alder of western North America, *Alnus viridis* subsp. *sinuata* (Regel) Á. Löve & D. Löve, and *A. viridis* subsp. *fruticosa* (Ruprecht) Nyman, respectively.

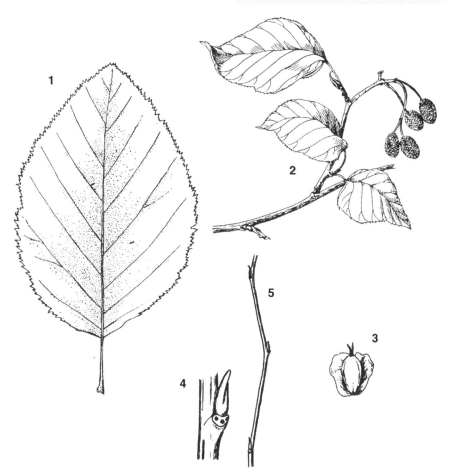

1. Leaf, × 1
2. Fruiting shoot with fruit-bearing catkins, × 1/2
3. Fruit, nutlet, × 3
4. Winter lateral bud and leaf scar, twig, enlarged
5. Winter twig, reduced enlarged

KEY CHARACTERS

- bushy shrub to 3 m, multiple stems in clonal clumps
- leaf blades glutinous beneath and when young; margins sharply serrulate, often "crispy-wavy" but not appearing coarsely dentate
- buds short-stalked to sessile, resinous-gummy
- fruit a small nut with a broad wing, borne in a conelike, woody catkin persistent into the next growing season
- habitat open, harsh, wet- to dry-mesic, sandy to rocky sites, especially along the shores of Lake Superior; absent in lower Michigan

Distinguished from speckled alder, *Alnus incana*, by the presence of both long and short shoots; leaf blade margins sharply serrulate, curled; blades distinctly glutinous-resinous beneath; nutlets with a broad membranous wing at maturity; occurs only in upper Michigan and almost always growing on drier sites.

THE SERVICEBERRIES—*AMELANCHIER*

Thirty-three species are recognized, primarily in North America, with few in Europe and Asia. Several are noted for their ornamental value of showy flowers and sweet, juicy fruit, especially *Amelanchier laevis*. The first species to be recorded, *A. ovalis* Medicus (syn. *A. vulgaris* Moench), was found "growing naturally upon the Alpes" and was named by Linnaeus *Mespilus amelanchier* (Coates 1992). In 1789 Medicus gave the name *Amelanchier* to the genus, based on the writings of the Flemish doctor and botanist Clusius (1526–1606), who noted that the natives of Savoy used the French common name amelanche because the fruit tasted of honey.

The shrubs and small understory trees in *Amelanchier* are characterized by stems with smooth, thin, grayish bark when young, very early spring flowering, and widely bird-dispersed fruits. The alternate, simple leaves have serrate margins of coarse to fine, sharp-pointed teeth. The slender stems when bruised have the odor of bitter almond, not as strong as in *Prunus*, and with bitter taste. Buds of most species are long, slender, and sharply pointed but with fewer exposed scales than American beech. Most species reproduce vegetatively by means of basal sprouting or rhizomes.

The name *serviceberry* is a corruption of *sarvissberry. Sarviss* is a transformation of the word *sorbus,* given by the Romans to a related kind of fruit (Peattie 1991). Peattie notes, "Sarviss is a good Shakespearean English form of the most classic Latin, whereas Serviceberry is meaningless as a name, or is at least a genteel corruption of an older and more scholarly form." The name *serviceberry* has its own folklore, dating back to pioneer days when mountain people used the showy flowers marking the end of winter for weddings, and for memorial services for people who had died during the winter season. According to some, services were deferred because travel conditions in the mountains were too difficult for all those who wished to pay their respects. According to others, services were deferred until the blooming signaled the time when the soil could be turned and the bodies buried in the ground. The name *Juneberry* is derived from the early ripening of the fruit in some areas. The name *shadbush* is also used; it is an eastern name, which refers to the coincidence of the flowering time with the spawning runs of the shad fish. Voss (1985) notes that more than eighty common names are reported for the group.

The serviceberries range in shade tolerance from intolerant to moderately tolerant and are habitat generalists and not well site differentiated by species as is typical in other groups. The fruits are consumed by many bird and mammal species and seeds very widely dispersed. Throughout diverse forest ecosystems of lower Michigan, their seedlings, <10 cm, probably are more common in the ground-cover layer than those of any other species except black cherry (*Prunus serotina*). In western North America, the dominant species is Saskatoon or western serviceberry, *A. alnifolia*. It has several varieties or subspecies and ranges from Ontario west to Alaska and south into the Pacific Northwest (subsp. *florida*), California, and Utah (var. *utahensis*). Other related species occur in Europe (*Amelanchier lamarckii, A. ovalis*) and East Asia (*A. asiatica*).

The serviceberries are a confusing group, characterized by hybridization, polyploidy, and seed apomixis and thus leading to identification difficulty. Especially for Michigan, Voss and Reznicek (2012) provide a succinct description of this difficult group. For other treatments, see Soper and Heimburger 1982; Gleason and Cronquist 1991; and Smith 2008. We have included the *Amelanchier spicata* complex in the key below, although it is not illustrated or described in the text. The *A. sanguinea* complex is included as a low shrub because some authors recognize it as such, but it does not spread by rhizomes to form clones. See Voss and Reznicek 2012 for a description this and other species.

KEY TO SPECIES OF *AMELANCHIER*

1. Tall shrubs or small trees 3–10 m; stems solitary or in few-stemmed clumps; flowers usually in long, lax, drooping racemes.
    2. Summit of ovary glabrous; branchlets grayish.
        3. Leaves young and folded or just beginning to unfold at flowering time, downy-white pubescent; lower surface remaining more or less pubescent at maturity; fruiting pedicle relatively short, 0.8–1.2 cm long . . . *A. arborea*, p. 64
        3. Leaves mature or at least half-grown at flowering time, sparsely pubescent to glabrous, coppery-red; glabrous both sides at maturity; longer fruiting pedicle, 2.5–5 cm . . . *A. laevis*, p. 66
    2. Summit of ovary tomentose, sparsely in fruit; branchlets reddish . . . *A. sanguinea* complex, p. 68
1. Low to medium shrubs, 0.3–3 m.
    4. Stems usually few, not spreading from base by rhizomes; early leaves relatively coarsely serrate-dentate (3–5 teeth per cm when mature); the veins conspicuous and running to tips of the teeth (or a principal fork into a tooth), at least toward apex of blade; petals 11–18 mm long . . . *A. sanguinea* complex, p. 68
    4. Stems usually numerous, growing in spreading clones from rhizomes; leaves more finely toothed, at least near the apex (5–8 teeth per cm when mature); the veins not prominent and anastomosing and becoming indistinct near the margin, at most with weak veinlets ending in the teeth; petals 5–9 mm long . . . *A. spicata* complex.

ROSACEAE
Amelanchier arborea (Michaux f.) Fernald
Downy Serviceberry, Juneberry

Size and Form. Tall, erect shrub or small tree, 4–10 m and 10–30 cm in diameter. Trunk slender; forming a narrow, rounded crown of many small branches and slender twigs; forming clonal clumps by basal sprouting. Michigan Big Tree: girth 200.7 cm (79 in), diameter 63.9 cm (25.1 in), height 19.2 m (63 ft), Barry Co.

Bark. Thin; pale, smooth on young trees; becoming grayish to reddish-brown and divided by dark, shallow fissures into narrow, longitudinal ridges.

Leaves. Alternate, simple, deciduous, blades 7–10 cm long, about half as wide; ovate to obovate; folded or just beginning to unfold at flowering time; acuminate or acute, rounded or subcordate, finely and sharply serrate; downy-white pubescent when young, becoming glabrous and dark green above, paler beneath, with pubescent midrib and main veins; petioles slender, pubescent at first, becoming glabrous, 2–3 cm long.

Stems-Twigs. Very slender, smooth, slightly to moderately zigzag, light green, becoming reddish-brown; leaf scars crescent-shaped, bundle scars 3; pith more or less 5-angled, continuous, pale.

Winter Buds. Terminal bud slender, sharp-pointed, narrow-ovoid to conical, 0.6–1.4 cm long, 5–6 visible, often twisted and ending in a short-pointed tip, greenish to reddish-brown, slightly pubescent; lateral buds similar, recurved toward twig.

Flowers. April–May, when the leaves are small, downy, and folded; bisexual; large, white, borne in lax or erect racemes 7–12 cm long; calyx 5-cleft, campanulate, villous on the inner surface; petals 5, narrow, strap-shaped, 2–3 cm long; stamens numerous; styles 5, united below; summit of ovary glabrous. Insect pollinated.

Fruit. Pome; June–August; globose, 0.8–1.2 cm long; the lowest in a cluster borne on a pedicel about 1.2 cm long; turning from bright red to dark purple with slight bloom; dry and tasteless; seeds 80,000/lb (176,000/kg).

Distribution. Frequent to locally common throughout the state. County occurrence (%) by ecosystem region (Fig. 19): I, 79; II, 67; III, 57; IV, 75. Entire state 72%. Widely distributed throughout e. North America, ME to s. QC, ON, and e. MN s. to MO, e. OK, TX and e. to LA and n. FL.

Site-Habitat. Moderately-shaded to open areas, dry to mesic, intolerant of wetlands; dry-mesic oak and oak-hickory and mesic beech–sugar maple or hemlock–northern hardwoods forests. Associates many, including trees and shrubs associated with these forest types and those of forest edges and disturbed areas.

Notes. Shade-tolerant as a juvenile, becoming moderately tolerant; slow-growing. Warm, moist, late autumn weather will cause the bud scales to swell, and sometimes the leaves flush part-way out. Widely dispersed by birds, downy serviceberry, together with smooth serviceberry, *Amelanchier laevis*, is one of the most commonly occurring seedlings in upland forests throughout the state, including in ecosystems where it is absent as an adult (very dry, excessively drained, sandy soils; slight rises or mounds in wetlands). Even seedlings <20 cm can be easily identified to genus by their finely and sharply serrate leaves, subcordate base, and relatively long, sharply pointed terminal bud.

Chromosome No. $2n = 34$, $n = 17$; $x = 17$

Similar Species. Most similar in habit and other characteristics is the smooth serviceberry, *Amelanchier laevis* Wiegand, which is found in similar soil-site conditions and distributions in Michigan, although it is more prevalent in southern lower Michigan. In western North America, the dominant species is the Saskatoon or western serviceberry, *A. alnifolia* (Nutt.) Nutt. Other related species occur in Europe (*A. lamarckii* Schroeder) and East Asia (*A. asiatica* [Siebold & Zucc.] Endl. ex Walp.).

1. Winter twig, × 1
2. Portion of twig, enlarged
3. Leaf, × 1
4. Flowering shoot, × 1/2
5. Vertical section of flower, enlarged
6. Fruiting shoot, × 1/2

KEY CHARACTERS

- tall shrub or small tree to 10 m
- bark smooth, grayish, thin; conspicuous network of dark vertical lines
- winter buds long and narrow, sharp-pointed, scales often twisted
- leaf blades finely and sharply serrate, folded or just unfolding at flowering time, downy-white-pubescent when young, lower surface retaining pubescence into maturity
- fruit a globose pome, glabrous, dark purple when mature with slight bloom; dry and tasteless.

Distinguished from smooth serviceberry, *Amelanchier laevis*, by leaves folded or just beginning to unfold at flowering time, young leaves downy-pubescent when young, lower surface retaining pubescence into maturity; shorter fruiting pedicle, 0.8–1.2 cm long; and fruit dry and tasteless.

ROSACEAE
Amelanchier laevis Wiegand

Smooth Serviceberry

Size and Form. Tall, erect, slender shrub with several stems by basal sprouting or a small tree to 10 m. Trunk and crown similar to that of downy serviceberry, *Amelanchier arborea*. Michigan Big Shrub: girth 1.75 m (69 in); diameter, 55.8 cm (22.0 in); height, 12.8 m (42 ft); Leelanau Co.

Bark. Thin, smooth, pale, mottled gray.

Leaves. Alternate, simple, deciduous; blades thin, 3–8 cm long and half as wide; elliptic-ovate to ovate-oblong; short-acuminate or acute; subcordate to rounded; margin finely and sharply serrate nearly to the base, teeth >20 per side, 6–8 per cm, and more than twice as many as the veins; young leaves usually at least half grown at flowering time, sparsely pubescent to glabrous, strongly coppery-red becoming dark green and glabrous at maturity, paler beneath, glabrous; petiole long, 1–3 cm, glabrous.

Stems-Twigs. Slender, rounded, reddish-brown, glabrous; more or less 5-angled, continuous; leaf scars crescent-shaped, bundle scars 3; pith more or less 5-angled, continuous, pale.

Winter Buds. Terminal bud long, slender, sharp pointed, ovate, acute, 0.9–1.7 cm long; 5–6 often twisted scales visible, greenish-brown or greenish-yellow; lateral buds smaller, sessile, solitary, elongated and recurved toward stem.

Flowers. March–May, bisexual, when the leaves are at least half grown; large, white or pinkish, showy, borne on erect or drooping racemes 7–12 cm long, glabrous or nearly so at flowering time, pedicels long, 2.5–3 cm long; calyx 5-cleft, campanulate, villous on inner surface; petals 5, narrow, linear-oblong, strap-shaped, 1–2 cm long; summit of ovary glabrous; stamens numerous. Insect pollinated.

Fruit. Pome; June–August, globose, 6–8 mm wide, dark purple to blackish with bloom; long fruiting pedicle, 2.5–5 cm long; sweet and juicy.

Distribution. Common throughout the state. County occurrence (%) by ecosystem region (Fig. 19): I, 71; II, 73; III, 86; IV, 88. Entire state 75%. E. North America, NL w. to MN, se. through n. IL, OH, PA, s. in Appalachian mts. to n. GA.

Site-Habitat. Wide tolerances in open to moderately shaded areas, including dry to mesic sites; intolerant of wetlands. It occurs as an understory tree in dry-mesic oak and oak-hickory and mesic beech–sugar maple or hemlock–northern hardwoods forests. Associates include trees and shrubs associated with these forest types and those of forest edges and disturbed areas.

Notes. Shade-tolerant as a juvenile, becoming moderately tolerant; slow-growing. One of the earliest shrubs or trees to flower in the spring. Warm and moist autumn conditions may cause buds to swell and leaves to initiate flushing. Widespread bird dispersal. It is one of the most commonly occurring seedlings in upland forests throughout the state, including ecosystems where it is absent as an adult (very dry, excessively drained sandy soils; slight rises or mounds in wetlands). Even seedlings <20 cm can be easily identified to genus by their finely and sharply serrate leaves, subcordate base, and relatively long, sharply pointed terminal bud. The name *laevis* means "smooth, lacking hairs."

Chromosome No. $2n = 34$, $n = 17$; $x = 17$

Similar Species. Most similar in form and other characteristics is downy serviceberry, *Amelanchier arborea*. Also it is similar in site conditions and geographic occurrence. A main difference is reflected in the respective common names, downy and smooth. Young, preformed leaves in the winter buds of *A. arborea* are downy-pubescent, whereas in *A. laevis* they are glabrous or nearly so.

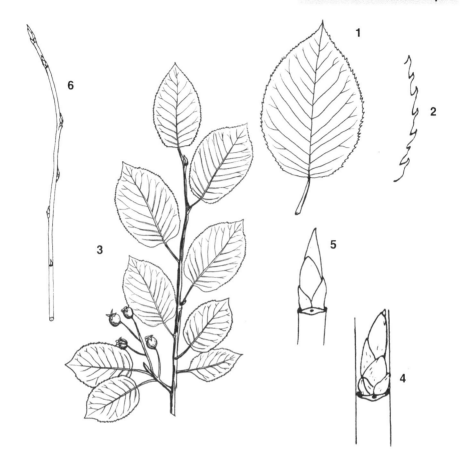

1. Leaf, reduced, × 1/2
2. Blade margin, enlarged
3. Vegetative twig with fruiting shoot at base, pome, × 1/3
4. Winter lateral bud and leaf scar, enlarged
5. Winter terminal bud and leaf scar, × 1 1/2
6. Winter twig, reduced

KEY CHARACTERS

- tall shrub with several slender stems or small tree to 12 m
- leaves with blades elliptic-ovate to ovate-oblong, glabrous, margins finely and sharply serrate; young leaves usually at least half grown at flowering time and coppery-red
- flowers, white, showy, early blooming, March–May; fruiting pedicle long, 2.5–5 cm
- fruit a pome, globose, dark purple to blackish with bloom, sweet and juicy

Distinguished from downy serviceberry, *Amelanchier arborea*, by young leaves glabrous or nearly so, flowering when the leaves are at least half grown and coppery-red; longer fruiting pedicle, 2.5–5 cm; and fruit sweet and juicy.

ROSACEAE
Amelanchier sanguinea (Pursh) DC.
Round-leaved Serviceberry

Size and Form. Low to tall, often arching shrub to 8 m; scattered solitary individuals or usually with several slender stems in a tight clump; not spreading by sprouts from rhizomes or roots.

Bark. Thin, grayish, smooth.

Leaves. Alternate, simple, deciduous; blades thin, 3–7 cm long and 2–4 cm wide; elliptic-oblong to suborbicular; blunt to rounded or acute; rounded or subcordate; margin of early (preformed) leaves relatively coarsely serrate-dentate nearly to the base, teeth sharp, 3–5 per cm toward the apex, lateral veins prominent, 9–13 per side; leaves nearly or fully unfolded at flowering time, at first grayish-green to whitish-tomentose beneath, green and glabrous when mature except for sparse pubescence on midrib beneath and on the petiole; petioles pubescent, 1–2 cm long.

Stems-Twigs. Very slender, glabrous, red to reddish-brown when young, becoming grayish; leaf scars raised, crescent-shaped, bundle scars 3; pith more or less 5-angled, continuous, pale.

Winter Buds. Terminal bud slender, sharp-pointed, 0.6–1 cm long; lateral buds smaller, 6–7 mm, narrowly ovate, acute, pink to reddish-brown, curved toward stem.

Flowers. Early April–May, with the unfolding and expanding leaves; bisexual; large, white, borne in loose, arching or drooping racemes 4–7 cm long, rachis and pedicels pubescent at first, becoming glabrate, pedicels to 1–2 cm long; sepals ovate-lanceolate, silky-hairy, deciduous; petals 5, narrowly spatulate to linear, 1–2 cm long, summit of ovary densely tomentose. Insect pollinated.

Fruit. Pome, June–August, globose, 5–10 mm wide, dark purple to black with bloom; pedicels 1–3 cm long; sweet and juicy; seeds 84,000/lb (184,800/kg).

Distribution. Occasional in southern lower Michigan; common in northern lower Michigan and upper Michigan. County occurrence (%) by ecosystem region (Fig. 19): I, 26; II, 57; III, 86; IV, 88. Entire state 48%. E. North America, ME w. to s. QC, ON; MN se. through n. IL, OH, PA, s. in Appalachian mts. to n. GA.

Site-Habitat. Wide tolerance of open or lightly shaded, dry and dry-mesic, sandy and gravely sites, roots typically reaching mildly basic to calcareous substrates; edges of forests and bogs; more often found along shores, low dunes, and calcareous gravels than other *Amelanchier* species (Voss and Reznicek 2012).

Notes. Shade-intolerant, tolerating light shade; slow-growing, short-lived. One of the earliest shrubs or trees to flower in the spring. Warm and moist autumn conditions may cause buds to swell and leaves to initiate flushing. Widespread bird dispersal. Recognized by Voss and Reznicek (2012) as the *A. sanguinea* complex, a polyploid mixture of triploid and tetraploid plants.

Chromosome No. A polyploid mixture, including many triploid and tetraploid plants.

Similar Species. This complex of species, varieties, and hybrids naturally has several related taxa that have been associated with it. These include *A. huronensis* Wiegand and species associated with the *A. spicata* complex (Voss and Reznicek 2012). In Europe the rock pear, *A. ovalis* Medikus, which grows on cliffs and rocky slopes, is the counterpart species.

1. Twigs with fruiting shoots, pome, × 1/2
2. Winter lateral bud and leaf scar, × 2
3. Winter 2-year-old lateral twig with terminal bud, × 2

KEY CHARACTERS

- low, straggling, arching, or erect shrub, 1–3 m
- leaves with blades elliptic-oblong to suborbicular; margin coarsely serrate-dentate nearly to the base, young leaves grayish-green to whitish-tomentose beneath
- flowers white, showy, early blooming, April–May, top of ovary tomentose
- fruit a pome, globose, dark purple to blackish with bloom, sweet and juicy.

ERICACEAE
Andromeda glaucophylla Link
Bog-Rosemary

Size and Form. Low, trailing to upright, evergreen bog shrub, 5–40 cm, little-branched, branches very slender, ascending from a trailing or decumbent stem; clone-forming by layering.

Leaves. Alternate, simple, evergreen, blades 2–6 cm long, 3–8 mm wide; linear to narrowly oblong; abruptly acute and mucronate; long-tapered; entire, strongly, often completely revolute; thick and leathery; glabrous, young leaves pale blue-green, glaucous, becoming lustrous, dark bluish-green above, very finely and densely white-pubescent beneath, midvein impressed; leaves sessile or with very short petioles, 1–2 mm long.

Stems-Twigs. Very slender, somewhat 3-sided, glaucous, becoming gray and reddish-brown with age; leaf scars small, half round or triangular; bundle scars 1; pith small, 3-sided, continuous.

Winter Buds. Terminal bud absent; lateral buds small, 0.8–1.2 mm, solitary, sessile, ovoid, subglobose, or conical.

Flowers. May–June; bisexual; borne in small, terminal, 5–10-flowered, umbel-like clusters at ends of branchlets; pedicels recurved, <1 cm long, not more than twice as long as the corolla; small, about 6 mm long; calyx white, usually spreading; corolla globose, urn-shaped, pinkish (rarely white), with 5 short, recurved lobes; stamens 10. Insect pollinated.

Fruit. Capsule; August–September; 5-celled, glaucous, turban-shaped, 4–5 mm across, much depressed at apex, bluish becoming brownish, persistent in winter; seeds many, lustrous, light brown.

Distribution. Common throughout northern lower and upper Michigan; less common in southern lower Michigan. County occurrence (%) by ecosystem region (Fig. 19): I, 58; II, 90; III, 100; IV, 100. Entire state 77%. Boreal and northern distribution, Greenland and NU, NL to SK, s. to New England, WV, w. to IL, MN.

Site-Habitat. Open, wet, cold, acidic, and nutrient-poor surface of bog and diverse fens, floating mats, occasional in interdunal swales and rock crevices; growing and layering in sphagnum moss where it is deep and wet (Smith 2008). Associates include a wide range species of sphagnum, tamarack, black spruce, northern white-cedar, leatherleaf, Labrador-tea, bog-laurel, large and small cranberries, black chokeberry, bog willow, shrubby cinquefoil, red-osier, and bog birch.

Notes. Shade-intolerant. Tolerant of low oxygen availability; plant indicator of acidic surface-soil conditions although the substrate conditions of the ecosystem, especially fens, is basic. Thus it is associated not only with acid-indicating shrubs but with northern white-cedar and shrubs such as shrubby cinquefoil and bog birch typical of basic soils and substrates.

Chromosome No. $2n = 48$, $n = 24$; $x = 12$

Similar Species. Closely related to the far northern and circumpolar wild rosemary, *Andromeda polifolia* L., which is distinguished by a glabrous but glaucous lower surface of leaf blades; reddish calyx; and longer flower pedicels, 1–2 cm long and 2–4 times as long as the corolla.

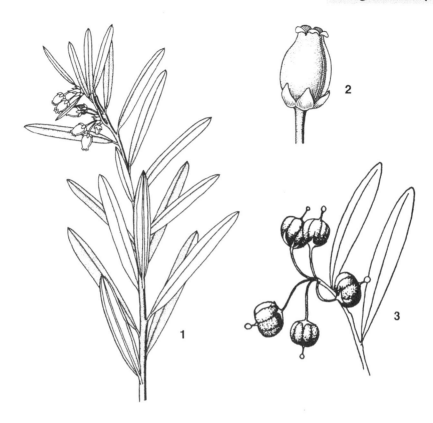

1. Flowering shoot, × 1/2
2. Flower, enlarged, × 2 1/2
3. Fruit, pome, × 2

KEY CHARACTERS

- low, evergreen bog shrub, ascending from a trailing or decumbent stem
- leaves linear, thick, strongly revolute, dark bluish-green above, densely white-pubescent beneath
- fruit a small bluish capsule becoming brownish and persistent in winter
- habitat open, cold, wet, sites and ecosystems with acidic surface soils

ARALIACEAE
Aralia hispida Ventenat
Bristly Sarsaparilla

Size and Form. Low, leafy, perennial shrub with a single erect, slender stem, 0.5–1 m; branches occasional near the base; woody at base and sometimes 5–20 cm higher; dying back in winter to the ground or woody base; dense bristlelike prickles, especially near the base; clones develop and spread by root suckers.

Bark. At base brownish-gray to brown, exfoliating, nodes enlarged.

Leaves. Alternate, bipinnately compound, deciduous; 10–30 cm long, the largest clustered near the base, rachis usually bristly; leaflets 2.5–8 cm long, 0.8–4 cm wide, mostly sessile; lanceolate, lance-ovate to elliptic; acute to acuminate; rounded or cuneate; margin sharply and irregularly serrate, occasionally doubly serrate, 12–30 teeth per side; dark green and glabrous or rarely bristly above, pale beneath, somewhat bristly on midrib and major veins; petioles of lower leaves 4–10 cm long, somewhat bristly, upper leaves nearly sessile.

Stems. Several at woody base, often bristly especially at nodes; at first greenish then tan to brown or reddish-brown with exposure; woody base 6–18 mm in diameter, often densely prickly; nodes conspicuous and enlarged on large woody bases; buds inconspicuous.

Flowers. June–July; bisexual and male flowers on the same plant (andromonoecious); white, about 2 mm in diameter, borne in umbels (3–13) about 2 cm in diameter, the lower solitary, long peduncled, the upper ones forming a loose spreading cluster; stamens 5; ovary 5-celled.

Fruit. Drupe, tipped by a persistent style; August–September; 6–8 mm across, purplish-black; strongly 5-lobed when dry; seeds 92,000/lb (202,400/kg).

Distribution. Common to locally abundant in upper Michigan and northern lower Michigan; in southern lower Michigan common in local sites in counties adjacent to lakes Michigan and Huron. County occurrence (%) by ecosystem region (Fig. 19): I, 45; II, 67; III, 100; IV, 100. Entire state 63%. Occurring in 82% of the counties bordering lakes Huron, Michigan, and Superior. A northern species ranging from New England n. to NL, QC, w. to MB, s.e. to n. MN, n. IL, IN, OH to WV, NJ, and n. VA.

Site-Habitat. Very dry to wet-mesic sites that are open due to natural and human disturbance, particularly sand dunes and open, sandy sites of diverse kinds near the Great Lakes but also on interior sandy plains, rock outcrops, and occasionally disturbed borders of swamps and bogs.

Notes. Very shade-intolerant, persisting in light shade; a pioneer opportunist of disturbances of all kinds. Dying back each winter to a woody base and losing its shrub form. Following winter dieback, sprouting vigorously in spring. It often forms large clones from an extensive horizontal root system in newly disturbed sandy soils, persisting for some years after disturbance (Voss and Reznicek 2012).

Chromosome number. $2n = 24$, $n = 12$; $x = 12$

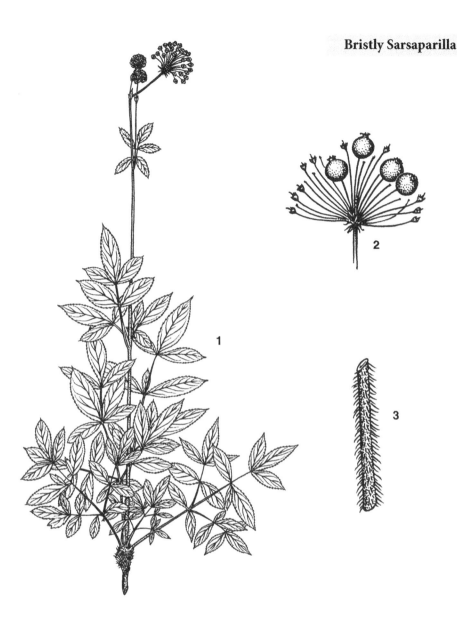

1. Individual with vegetative and flowering shoots
2. Fruiting shoot, drupe, × 1
3. Lower stem with prickles, × 1/2

KEY CHARACTERS

- low perennial shrub usually less than 1 m tall with an erect main stem during the growing season; dying back in winter to the ground or a short woody base
- lower stem supporting several leafy branches
- leaves bipinnately compound, lance-ovate leaflets sharply and irregularly serrate
- usually bristly on lower stem, leaf rachis and petiole, and major veins of lower blade surface
- fruit a blue-black drupe borne in loose terminal umbels
- characteristic of diverse, open, sandy to rocky sites but especially sand dunes along the Great Lakes.

ERICACEAE
Arctostaphylos uva-ursi
(Linnaeus) Sprengel

Bearberry, Kinnikinnick

Size and Form. Decumbent, trailing evergreen shrub with ascending branches to 30 cm and several meters long; branches flexible; clone-forming by stems rooting at nodes, often in large mats.

Bark. Main stems reddish-brown to ashy, exfoliating and appearing papery or flaky.

Leaves. Alternate, simple, evergreen, persistent 1–2 years; 1–3 cm long and about half as wide; obovate to spatulate; rounded to obtuse; cuneate; entire; leathery and coriaceous; bright green above, paler beneath, becoming bronze to reddish in autumn; glabrous; petioles short, often pubescent, 2–4 mm long.

Stems-Twigs. Main stems 3–5-sided, decumbent with slender ascending branches, brown, minutely tomentose-viscid, becoming glabrate; leaf scars small, crescent-shaped, bundle scar 1; pith small, slightly angled, continuous.

Winter Buds. Terminal bud absent; lateral buds solitary, sessile, ovoid, typically 3 scales exposed.

Flowers. May–early June with the leaves; bisexual; borne in 1–6-flowered racemes or panicles; pedicels 2–4 mm long; white to pale pink; calyx 4–5 parted; corolla urn- or bell-shaped with very short, rounded lobes, 4–6 mm long; stamens 8–10. Insect pollinated.

Fruit. Drupe; July–September; globose, lustrous, bright red, 6–8 mm across, dry or mealy, persistent through winter; pit 3–7 mm across, composed of 5–10 wholly or partially fused pits; seeds 58,000/lb (127,600/kg).

Distribution. Common to locally abundant in northern lower and upper Michigan; infrequent in southern lower Michigan. County occurrence (%) by ecosystem region (Fig. 19): I, 32; II, 87; III, 86; IV, 100. Entire state 63%. Circumboreal; transcontinental in northern temperate climates; lacking in much of the interior West, Southwest, Great Plains, and the mid- and deep South, w., to TX, OK, n. to NE.

Site-Habitat. Open to lightly shaded sites in diverse ecosystems usually in dry to dry-mesic, well-drained, sandy and rocky sites with either acidic or basic soils; pine and oak barrens, cobble and bedrock lakeshores, sandy Great Lakes shorelines, limestone pavements, beaches, wooded dunes, and swale complexes.

Associates include a great many tree and shrub species present in the above ecosystems and communities; reported from 18 near-natural communities of Michigan (Kost et al. 2007); occurs in more communities than any of the 132 species we describe.

Notes. Very shade-intolerant, persistent in light shade. Tolerant of drought, low nutrient availability, and cold temperatures. Distributed worldwide in northern, boreal, and alpine regions. The scientific name from the Greek *arcto*, "bear," and *staphyle*, "bunch of grapes of the bear," hence bearberry. It is known as kinnikinnick in the Pacific Northwest. Leaves are reported used for a variety of medical issues, including kidney and urinary problems (Moore 1979; Kershaw, MacKinnon, and Pojar 1998). Describing the uses of kinnikinnick in the Rocky Mountains, Kershaw, MacKinnon, and Pojar report that tea is widely used, but extended use may lead to "stomach and liver problems (especially in children)." They also note that bears are very fond of the fruits. Leaves were also used as a substitute for tobacco or a tobacco extender by Native Americans in Michigan and the Rocky Mountains (Moore 1979; Voss 1996; Kershaw, MacKinnon, and Pojar 1998). It is difficult to transplant but may make a superb ground cover in landscape plantings.

Chromosome No. $2n = 52$, $n = 26$; $x = 13$

Similar Species. The genus *Arctostaphylos* includes many western species. Throughout the interior West and California the common name for *Arctostaphylos* is manzanita (Spanish word meaning "little apples" for the fruit). Stuart and Sawyer (2001) contrast 12 species of manzanita (including *A. uva-ursi*) typical of chaparral vegetation of the Upper Sonoran life

1. Flowering shoot, × 1
2. Prostrate stem bearing vegetative and fruiting shoots, drupe, reduced

KEY CHARACTERS

- procumbent or trailing evergreen shrub with ascending branches
- leaves alternate, evergreen, leathery, and coriaceous, bright green above, short, obovate to spatulate
- fruit a globose drupe, lustrous, bright red
- habitat open, usually dry to dry-mesic sites with either acidic or basic soils

zone of California. Chaparral includes species of 3 major genera, *Ceanothus, Arctostaphylos*, and *Cercocarpus*. Chaparral forms extensive brushlands on arid lands, and the species are characterized by marked xerophytic structures such as small leaves, entire leaves, thickened epidermis, hard and very dense wood, permanently vertically oriented leaves, small flowers, and seeds adapted to very dry conditions and fire. Seeds may remain viable in the soil for decades and are simulated to germinate by wildfire.

ROSACEAE
Aronia prunifolia (Marsh.) Rehder
Black Chokeberry

Size and Form. Erect, often weakly spreading shrub, 0.5–2 (4 ft.) m; multistemmed, open, and round-topped; forming clones by rhizomes. Michigan Big Tree: girth 12.7 cm (5 in), diameter 4.0 cm (1.6 in), height 5.5 m (18 ft), Oakland Co.

Bark. Smooth to somewhat roughened, reddish-brown to grayish.

Leaves. Alternate, simple, deciduous; blades 2–8 cm long, 1.5–3 cm wide; elliptic-obovate to oval or oblanceolate; acuminate or abruptly long-pointed, mucronate; rounded to broadly cuneate; finely crenate-serrate, serrations incurved; bright green and somewhat shining above with dark reddish-black, hair-like glands along the midrib, glabrous or soon glabrate above, paler beneath and glabrous to hairy; petioles short, pubescent to glabrate, 0.2–1 cm long.

Stems-Twigs. Slender, rounded to somewhat flattened; red or brownish-red; glabrous or with a few scattered hairs; leaf scars U-shaped, low; bundle scars 3; pith moderate, rounded, pale, more or less continuous.

Winter Buds. Terminal bud present, 5–9 mm long, oblong, sharply pointed and often slightly curved or hooked at tip, sessile, strongly flattened, appressed, glabrous, dark red, bud scales about 4, keeled, the lower scales double notched and irregularly toothed; lateral buds usually smaller, often abruptly short-pointed.

Flowers. June; bisexual; white or pinkish; borne 15 or fewer in clusters at the ends of leafy shoots; peduncles and pedicels at first loosely pubescent, becoming glabrous or nearly so; about 1 cm across; calyx urn-shaped, 5-lobed; petals 5; styles 5; stamens numerous. Insect pollinated.

Fruit. Pome; September–October; small, 6–10 mm wide; globose; lustrous black to purplish-black; sometimes persistent into winter; seeds 276,000/lb (607,200/kg).

Distribution. Frequent throughout the state; widespread in acidic wetlands. County occurrence (%) by ecosystem region (Fig. 19): I, 87; II, 93; III, 86; IV, 75. Entire state 88%. Distributed from NL s. to n. GA, and n. to MN.

Site-Habitat. Open to lightly shaded, wet to wet-mesic, extremely acidic to slightly acidic sites of low nutrient and oxygen availability;

occurs in a diverse array of communities including coastal marsh, sand prairie, nutrient-poor fen, bog, muskeg, shrub thicket and carr, various swamps, interdunal swales, and lake plain oak openings. Because it occurs from extremely acidic sites to a few communities where acidic microsites are underlain by basic and calcareous substrates, it has many tree and shrub associates—at least 98 individual tree and shrub species are reported in the 13 natural communities where it occurs in Michigan (Kost et al. 2007).

Notes. Shade-intolerant, persisting in light shade. Voss and Reznicek (2012) note, "Completely glabrous plants have often been recognized as *A. melanocarpa* (Michx.) Elliot, which in our area seems quite impossible to separate consistently from the somewhat pubescent *A. prunifolia*." For various viewpoints, including interspecific hybridization, see Hardin 1973; Gleason and Cronquist 1991; and Voss 1985.

Chromosome No. $2n = 34$, $n = 17$; $x = 17$

Similar Species. The closely related red chokeberry, *Aronia arbutifolia* (L.) Elliott, of southern New England, s. to FL and w. to TX, is pubescent in twigs, leaves, and distinguished by bright red pomes. It is sometimes used in landscape plantings because of its climatic and soil adaptability and the brilliant red fruit display in autumn and winter. In November 2011, eight individuals of it, possibly one clone, were found in Kent Co., a new state record (Ryskamp and Warners 2012).

1. Leaf, × 1
2. Flowering shoot, × 1/2
3. Fruiting shoot, pome, × 1/2
4. Winter lateral bud and leaf scar, enlarged
5. Winter twig, × 1

KEY CHARACTERS

- an erect, loosely spreading clonal shrub usually <2 m
- leaf blades elliptic-obovate or oblanceolate; tiny blackish, hairlike glands along the midrib of the upper surface
- buds dark red, sharply pointed and slightly curved or hooked at tip, keeled and strongly flattened
- fruit a small lustrous blackish pome, sometimes persistent
- habitat wet to wet-mesic, open or lightly shaded acidic sites

BERBERIDACEAE
Berberis thunbergii de Candolle

Japanese Barberry

Size and Form. Low, erect, compact, spiny shrub, usually wider than tall, 0.5–1.5 m high; densely branched, branches stiff.

Leaves. Alternate, simple, deciduous, blades 1–3 cm long, 0.2–1 cm wide; borne singly on long shoots or in clusters on short shoots in axils of spines; highly variable in size and shape; spatulate to narrowly obovate; obtuse, rarely acute; tapering gradually; entire; glabrous, bright green above, and paler with slight bloom beneath, turning reddish to dark purple in autumn; petioles 0.2–1 cm long.

Stems-Twigs. Slender, strongly 3-grooved downward from the nodes, purplish-red in first year, turning purplish-brown the second year; glabrous; spines usually simple, 4–15 mm long, straight and slender, sharp, borne singly at a node; inner bark and wood bright yellow; pith relatively large, continuous.

Winter Buds. Terminal bud absent; lateral buds small, ovoid, 6–8 overlapping scales.

Flowers. April; bisexual; borne solitary or 2–5 in umbels under the foliage; small; yellowish; sepals and petals 6; stamens 6; plants dioecious. Insect pollinated.

Fruit. Berry; September–October; persistent into winter; bright red, narrowly ellipsoid, about 1 cm long; seeds small, brown, ca. 29,000/lb (63,800/kg).

Distribution. Relatively common locally in southern lower Michigan, infrequent to rare elsewhere. County occurrence (%) by ecosystem region (Fig. 19): I, 68; II, 23; III, 14; IV, 38. Entire state 45%. Planted or escaped and naturalized, NL, QC, ON, s. to TN, GA; w. of the Mississippi River MN s. to MO; Great Plains ND s. to WY, KS.

Site-Habitat. Very adaptable to sandy loam to clayey soils; spreads vigorously in fertile mesic to wet-mesic sites, including swamp edges; grows its bushy best in full sunlight. Associates include a wide range of species, especially in moist, disturbed sites.

Notes. Moderately shade-tolerant; moderately slow-growing. Native of Japan, introduced in North America ca. 1864 and widely planted for hedges. Becoming naturalized and invasive in a wide range of sites, especially in mesic to wet-mesic, fertile, open areas and lightly shaded forests. Seeds widely dispersed by birds. Flowers are sensitive to touch, if the organs are touched the whole flower snaps shut. Not recommended for ornamental planting; cultivation illegal in Canada.

Chromosome No. $2n = 28$, $n = 14$, x = 7.

Similar Species. The American or Allegheny barberry, *Berberis canadensis* Miller, occurs in mountains from VA to GA and in the Mississippi Valley. The European or common barberry, *B. vulgaris* L., was once widely planted but is now largely eradicated since it is thought to be an alternate host of the common stem rust of wheat (*Puccinia graminis*).

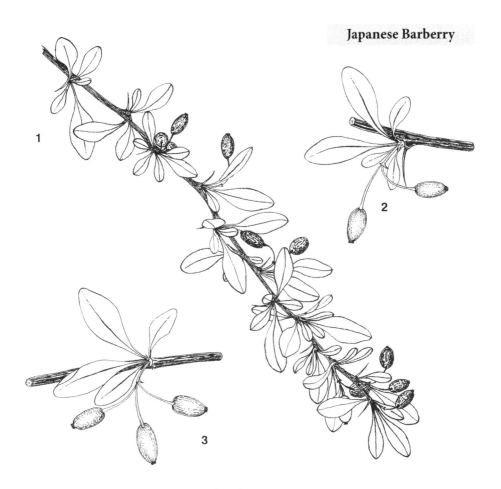

1. Stem with short-shoot leaves, spines, fruit, berry, × 1
2. Stem node with short-shoot, spine, and fruit, × 1
3. Stem node with short-shoot leaves and fruit, × 1

KEY CHARACTERS

- low shrub to 1.5 m, densely branched
- leaves spatulate to obovate, bright green above, entire
- twigs with conspicuous spines at nodes, strongly 3-grooved downward from nodes
- fruit a bright red berry persisting into winter
- widely cultivated and naturalized in diverse open or lightly shaded sites

BETULACEAE
Betula pumila Linnaeus

Bog Birch

Size and Form. Low, erect shrub, 0.5–3 m; few to many ascending branches from sprouting at the root collar.

Bark. Smooth, reddish-brown, light-colored lenticels prominent.

Leaves. Alternate, simple, deciduous, leaves on short shoots borne in 2s or occasionally 3s, leaves on long shoots borne singly; blades 1–5 cm long, 1–3 cm wide, somewhat leathery; suborbicular, obovate, or broad-elliptic; rounded or obtuse; rounded or cuneate; coarsely crenate-dentate; 7–11 teeth per side; dull green above, paler to whitish beneath; pubescent when young, becoming more or less glabrous; veins finely reticulated, 4–6 main pairs; petioles short, 0.3–1 cm long.

Stems-Twigs. Slender, reddish-brown, becoming grayish, pubescent, usually not resinous-glandular.

Winter Buds. Terminal bud absent, laterals small, 2–4 mm long, ovoid, acute, usually divergent.

Flowers. May–June; unisexual, plants monoecious; borne in catkins; male catkins clustered or in pairs, 1.5–3 cm long, slender, brownish; female catkins erect, 1–3 cm long, 6–10 mm in diameter, slender, peduncles very short, greenish; bracts pubescent, lateral lobes spreading; plants monoecious. Wind pollinated.

Fruit. Samara; August–September; small, bracts cross-shaped, wings narrow; seeds very small, 2,422,000/lb (5,328,400/kg).

Distribution. Frequent to common throughout the state. County occurrence (%) by ecosystem region (Fig. 19): I, 76; II, 57; III, 100; IV, 75. Entire state 71%. Primarily northern transcontinental range, NL to NT, s. to CA; New England and upper Great Lakes Region and IA, IL, IN, OH.

Site-Habitat. Wet to wet-mesic, circumneutral, nutrient-rich sites: fens, bogs, moderately open swamps, marsh edges. Also on lake edges, in coastal wet-mesic habitats, and in wet, open backswamps of river floodplains. Associates include American elm, red maple, black ash, yellow birch, northern white-cedar, tamarack, speckled alder, mountain holly, winterberry, common elder, poison-sumac, sage willow, sweet gale, red raspberry.

Notes. Shade-intolerant. Also called swamp birch. Although associates include plants regarded as acid indicators, bog birch roots in somewhat deeper layers that are circumneutral to basic, thus indicating a basic substrate. Contrary to its common name, it is absent in true bogs with deep acidic soils. Hybrids with yellow birch, *B.* × *purpusii* Schneid., are more frequent in southern Michigan than northern Michigan. See Dancik and Barnes 1972 for details of hybridization with yellow birch. Hybrid with white birch, *B.* × *sandbergii* Britt., occasional in the state.

Chromosome No. $2n = 56$, $n = 28$; $x = 14$

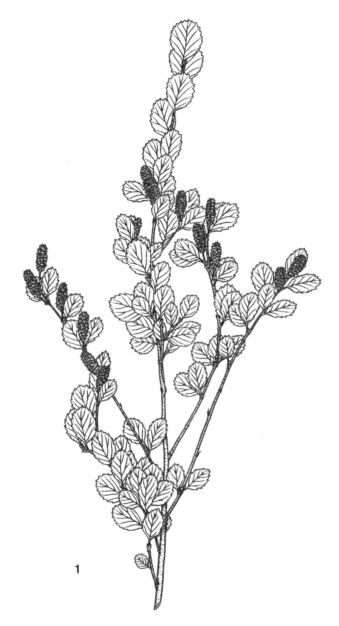

1. Individual with short and long vegetative shoots and fruiting shoots with catkins, fruit a samara

KEY CHARACTERS
- low, erect clonal shrub, 0.5–3 m with ascending branches
- leaves borne in 2s or occasionally 3s on short shoots and singly on long shoots, blades obovate with crenate-dentate margin and few teeth
- bracts cross-shaped, subtending seeds
- habitat diverse wetland ecosystems with circumneutral or basic substrates

BIGNONIACEAE
Campsis radicans (L.) Bureau

Trumpet-Creeper, Trumpet-Vine

Size and Form. Large, vigorous vine, creeping but usually climbing trees, fence posts, walls, old buildings, and other objects to 15 m or more; attachment by aerial roots; clone-forming by rooting at nodes.

Leaves. Opposite, deciduous, pinnately compound, up to 50 cm long; leaflets 7–13 (usually 9 or 11), the terminal leaflet on a stalk 10–20 mm long, the upper pair usually sessile or on short stalks; stalks of the lowest pair the longest, 0.5–1.5 cm; leaflets 3–8 cm long and 1–4 cm wide; ovate to lanceolate; acuminate; rounded or cuneate; sharply and coarsely toothed; glabrate and yellowish-green to bright green above, paler and usually more or less pubescent beneath, especially along the veins and midrib; petioles 2–5 cm long.

Stems-Twigs. Moderate, greenish becoming pale yellowish or straw colored, puberulent or scabrous; leaf scars opposite, low, connected by transverse hairy ridges, shield-shaped or C-shaped open side up; bundle scar 1, compound and crescent-shaped; aerial roots in 2 rows or patches below the nodes.

Winter Buds. Small, solitary, sessile, glabrous, yellowish, with 2 or 3 pairs of visible scales.

Flowers. June–September; bisexual; showy, reddish-orange; borne on short pedicels in conspicuous crowded clusters at the ends of curved branches; calyx about 1.5 cm long, glabrous; corolla trumpet-shaped, glabrous, generally 6–9 cm long; pollinated to a large extent by birds; besides the nectar-secreting ring in the flower, there are 4 sets of nectaries outside the flower: minute glands on the petioles, on the calyx, on the corolla lobes, and over the developing fruit (Elias and Gelband 1975). Insect and hummingbird pollinated.

Fruit. Capsule; August–October; 2-celled, thick, 10–20 cm long, green, turning grayish-brown as they mature; compressed at right angles to its internal partition and forming a longitudinal ridge on each side; seeds many, flat, with tissuelike wings; seeds 136,000/lb (299,200/kg).

Distribution. Locally frequent in southern lower Michigan. County occurrence (%) by ecosystem region (Fig. 19): I, 37; II, 10; III, 0; IV, 0. Entire state 20%. Widely distributed throughout e. North America, the Great Plains, w. to UT, CA, WA.

Site-Habitat. In the South its native habitats include alluvial bottomlands, stream banks, and open forests with moist soil; also occurring in the same sites in Michigan. Tolerates drought, flooding, low nutrient availability, and virtually any soil, including that of sidewalk cracks.

Notes. Shade-intolerant, tolerating moderate shade in juvenile phase. Native mostly south of the Ohio River, but widely escaped from cultivation in Michigan and other parts of North America. Deam (1924) described it as a weed and a nuisance in fields and open woodlands in the valleys of the Ohio, Wabash, and White rivers; climbing trees to 50 m or more. Described by Dirr (1990) as a rampant, clinging, strangling vine; at its best on fence posts; when it reaches the top of the post, it forms an immense whorl of stems and flowers, making the post appear as if it is about to fly. Once popular and attractive in landscaping for screening: "If you can't grow this plant, you should give up gardening." However, it is not recommended for planting due to its potential invasive harm in strangling and outcompeting native plants. Described as *Bignonia radicans* L., the name *Campsis* (from the Greek, *kampsis*, referring to curved stamens) is used to distinguish its aerial rooting habit from the that of other vines of the Bignoniaceae that attach by tendrils, such as cross vine, *Bignonia capreolata* L.

Chromosome No. $2n = 40$, $n = 20$; $x = 20$

Similar Species. Native in China and planted in Japan is *Campsis grandiflora* (Thunb.) Loisel, Chinese trumpet-creeper, noted for less climbing vigor and its early production of very large, scarlet to orange-pink flowers.

1. Leaf, × 1/5
2. Single leaf and flowering shoot, reduced
3. Fruit, capsule, × 1/5
4. Winged seeds, reduced
5. Winter lateral bud and leaf scar, enlarged
6. Winter twig with aerial roots, reduced

KEY CHARACTERS

- vigorous vine, climbing trees and other objects
- leaves opposite, pinnately compound, with 9–11 sharply and coarsely toothed leaflets
- stems straw-colored, with 2 sets or patches of aerial roots below the nodes
- flowers conspicuous, trumpet-shaped, orange-red, clustered at the ends of curved, upturned branches
- fruit a thick capsule, 10–20 cm long, splitting along 2 sutures to release many small winged seeds
- habitat open alluvial floodplains or woodlands of any soil type in southern lower Michigan and the southern part of the western Great Lakes Region

RHAMNACEAE
Ceanothus americanus Linnaeus

New Jersey Tea

Size and Form. Low, erect shrub, to 1 m; branches typically upright, often dense and spreading, usually many stems arising from a large, deep, dark red, root collar, plate, or grub (see "Notes").

Bark. Thin, grayish to reddish-brown.

Leaves. Alternate, simple, deciduous, blades 3–10 cm long, 2–6 cm wide; ovate to ovate-oblong; acute or acuminate; rounded or broadly cuneate; irregularly serrate; dull green above, slightly pubescent, becoming glabrous, grayish-green, and pubescent or nearly glabrous beneath; 3 prominent veins from the base, arcuate; petioles short, 0.6–1.2 cm long.

Stems-Twigs. Slender, rounded, yellowish-green to brownish, puberulent when young, becoming glabrate; leaf scars small, half elliptical, bundle scar 1; stipule scars present: pith large, white, continuous.

Winter Buds. Terminal bud present; lateral buds small, ovoid, sessile, with about 4 scales exposed, the first 2 nearly as long as the bud.

Flowers. July; bisexual; white, borne in dense, showy, round-topped clusters terminating from long stout, glabrous peduncles, which are usually axillary, the lower peduncles progressively longer, from 3–15 mm; small, about 4 mm across; pedicels 4–5 mm long; calyx 5-lobed; petals 5; stamens 5. Insect pollinated.

Fruit. Capsule, borne on a stalked, saucerlike base; September–October; 3-celled, brown, roughened or crested on the angles; globose, depressed, 3–6 mm wide; fruit falling and leaving the silvery-lined base, persisting into winter; 1 seed per cell; seeds 112,000/lb (246,400/kg).

Distribution. Locally frequent in the lower Michigan; occasional in upper Michigan. County occurrence (%) by ecosystem region (Fig. 19): I, 76; II, 60; III, 14; IV, 38. Entire state 61%. Widely distributed throughout e. North America, QC, ON s. to FL, LA; Great Plains from NE to TX.

Site-Habitat. Open, dry to dry-mesic, fire-prone sites; primarily in ecosystems with circumneutral to basic substrates, although topsoil may be acidic; oak and pine barrens, prairies on steep slopes, oak openings, dry oak forests. Associates include jack pine; eastern redcedar; northern pine; black, white, and bur oak; pignut hickory; sassafras; wild plum; prairie willow; hazelnuts; bearberry; and fragrant sumac.

Notes. Shade-intolerant. Useful as an indicator species of dry, fire-prone sites usually with calcareous substrates. Through its relationship with nitrogen-fixing bacteria, it can provide nitrogen to the soil from the atmosphere. Most of the >50 species in the genus occur in western North America. Leaves and roots used medicinally, and leaves were reported a good substitute for oriental tea during the American Revolutionary War, but they lacked caffeine. Species of *Ceanothus* sprout readily following a fire, and seeds may survive for years in the soil awaiting fire to break their dormancy and germinate. Seeds of snowbrush, *C. velutinus* Hook., of the central and northern Rocky Mountains and west to California, are reported to remain dormant for at least 200 years (Kershaw, MacKinnon, and Pojar 1998).

A *grub* is a local expression for the woody root-collar plate of certain oaks and both *Ceanothus* species growing on dry, fire-prone sites (Curtis 1959, 336). When a *Ceanothus* sprout has been killed by fire, the root collar forms callus tissue, and the plant continues to live. As a result of repeated fire and resprouting, a large, extremely tough, burllike root stock builds up. Settlers had to dig out such surface root plates, which had built up over decades of fire. The word *grub* comes from the German *grubben*, "to dig."

Chromosome No. $2n = 24$, $n = 12$; $x = 12$

Similar Species. The prairie redroot, *Ceanothus herbaceous* Raf, with smaller leaves, has a narrower geographic and ecological distribution. Several other species are native in the Southwest, Mexico, the Rocky Mountains, and the Pacific Northwest, including redstem ceanothus, *C. sanguineus* Pursh, a disjunct outlier of which occurs in northern Keweenaw Co., Michigan (Voss and Reznicek 2012).

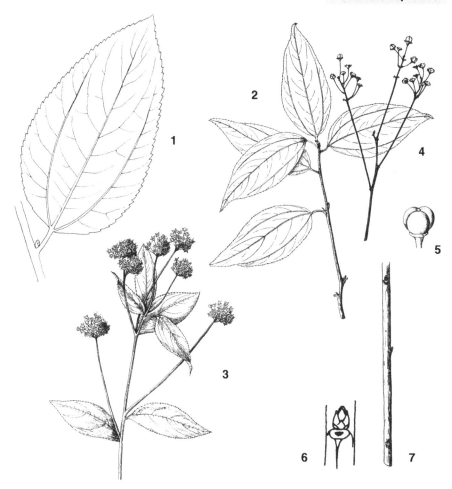

1. Leaf, × 1
2. Vegetative shoot, × 1/2
3. Flowering shoot, × 1/2
4. Fruiting shoot, capsule, reduced
5. Single fruit, capsule, × 2
6. Winter lateral bud and leaf scar, × 1
7. Winter twig, × 1

KEY CHARACTERS

- low, erect, many-branched shrub to 1 m, stems arising from a large, deep, dark-red root plate or grub
- leaves with blades 3–10 cm long, 2–6 cm wide; ovate to ovate-oblong with 3 prominent veins from the base
- flower and fruit peduncles usually axillary
- fruit a capsule borne on a stalked, saucerlike base, persisting into winter

Distinguished from prairie redroot, *Ceanothus herbaceous,* by longer and wider leaves, ovate to ovate-oblong leaf shape, flower and fruit peduncles much longer and usually axillary, capsule roughened or crested on angles.

RHAMNACEAE
Ceanothus herbaceus
Rafinesque-Schmaltz

Prairie Redroot,
Narrow-leaved New Jersey Tea

Size and Form. Low, erect, many-branched shrub, to 0.8 m; usually many stems arising from a large, deep, dark red root collar, plate, or grub (see "Notes" for *Ceanothus americanus*); also clone-forming by means of rhizomes.

Bark. Thin, grayish to reddish-brown, becoming purplish with age.

Leaves. Alternate, simple, deciduous, blades 1.5–5 cm long, 0.8–2 cm wide; elliptic-lanceolate, obtuse or subacute; rounded or broadly cuneate; crenate-serrulate; glabrous, lustrous green above, paler, nearly glabrous beneath, with hairs along the veins; 3 prominent veins from the base, arcuate; petioles very short, 4–6 mm long, hairy.

Stems-Twigs. Slender, rounded, yellowish-green to brownish, puberulent when young, becoming glabrate; branches from the central woody stalk appearing herbaceous, dying back in winter; leaf scars small, half elliptical, bundle scar 1; stipule scars present; pith large, white, continuous.

Winter Buds. Terminal bud present; lateral buds small, ovoid, sessile, with about 4 scales showing, the first 2 nearly as long as the bud.

Flowers. June; bisexual, white, borne in dense, showy, round-topped clusters on slender, short, terminal peduncles <5 cm long; small, about 3 mm across; white; calyx 5-lobed; petals 5; stamens 5. Insect pollinated.

Fruit. Capsule, borne on a stalked, saucerlike base; August–September; 3-celled, dark brown, not roughened or crested on the angles; globose, depressed, 5–6 mm wide; fruit falling and leaving a saucer-shaped, silvery-lined base, persisting into winter; 1 seed per cell.

Distribution. Rare in southern lower Michigan. Locally frequent in northern lower Michigan; occasional in the western part of upper Michigan. County occurrence (%) by ecosystem region (Fig. 19): I, 5; II, 43; III, 29; IV, 50. Entire state 25%. Widely distributed in e. North America (QC to MN, s. to VA, TN), Great Plains, and Rocky Mountains (ND, MT s. to TX, NM).

Site-Habitat. Open, dry to dry-mesic, fire-prone sites; primarily in ecosystems with circumneutral to basic substrates, although topsoil may be acidic; prairies, pine and oak barrens, and rocky limestone barrens and limestone bedrock lakeshores. Frequently found in the High Plains District (Fig. 19) of northern lower Michigan, characterized by rapid increase in heat sum in early spring and concomitant "late frosts" (i.e., severe freezing) in late May and early June. Associates include jack pine, northern pin and black oak, black cherry, sand cherry, sweetfern, low sweet blueberry, and bearberry.

Notes. Shade-intolerant. Useful as an indicator species of dry, fire-prone sites, usually with calcareous substrate. It also develops a large, belowground, woody root-collar plate or grub as in *Ceanothus americanus*. Through its relationship with nitrogen-fixing bacteria, it can provide nitrogen to the soil from the atmosphere.

Chromosome No. $2n = 24$, $n = 12$; $x = 12$

Similar Species. New Jersey tea, *Ceanothus americanus* L, is similar but has wider leaf blades and is much more widely distributed in the southern half of lower Michigan. The littleleaf redroot, *C. microphyllus* Michaux, of the dry pinelands and sand hills of the southeastern coastal plain has leaves only 3–8 mm long. See *C. americanus* for other species of the genus.

1. Leaf, × 1 1/2
2. Flowering shoot, × 1/2
3. Fruiting shoot, capsule, × 1/2

KEY CHARACTERS

- low, erect, many-branched shrub to 0.8 m, stems arising from a large, deep, dark-red root plate or grub
- leaves with blades 1.5–5 cm long, 0.8–2 cm wide; elliptic-lanceolate with 3 prominent, veins from the base
- flowers and fruit borne on terminal peduncles
- fruit a capsule borne in clusters, not roughened or crested on angles; peduncles terminal, <5 cm long, persisting into winter

Distinguished from New Jersey tea, *Ceanothus americanus*, by leaf blades shorter and narrower, elliptic-lanceolate shape; flower and fruit peduncles shorter and terminal; and capsule not roughened or crested on angles.

CELASTRACEAE
Celastrus scandens Linnaeus

American Bittersweet

Size and Form. High-climbing, twining vine to >15 m, climbing on trees, shrubs, and other objects, sometimes trailing or sprawling on the ground and over low vegetation, attaching by spiraling around host stems; forming clones by suckering from roots.

Bark. Large branches smooth, becoming rough and flaking with age, gray or brown.

Leaves. Alternate, simple, deciduous blades 5–12 cm long and about half as wide; ovate to oblong-ovate; acuminate or acute; broad-cuneate, finely serrate or crenulate; moderately thick; deep glossy green above, becoming yellowish-green in autumn; glabrous; petioles short, 0.5–1.5 cm long.

Stems-Twigs. Twining, moderately slender, main stem to 4 cm in diameter, flexible, circular in cross section, glabrous, at first green, becoming brownish-gray; no tendrils or aerial roots present; leaf scars slightly raised, half elliptical or broadly crescent-shaped; bundle scar 1; pith large, white, continuous.

Winter Buds. Terminal bud absent; laterals small, sessile, subglobose, pointing outward at nearly a right angle to the stem, glabrous, brownish-gray; 6 or more mucronate scales, outer scales long-pointed; stipule scars minute.

Flowers. April–May; unisexual, plants dioecious; borne in terminal panicles, 3–5 cm long; small; greenish or whitish; calyx 5-lobed; petals 5, male flowers with 5 stamens and a rudimentary pistil, female flowers with 1 compound pistil and rudimentary stamens; plants dioecious or polygamodioecious. Insect pollinated, especially by bees.

Fruit. Capsule, borne in terminal clusters; September–October; valves orange, splitting and displaying scarlet arils covering the seeds, 1–1.2 cm across, persistent in winter; seeds 3–8 per fruit, reddish brown, elliptical, 3–6 mm long; seeds 26,000/lb (57,200/kg).

Distribution. Common to infrequent in the southern half of lower Michigan and lake-moderated counties of northern lower Michigan and Upper Peninsula. County occurrence (%) by ecosystem region (Fig. x19: I, 84; II, 47; III, 71; IV, 75. Entire state 69%. Widely distributed throughout e. North America and the Great Plains, QC to MB, s. to TX.

Site-Habitat. Open to moderately shaded diverse upland sites of moderate to high fertility; stream banks, floodplains, lakeshores, disturbed forests and edges, fencerows, sand dunes, roadsides. Associates include species in disturbed oak-hickory, oak savanna, dry-mesic to mesic forest edges, and floodplain forests.

Notes. Shade-intolerant, establishing and persisting in light to moderate shade. Climbs by twining with an open, spiraling pattern from left to right; by twining it may girdle and constrict or occasionally kill small trees and shrubs. Fruit production favored by well-lighted environments, which it attains by climbing or sprawling. Vigorous growth on circumneutral and basic soils. Widely planted as an ornamental vine because of its yellowish-green leaves and orange to scarlet fruits. The bark was formerly used in medicine. The genus includes 45–60 species, only one of which occurs in North America. Others are found in East Asia, Australia, and Madagascar. It is a species of special concern, protected from collection on public lands under Michigan's Christmas Greens and Wildflower Protection Law, Public Act 182, 1962.

Chromosome No. $2n = 46$, $n = 23$; $x = 23$

Similar Species. The closely related oriental bittersweet, *Celastrus orbiculatus* Thunberg, of East Asia has become locally naturalized in urban-suburban areas of 11 counties of southern Michigan—a locally serious pest. Its flowers and orange fruits are borne in axillary cymes in small clusters in the axils of nearly round or broadly ovate leaves with rounded teeth and short-pointed apex. An increasingly serious invasive in domesticated landscapes, woodland edges, and cutover forests. It should be eliminated when found. Its introduction has brought about extensive hybridization, and Leicht-Young et al. (2007) report that the two species are sometimes difficult to distinguish using vegetative characters.

1. Flowering shoot, reduced
2. Female flower, × 4
3. Male flower, × 4
4. Fruiting shoot, capsule, reduced
5. Fruit, capsule, closed and open, × 1 1/2
6. Winter lateral bud and leaf scar, front view, enlarged
7. Winter lateral bud with long-pointed outer scales, side view, enlarged
8. Winter twig, reduced

KEY CHARACTERS

- twining vine, climbing trees, shrubs, and other objects or trailing on the ground
- stems slender, flexible, grayish-brown, with large, white, continuous pith
- leaves alternate, simple, blade ovate to oblong-ovate, finely serrate to crenulate; deep green, turning greenish-yellow in autumn
- fruit a capsule, orange, splitting and displaying bright red arils covering the seeds, persistent in winter, borne in terminal clusters

Distinguished from oriental bittersweet, *Celastrus orbiculatus*, by leaves ovate to oblong-ovate, finely serrate or crenate, apex acuminate to acute; flowers and fruit in terminal clusters; fruit with scarlet arils; occurring primarily in sites distant from urban areas.

CANNABACEAE
Celtis tenuifolia Linnaeus

Dwarf Hackberry

Size and Form. Straggly shrub or small tree, 2–6 m (8–20 ft) high and 8–20 cm (3–8 in) in diameter. Trunk commonly forked; irregular and asymmetrical crown; branches numerous, stiff, twisted, and intergrown. Michigan Big Tree: girth 109.2 cm (43 in), diameter 34.8 cm (13.7 in), height 7.9 m (26 ft), Washtenaw Co.

Bark. Thin, smooth, pale gray, developing corky ridges with age.

Leaves. Alternate, simple, blades 3–6 cm long, 2.5–4 cm wide; obliquely ovate; acute to short-acuminate; rounded to moderately asymmetrical; entire to irregularly shallowly toothed above the middle, teeth absent on leaves of flowering shoots; gray-green above, paler beneath; glabrous to scabrous or finely hairy above, pubescent beneath; 3 primary veins from the base; turning yellow in autumn; petiole pubescent, short, 0.5–1 cm long.

Stems-Twigs. Slender, narrow, zigzag, stiff; pubescent; grayish to brown; leaf scars two-ranked, oval to crescent-shaped; bundle scars 3; pith chambered, especially at the nodes; pith white.

Winter Buds. Terminal bud absent; end and lateral buds ovoid, light brown, 1–2 mm long; ovoid, acute, closely appressed.

Flowers. May, monoecious or polygamous; well ahead of the vegetative shoots; greenish, inconspicuous; on slender pedicels; male flowers in clusters at the base of the shoot; calyx greenish, deeply 5-lobed; corolla 0; stamens 5; female flowers usually solitary in the axils of the upper leaves; ovary 1-celled. Wind pollinated.

Fruit. Drupe, September; borne on a short pedicel, 3–6 mm, subglobose, 5–9 mm in diameter; distinctive coloration includes orange-red, salmon-colored, and orange to light cherry in late summer, becoming reddish-purple in late autumn; thick-skinned, thin-fleshed; mature fruit plump, with smooth surface, persistent on a wiry stalk through the winter, insipid, 5–8 mm; pit smooth to shallowly and obscurely pitted.

Distribution. Rare in the southern Lower Peninsula, highly localized; known from Jackson, Lenawee, Livingston, St. Joseph, and Washtenaw Co.; reaching one part of its northernmost distribution in southern Michigan. County occurrence (%) by ecosystem region

(Fig. 19): I, 13; II, 0; III, 0; IV, 0. Entire state 6%. Primarily a southerly species, FL w. to TX, OK, MO, n. and e. to southern MI and ON, NJ.

Habitat. Characteristic of dry-mesic to dry sites, including wooded sand dunes; sandy, calcareous glacial deposits; and soils with limestone near the surface. Associates include dwarf chinkapin oak, pignut hickory, sassafras, black cherry, black oak, ground juniper, New Jersey tea, fragrant sumac, gray dogwood, northern dewberry.

Notes. Shade-intolerant; slow-growing; short-lived. Germination epigeal. Usually bearing fruit when only 1.5–2 m tall. A rare but easily overlooked species, becoming more frequent south of Michigan. Found on disturbed sites, often eliminated by developing oak-hickory forests. It rarely occurs together with the more common northern hackberry. First discovered in Michigan in 1971 and described in detail by Wagner (1974).

Chromosome No. $2n = 20$, $n = 10$; $x = 10$

Similar Species. Desert hackberry, *Celtis pallida* Torrey, is a small shrub ranging from central and southern Texas to New Mexico and Arizona into Mexico and Central and South America.

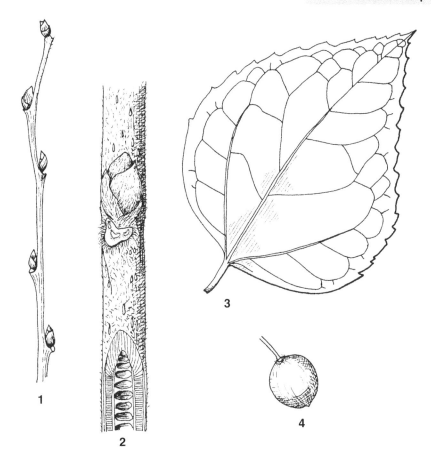

1. Winter twig, × 1
2. Portion of twig, × 4
3. Leaf, × 2
4. Fruit, drupe, × 1-1/2

KEY CHARACTERS

- scraggly shrub or small tree with asymmetrical crown
- bark gray with corky warts or ridges
- leaf blade entire or irregularly shallowly toothed above the middle, tip acute to short-acuminate, base rounded or inequilateral, three primary veins from the base
- drupe plump and smooth when mature; orange or salmon to reddish purple; persistent; pit smooth
- rare, reported from only five counties in southern lower Michigan

Distinguished from northern hackberry, *Celtis occidentalis*, by scraggly shrub or small tree form; leaves smaller and sparsely veined; acute to short-acuminate tips; fruits with short pedicels, plump with smooth surface when mature, orange-red becoming dark cherry-red in late autumn.

RUBIACEAE
Cephalanthus occidentalis Linnaeus

Buttonbush

Size and Form. Tall, erect, spreading shrub, 2–5 m; many stems branch vigorously and form a densely foliated shrub or clonal thicket in full sun; clonal by layering or basal sprouting. Michigan Big Tree: girth 88.9 cm (35 in), diameter 28.3 cm (11.1 in), height 10.7 m (35 ft), Oakland Co.

Bark. Thin, yellowish-brown to dark gray or brown, larger basal stems with flat-topped brown ridges and tan fissures, flaking with age.

Leaves. Opposite or in whorls of 3, rarely 4 (see Notes); simple, deciduous, blades 6–18 cm long and about half as wide; ovate to elliptic-lanceolate, acute to acuminate; broadly cuneate or rounded; entire but often wavy; thick and leathery; glabrous, lustrous, dark green above, glabrous to somewhat pubescent and paler beneath; petioles stout, grooved, 1–2 cm long; stipules small, triangular, occurring singly between the opposite or whorled leaf bases.

Stems-Twigs. Slender to moderately stout, with prominent, vertical, raised lenticels; current shoots and young stems slightly ridged or angled, becoming rounded on older twigs; glabrous and glossy, young stems light brown to yellowish, becoming grayish-brown; larger stems usually swollen at the base; leaf scars half round, bundle scars in a half circle, indistinct; stipule scars or persistent stipules connecting the leaf scars; pith light brown, continuous.

Winter Buds. Terminal bud absent, lateral buds very small, somewhat raised, dome-shaped and appearing sunken in bark, brownish, ca. 1–1.5 mm above top of leaf scar; bud scales indistinct.

Flowers. July–August; bisexual; creamy white, very fragrant, borne in dense, globose heads 2–3 cm across, on peduncles 3–7.5 cm long, 100–200 flowers per head; heads terminal, borne on long peduncles in axils of upper leaves; calyx with 4–5 rounded sepals; corolla 0.6–1.3 cm long, tubular, with 4 short, spreading lobes; stamens 4; ovary 2-celled. Insect pollinated.

Fruit. Achene, borne as a multiple of achenes in a dense, round head; September–October; heads becoming reddish-brown as they ripen; achenes reddish-brown, conical, 6–8 mm long; seeds ca. 2,250/lb (4,950/kg).

Distribution. Abundant in the southern half of lower Michigan; frequent in the west-ern part and northern tip of northern lower Michigan; absent in upper Michigan. County occurrence (%) by ecosystem region (Fig. 19): I, 92 ; II, 53; III, 0; IV, 0. Entire state 61%. Distributed throughout e. North America, the Great Plains, KA to TX; AZ, CA.

Site-Habitat. Open, wet, seasonally inundated, primarily circumneutral to basic sites, tolerant of inundation and sedimentation; river and stream margins, marsh edges, dunal depressions, bogs, lakeshores; wet microsites in hardwood swamps, likely in any low area where seed source is available and water flows or stands part of the growing season. Associates include black ash, red maple, American elm, swamp white oak, speckled alder, willows, dogwood, poison-sumac, shrubby cinquefoil, winterberry.

Notes. Very shade-intolerant. Early leaves on indeterminate shoots are opposite, whereas late leaves on such vigorous shoots are often whorled. Establishing in the open sites that are inundated by high groundwater levels or river flooding; persisting with fewer and fewer stems in light to moderate shade. High tolerance of seasonally high water table and very slow or stagnant-appearing water. Among the few woody plant species to survive and grow when roots and lower stems are covered with water for part or much of the growing season. Forming adventitious roots in moving water. *Cephalanthus* was one of only two shrubs reported "most tolerant" of woody species of the South to waterlogging (i.e., inundation) and flooding; in the same category as black willow (*Salix nigra*), water tupelo (*Nyssa aquatica*), and bald cypress (*Taxodium distichum*) (Hook 1984). *Most tolerant* is defined as "Species capable of living from seedling to maturity

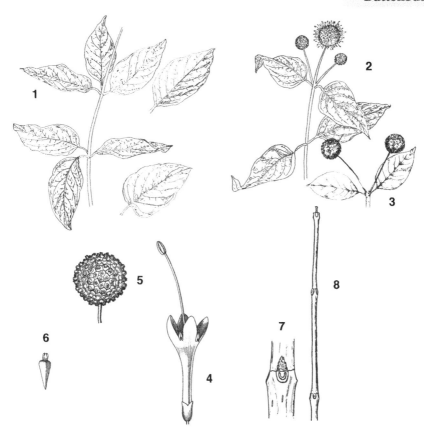

1. Leaves and vegetative shoot, reduced
2. Flowering shoot, reduced
3. Fruiting shoot, with head of multiple achenes, reduced
4. Flower, × 5
5. Head of achenes, × 1
6. Fruit, achene, × 2
7. Winter lateral bud and leaf scar, enlarged
8. Winter twig, reduced

KEY CHARACTERS

- tall, erect, spreading shrub
- leaves opposite or in whorls of 3, glabrous, lustrous, dark green above
- stems moderately stout with prominent vertical, raised lenticels
- fruit an achene, borne as a multiple of achenes in a dense, round head
- habitat open, circumneutral to basic sites with standing water throughout much of the growing season

in soils that are waterlogged almost continually year after year except for short durations during droughts. The name *Cephalanthus* is from the Greek, meaning "head-flower." Flowers are very attractive to small butterflies.

Chromosome No. $2n = 44$, $n = 22$; $x = 11$
Similar Species. The southern variety is

Cephalanthus occidentalis var. *pubescens* Raf.; it has pubescent twigs and leaf blades with soft pubescence beneath. A western variety, *californicus* Benth., occurs in California, and *C. salicifolius* Humb. & Bonpl. occurs in Texas and Mexico.

ERICACEAE
Chamaedaphne calyculata
(Linnaeus) Moench

Leatherleaf, Cassandra

Size and Form. Upright, evergreen bog shrub with many slender, wiry stems, 0.3–1 m; clone-forming by rhizomes and layering, noted for forming large clones.

Leaves. Alternate, simple, evergreen, blades 1–4 cm long, 0.5–1.5 cm wide; oblong, elliptic, or oblanceolate; acute or rounded; slightly rounded or cuneate; entire or obscurely finely toothed, slightly revolute; thick and leathery; dull green, dotted with silvery scales above, pale green becoming brownish, coriaeous and scurfy beneath, with orange-brown scales turning brownish or reddish in winter; petioles pubescent or glabrous, short, about 2 mm long.

Stems-Twigs. Slender, wiry, puberulent and scurfy at first, becoming gray and shreddy, finally smooth and dark reddish-brown; leaf scars minute, crescent-shaped, bundle scar 1; pith small, round, continuous.

Winter Buds. End bud when present small, 1.5–3 mm long, short-stalked, oblong to globose, with 4 exposed scales, basal scales ciliate; lateral buds very small, sessile, globose.

Flowers. April–June; bisexual; borne in many-flowered terminal, leafy racemes, 4–12 cm long; occurring singly in axils of the small leaves; small, 5–7 mm long; white; calyx 5-lobed; corolla oblong, bell- or urn-shaped, with 5 short lobes; stamens 10. Insect pollinated.

Fruit. Capsule; September; dry, roundish, somewhat flattened, 5-celled, 3–5 mm across, persistent in winter; seeds small, shiny, light brown, 1–2 mm long, wingless, numerous.

Distribution. Common to abundant locally throughout the state, increasing northward. County occurrence (%) by ecosystem region (Fig. 19): I, 66; II, 97; III, 100; IV, 100. Entire state 83%. Circumboreal but absent in w. North America; as far s. as n. IL, IN, OH, to NJ and mts. of NC.

Site-Habitat. Wet, open, acidic, nutrient-poor sites but also acidic microsites in ecosystems with circumneutral substrate below the rooting zone; wetland communities of many kinds, bog, muskeg, fens, acidic and circumneutral swamps, marsh borders, wooded dune and swale complexes. Associates are typical acidic sphagnum bog plants, including black spruce, tamarack, Labrador-tea, bog-rosemary,

bog-laurel, *Vaccinium* spp., mountain holly, and also calciphilic species such as wild black currant, *Ribes* spp., wild-raisin, bog birch, poison-ivy (Kost et al. 2007).

Notes. Very shade-intolerant. Occurs in single clumps, large clones, or large beds in pure stands. In Minnesota its presence defines the "muskeg" plant community (Smith 2008). Sprouts vigorously from rhizomes following fire. Circumboreal occurrence; present in Japanese bogs, northern Honshu and Hokkaido islands. The name *Chamaedaphne* is from Greek *chamae*, "on the ground," and *daphne*, "laurel."

Chromosome No. $2n = 22$, $n = 11$; $x = 11$

1. Leaf, × 1-1/2
2. Flowering shoot, × 1/2
3. Flower, × 2
4. Fruit, capsule, × 2
5. Fruiting shoot, × 1/2, and single capsule, ×1

KEY CHARACTERS

- low, upright, bog shrub, clone-forming
- leaves alternate, evergreen, elliptic, slightly revolute, thick, and leathery; green with silvery scales above; brownish, coriaeous, and scurfy beneath
- fruit a small capsule, persistent in winter
- habitat wet, open, acidic sites; especially bog and muskeg

RANUNCULACEAE
Clematis virginiana Linnaeus
Virgin's Bower

Size and Form. Slender, soft-wooded vine; trailing on the ground or climbing on shrubs, trees, and other objects by means of twining petioles to 5 m.

Leaves. Opposite, simple, deciduous; compound, mostly trifoliate; leaflets 3–10 cm long, 2.5–8 cm wide, all long-stalked; coarsely toothed, frequently deeply incised or 3-lobed; ovate or slightly heart-shaped; acuminate; subcordate to rounded; margins entire or with few, irregularly spaced, rounded to acute-tipped teeth; bright green above, paler beneath; glabrous on both sides or with sparse, pale hairs beneath when young; petioles 4–10 cm long, when leaf blades are shed, petioles remain, twisting on supports to pull the plant upward; usually persistent while blades wither, decay, and are removed piecemeal by winter winds.

Stems-Twigs. Slender, flexible, soft, the lower part persistent, becoming woody, dying back in winter; six prominent ridges; twining by petioles that are persistent after leaf blades are shed; at first brownish-green, becoming brown to reddish-brown; hairy or glabrous; leaf scars joined around stem by a transverse ridge, broadly V-shaped but concealed by persistent petiole; bundle scars 5–7; pith white, angled or star-shaped, continuous, with a thin partition at nodes.

Winter Buds. Terminal bud absent; lateral buds small, solitary, sessile, somewhat hairy; ovoid or flattened, with 2–4 mostly hairy scales.

Flowers. July–August; unisexual, plants dioecious; small; white to cream-colored, borne in large, open, axillary leafy panicles on pubescent peduncles 1–8 cm long; sepals 4, elliptical to obovate, 6–10 mm long; petals absent; male flowers have fertile stamens; female flowers have pistils and sterile stamens. Insect pollinated.

Fruit. Achene; late August–October; 3–5 mm long, < 2 mm across, flattened, brown, hairy, to which is attached a persistent feathery style, 2.5–4 cm long; achenes are borne in a head about 5 cm across, supporting a cluster of swirling feathery styles; seeds 192,000/lb (422,400/kg).

Distribution. Widely distributed throughout the state. County occurrence (%) by ecosystem region (Fig. 19): I, 84; II, 87; III, 86;

IV, 100. Entire state 87%. Widely distributed throughout e. North America, also the Great Plains from MN to TX.

Site-Habitat. Open to moderately shaded diverse upland sites of moderate to high fertility; stream banks, floodplains, lakeshores, edges of marshes, disturbed forests, fencerows, sand dunes, roadsides.

Notes. Shade-intolerant, establishing and persisting in moderate and rarely deep shade. Short-lived, it produces many wind-borne seeds and aggressively colonizes open, disturbed sites. The large genus *Clematis* includes about 300 species of woody and semiwoody vines and herbaceous perennials, chiefly occurring in temperate regions of both hemispheres. *Clematis* is derived from the Greek *clema*, "shoot," thus a slender vine. The common names virgin's bower and travelers joy derive from the fact that it forms bowers along fences and hedgerows that provide shade and shelter to maidens and travelers alike.

Chromosome number. $2n = 16$, $n = 8$; $x = 8$

Similar Species. A more northern relative (primarily in western upper Michigan), purple clematis, *Clematis occidentalis* (Hornem.) DC., is distinguished by early flowering as the leaves unfold (May–June), bisexual flowers that are solitary on elongated peduncles, and bluish to pink-purplish sepals. A variety, *grosseserrata* (Rydb.) J. Pringle, the western blue virgin's bower, is found in the Rocky Mountains and north and west to Oregon, Washington, and British Columbia. A perennial woody climber, the western white clematis, *C. ligusticiifolia* Nutt., occurs throughout western North America. In the Midwest and South, woody and semiwoody species include vasevine, *C. viona* L; the net-leaf clematis, *C. reticulata*

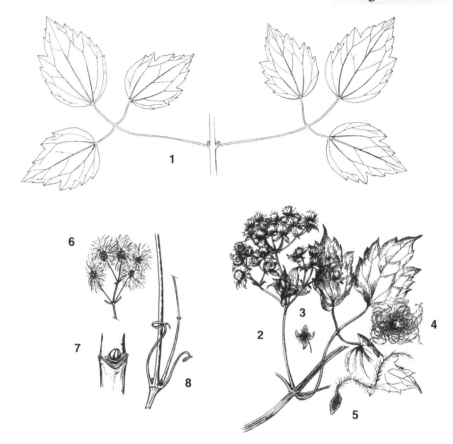

1. Leaves, × 1/2
2. Flowering shoot, × 1/2
3. Flower, × 1/2
4. Head of achenes, × 1/2
5. Achene, × 2
6. Fruiting shoot, × 1/2
7. Winter lateral bud and leaf scar, enlarged
8. Winter twig with persistent petioles at node, × 1/2

KEY CHARACTERS

- slender, soft-wooded vine; trailing on the ground or climbing by means of twining petioles
- leaves opposite, simple, deciduous; compound, mostly trifoliate, leaflets long-stalked, coarsely toothed or 3-lobed, ovate or slightly heart-shaped; petioles long
- stems soft, the lower part persistent, becoming woody, dying back in winter; six prominent ridges; pith white, angled or star-shaped
- flowers white to cream-colored, borne in profuse clusters
- fruit a flattened achene to which is attached a persistent feathery style, 2.5–4 cm long

Walter, in upland forests; and in swamps and bottomlands *C. crispa* L. In Central Europe, the woody climber is the common Waldrebe (forest vine), *C. vitalba* L. In East Asia, many herbaceous and semiwoody species occur, 147 in China alone.

MYRICACEAE
Comptonia peregrina
(Linnaeus) Coulter
Sweetfern

Size and Form. Low, many-branched shrub, 0.3–1 m; branches erect or spreading; forming dense clones by means of rhizomes.

Leaves. Alternate, simple, deciduous with a few withered ones persisting; fernlike, blades their entire length, 5–12 cm long, 1–1.5 cm wide; linear-oblong; acute or rounded lobes; base tapered or rounded; deeply pinnately lobed, 12–25 rounded or obtuse, often mucronate lobes per side, some sinuses nearly reaching the midrib; more or less pubescent, dark green above, pubescent and paler beneath; both sides covered with scattered yellow, shiny resin-dots; sweet-scented; petioles short, 1–5 mm long, with a pair of subcordate, long-pointed stipules at base, deciduous in winter, although some may persist.

Stems-Twigs. Slender, brown to purplish-brown, covered with white hairs, dotted with tiny yellowish resin glands when young; fragrant when bruised; leaf scars raised, 3-sided, bundle scars 3; stipule scars small; pith small.

Winter Buds. Terminal bud absent, lateral buds ovoid, 4 or more scales exposed; stipule scars small; fragrant when broken.

Flowers. April–May before the leaves appear; inflorescence unisexual with separate male and female catkins borne on the same or different plants; male catkins cylindrical, elongate, about 2–4 cm long, clustered at the ends of twigs; female flowers many, inconspicuous in globose catkins 1.5–2.5 cm long. Wind pollinated.

Fruit. Achene; July–August; small, boney; borne in a bristly burlike cluster, conical, 1–2 cm in diameter with 8–15, one-seeded achenes, each 3–4 mm long, light brown, smooth, fragrant, surrounded by 7–9 long, linear, glandular bristlelike bracts forming a bur; animal dispersed.

Distribution. Locally frequent to abundant in northern lower and upper Michigan; infrequent in southern Michigan. County occurrence (%) by ecosystem region (Fig. 19): I, 42; II, 87; III, 86; IV, 100. Entire state 67%. Primarily boreal and northern; distributed locally in e. North America from QC, ON, s. to IL, TN, GA.

Site-Habitat. Open, dry, sandy, acidic, infertile, fire-prone sites; typically in pine and oak barrens where it tolerates cold, frost-prone sites, especially in the Grayling Subdistrict (8.2, Fig. 19): also sand ridges in lake plain oak openings. Associates include jack and red pines, northern pin oak, black cherry, bigtooth aspen, black and white oaks, bearberry, sand cherry, low sweet blueberry, black huckleberry, wintergreen, and prairie willow.

Notes. Very shade-intolerant. Some brown, withered leaves usually persist through winter. Clones 110 m or more across are reported in Minnesota (Smith 2008). Heavily browsed by deer. All parts of the plant give off a fragrant, spicy odor. Root nodules contain symbiotic nitrogen-fixing bacteria, which may account in part for its competitive advantage in nutrient-poor sites. Seeds stored in forest soils are reported to have a viability of at least 70 years (Del Tredici 1977). Sweetfern is the alternate host for the rust fungus occurring on jack pine. Leaves formerly used in tonics, home remedies, and as a substitute for tea.

Chromosome No. $2n = 32$, $n = 16$; $x = 16$

Similar Species. Monotypic genus endemic to eastern North America.

1. Vegetative shoot, × 1/2
2. Flowering shoot with male catkins, overwintering form, × 1/2
3. Male catkin before expanding, × 1
4. Fruiting shoot with bur enclosing fruit, nut, × 1/2
5. Bur enclosing fruit, nutlet, × 2
6. Winter lateral bud and leaf scar, enlarged
7. Winter lateral bud, side view with raised leaf scar, enlarged

KEY CHARACTERS

- low, many-branched shrub; all parts fragrant
- stems slender, with white hairs and tiny yellow resin dots
- leaves fernlike their entire length, covered with tiny resin dots, stipules present
- fruit a small achene, enclosed in a bur of long, bristlelike bracts
- habitat dry, acidic, open, nutrient-poor, fire-prone sites

THE DOGWOODS—*CORNUS*

Widespread in the north temperate regions, the 60 species of *Cornus* are primarily deciduous, and 5 species are common in Michigan and the upper Great Lakes Region. Phylogenetically based biogeographic studies are available from Xiang et al. (2006). *Cornus* is one of the few genera of species we include that has large trees 20–30 m (66–100 ft) tall (*Cornus nuttallii*, Pacific dogwood of western North America, *C. controversa*, giant or pagoda dogwood of East Asia, and *C. macrophylla* of Japan), many shrubs, and two herbaceous species (*C. canadensis*, bunchberry of North America, and *C. suecica* of Siberia).

Shrub dogwoods are clone forming by sprouting from the root crown and roots, tip-rooting, and layering. *Cornus sericea*, red-osier, and *C. amomum*, silky dogwood, are indicators of open wet sites, whereas *C. alternifolia*, alternate-leaved dogwood, thrives in mesic, fertile, and shaded understories of beech-maple or northern hardwoods forests.

Cornus is characterized by opposite leaf arrangement, except in *C. alternifolia* and *C. controversa*, entire blade margins, and lateral veins strongly and evenly curving upward toward the tip as they near the margin (i.e., arcuate). When blades are slowly pulled apart, threadlike vascular strands still connect the parts up to 1 cm. Twigs are slender and often highly colored; leaf scars raised, bundle scars 3. The terminal bud is present and much larger than the laterals. The small, perfect flowers are white, greenish-white, or yellow and are borne in terminal clusters in the early spring. The fruit is a distinctively colored drupe with a 2-celled and 2-seeded pit. Fruits are widely dispersed by birds and mammals. The fruit, seeds, flowers, twigs, buds, bark, and leaves are utilized as food by various animals. The wood is hard and heavy. The name *Cornus* dates back to Theophrastus and signifies "horn," referring to extreme wood hardness.

The members of the genus *Cornus* are remarkably diverse in appearance and cultivated mainly for their bark, leaves, fruit, and bracts of the flowers. Large, showy bracts are typical of *C. florida*, *C. nuttallii*, *C. kousa*, and *C. canadensis*, whereas other species have small flowers lacking large bracts. Widely used in cultivation besides the North American species *Cornus florida* are *C. sericea* and *C. nuttallii*. European and Asian shrubs are also popular (*C. alba*, *C. kousa*, and *C. mas* L. (syn. *C. officinalis*). *Cornus mas*, the European cornelian cherry, was highly prized in England and Europe for its deep red to reddish-brown fruits (resembling the translucent mineral-gemstone for which it is named), early yellow flowers, and iron-hard wood.

1. Stems conspicuously bright or deep red.
    2. Stems bright to deep red and smooth usually nearly to the base of the plant except for conspicuous whitish lenticels increasing in size and abundance toward the base; current shoots slightly pubescent, soon becoming glabrous except for finely appressed hairs at the shoot apex; stems >8 mm with white pith; fruits whitish when ripe . . . *C. sericea*, p. 110
    2. Stems bright to deep red and smooth at first but with prominent reddish-brown fissures and rough bark soon increasingly abundant (and obscuring lenticels) toward the base; current shoots densely covered with fine rust-colored hairs to silky-downy pubescence, becoming glabrate; stems >8 mm with brown pith; fruit blue to bluish-white when ripe . . . *C. amomum*, p. 104
1. Stems not conspicuously bright or deep red.
    3. Leaves and leaf scars alternate; leaves crowded toward the end of the twig, sometimes appearing whorled; bud scales more than 2 . . . *C. alternifolia*, p. 102
    3. Leaves and leaf scars opposite; leaves not crowded toward the end of the twig; bud scales 2.
        4. Stems gray; leaf blades small, narrowly ovate to lanceolate; 3–4 veins per side; fruits white when ripe; borne in pyramidal, loose clusters . . . *C. foemina*, p. 106
        4. Stems mostly yellowish-green with dark purplish streaks; leaves large, broadly ovate to suborbicular; 6–9 veins per side; fruit pale blue to blue when ripe; borne in dense, flat-topped cymes . . . *C. rugosa*, p. 108

CORNACEAE
Cornus alternifolia Linnaeus f.

Alternate-leaved Dogwood, Pagoda Dogwood

Size and Form. Medium to tall, erect, understory few-stemmed shrub or small tree, 2–8 m, 4–15 cm in diameter with a short, slender, and straight bole; branches long, slender, arching and upswept at the tips; arranged in irregular horizontal whorls to form a storied or pagoda effect. Occasionally clonal by layering. Michigan Big Tree: girth 63.5 cm (25 in), diameter 20.2 cm (8.0 in), height 11.9 m (39 ft), Oakland Co.

Bark. Thin, dark reddish-brown, smooth, becoming shallowly fissured.

Leaves. Alternate, simple, deciduous; clustered at the ends of the current shoots, shoots often so short that the leaves appear to be opposite or whorled; blades 7–12 cm long, 6–8 cm wide; oval, ovate to obovate; acute or acuminate; cuneate to rounded; entire, slightly undulate or obscurely wavy-toothed; thin, dark green, nearly glabrous above, paler and covered with appressed hairs beneath; fall color varies from yellow to shades of red and purple; veins arcuate; petioles slender, grooved, hairy, with clasping bases, 2–7 cm long.

Stems-Twigs. Slender, greenish or reddish, becoming smooth, lustrous, dark green to dark purplish-red; upcurved branch tips give a pagodalike appearance; leaf scars crowded at the tip. Sylleptic growth often occurs on vigorous current shoots, the terminal part of shoot often exceeded in length by the sylleptic shoot just below it; leaf scars crowded toward the tip, crescent-shaped; bundle scars 3; pith white, continuous.

Winter Buds. Terminal bud present, small, narrowly ovoid, acute, 4–6 mm long with 2–4 keeled scales exposed; glossy, green, red, or purplish-brown; lateral buds much smaller; flower buds spherical or vertically flattened.

Flowers. May–June, after the leaves; bisexual; small, borne on slender pedicels in many flowered, irregular, open terminal cymes; calyx cup-shaped, obscurely 4-toothed, covered with fine, silky, white hairs, sepals minute; petals 4, cream-colored, oblong, 2.5–3.5 mm long; stamens 4; ovary 2-celled. Insect pollinated.

Fruit. Drupe; October; globular, bluish-black, 6–8 mm across, borne in loose, red-stemmed clusters, flesh bitter; seeds 8,000/lb (17,600/kg).

Distribution. Occasional to locally common throughout the state. County occurrence (%) by ecosystem region (Fig. 19): I, 84; II, 63; III, 100; IV, 88. Entire state 78%. Widely distributed, e. USA and adjacent Canada: NL to MN, s.e. MB, s to FL, AL, AR.

Site-Habitat. Mesic to wet-mesic, shaded, slightly acidic to basic, usually nutrient-rich sites; beech–sugar maple and hemlock-hardwood northern forests, floodplains, wooded dune and swale complexes, swamp edges. Associates include sugar, red, and striped maples; beech; basswood; bitternut hickory; hop-hornbeam; tulip tree, spicebush, elderberries, Canadian fly honeysuckle, maple-leaved viburnum, leatherwood. Tolerates moderately high water tables and organic soils; not found in dry, drought-prone sites.

Notes. Shade-tolerant; slow-growing, relatively short-lived. Shoots and leaves arranged to make the best use of low light conditions of the forest understory. The sylleptic growth habit is an adaptation for light foraging. Good indicator species for moist, nutrient-rich sites.

Chromosome No. $2n = 20$, $n = 10$; $x = 10$

Similar Species. Sister species of *Cornus controversa* Hemsl. of China and Japan (Xiang et al. 2006). It is similar in morphology and also occupies moist and fertile sites; it is a medium to large tree, to 25 m (80 ft) tall and 50 cm (20 in) in diameter.

1. Winter twig, × 1
2. Portion of twig, enlarged
3. Leaf, × 3/4
4. Flowering shoot, × 1/2
5. Flower, enlarged
6. Fruiting shoot with drupes, × 1/2

KEY CHARACTERS

- medium to tall, erect understory shrub or small tree to 8 m
- stems greenish or reddish, with leaf scars crowded near the tip
- leaves alternate, crowded near the ends of the current shoot, appearing opposite or whorled; blades long-pointed, entire, veins arcuate
- fruit a bluish-black drupe
- habitat mesic, nutrient-rich, shaded sites but best developed at forest edges

CORNACEAE
Cornus amomum Miller

Silky Dogwood

Size and Form. Upright, loosely-branched shrub with arching stems to 4 m; forming multistemmed clones in open sites by root-collar sprouting and to a lesser extent rhizomes and tip rooting.

Bark. Lower stems and those >0.8 cm with smooth, red epidermis breaking up to form narrow, longitudinally elongated, corky fissures, rough to touch.

Leaves. Opposite, simple, deciduous, blades 3–9 cm long, 2–6 cm wide; broadly ovate or elliptic; short-acuminate; rounded or broadly cuneate; entire; dark green to yellowish-green above, glabrous or with appressed pubescence, lighter green to glaucous and with appressed rust-colored pubescence beneath on immature blades; veins parallel-arcuate, 3–5 per side; petioles 0.5–1.5 cm long, often arching, causing appearance of drooping leaves.

Stems-Twigs. Slender to moderate; current shoots densely covered with fine rust-colored hairs to silky-downy pubescence, becoming glabrate; uppermost stems red, smooth, with scattered lenticels, soon becoming roughened in stems >0.5 cm with conspicuous, narrow, corky fissures; young shoots green becoming red to dark reddish-brown; bud scale scars narrow, crescent-shaped, bundle scars 3; pith small, brown, continuous.

Winter Buds. Terminal bud conical, 3–5 mm long, light brown, with dense grayish hairs; lateral buds sessile, small, 2–3 mm long, clearly visible, closely appressed, flattened triangular to oblong, densely hairy, 2 valvate scales exposed.

Flowers. May–July, after the leaves; bisexual; borne in flat to slightly convex, dense terminal cymes; small, yellowish-white; calyx lobes lanceolate, 0.8–1.3 mm long; petals 4, 4–5 mm long; stamens 4. Insect pollinated.

Fruit. Drupe; August–September; globose, dark blue with bluish-white or white patches when ripe, ca. 6–8 mm long and equally wide; pedicels yellow-brown to maroon; pit 2-celled and 2-seeded, seeds 12,200/lb (26,840/kg).

Distribution. Frequent to locally abundant in southern Lower Michigan; frequent in northern lower Michigan counties bordering the lakes; occasional to absent in the Upper Peninsula. County occurrence (%) by ecosystem region (Fig. 19): I, 95; II, 63; III, 57; IV, 0.

Entire state 71%. Distributed throughout most of e. USA and adjacent Canada; New England, s. to FL and w. to s. QC, ON, MN, s. to OK, MS.

Site-Habitat. Open, wet to wet-mesic sites, usually circumneutral to basic sites with moderate or high nutrient availability; swamps; marshes, bog and fen edges, lakeshores, along streams and first bottom of river floodplains; disturbed upland areas adjacent to wetland sites. Similar in habitat and associates as red-osier, *Cornus sericea* L.

Notes. Very shade-intolerant, persisting in light shade. On vigorous indeterminate shoots in full sun, the silky pubescence sometimes persists visibly >10 cm along the red stems after leaf shed. Easily transplanted. Good for landscape plantings on wet sites but not widely used. Latest to flower among all species of *Cornus*. Shedding leaves later than its common associate, *Cornus sericea*.

Chromosome No. $2n = 22$, $n = 11$; $x = 11$

Similar Species. Most Michigan individuals are *Cornus amomum* subsp. *obliqua* (Raf.) J. S. Wilson, whereas the closely related *C. amomum* subsp. *amomum* is reported only from Monroe Co. (Voss and Reznicek 2012). The subsp. *amomum* occurs primarily in the southeastern United States (Gleason and Cronquist 1991, p. 324).

1. Flowering shoot, × 1/2
2. Flower, × 2
3. Fruiting shoot, × 1/2
4. Fruit, drupe, × 4
5. Winter bud and leaf scar, enlarged
6. Winter twig, × 1

KEY CHARACTERS

- upright, loosely branched shrub to 3.5 m
- current shoots and upper stems reddish and smooth, rapidly forming longitudinally fissured epidermis lower on the branch; current shoots with fine, dense, silky-downy pubescence; pith of older stems brown
- leaves opposite with parallel-arcuate venation
- fruit when ripe a dark blue drupe with pale patches
- habitat open wetlands of many kinds

Distinguished from red-osier, *Cornus sericea*, by bark with rough and longitudinally fissured epidermis on lower part of medium to large stems (>0.5 cm); current shoots at first with fine silky-downy pubescence; pith brown, not white; fruit bluish when ripe.

Distinguished from gray dogwood, *Cornus foemina*, by red to reddish-brown, striated bark; current shoots at first with fine, silky-downy pubescence; inflorescence flat or slightly convex, fruit bluish when ripe, not white; occurring almost exclusively in wetlands.

CORNACEAE
Cornus foemina Miller

Gray Dogwood

Size and Form. Upright, much-branched shrub to 5 m high; forming dense, many-stemmed clones by sprouting roots. Michigan Big Tree: girth 35.6 cm (14.0 in), diameter 11.3 cm (4.5 in), height 6.7 m (22.0 ft), Oakland Co. Individual with larger diameter, 45.7 cm, Wayne Co.

Bark. Smooth, light gray; rough bark forms on older stems at base.

Leaves. Opposite, simple, deciduous, blades 2–10 cm long and half as wide; narrowly ovate to lanceolate, long-acuminate; cuneate; entire; grayish-green above, glaucous beneath, becoming reddish or purplish in autumn; appressed-pubescent both sides; usually 3–4 veins per side, parallel-arcuate; petioles 0.5–1.5 cm long.

Stems-Twigs. Slender, light reddish-brown to peach-colored, becoming gray, glabrous, lenticels prominent; bud scale scars narrowly V- to crescent-shaped, bundle scars 3; pith white on young twigs, becoming tan to brownish on larger stems, continuous.

Winter Buds. Terminal buds occasionally stalked, narrowly pyramidal, 2–5 mm long, sharp-pointed, brown with appressed silky hairs, 2–4 scales exposed; lateral buds sessile, 1–3 mm long, flattened-triangular, closely appressed, densely hairy.

Flowers. May–June; bisexual; borne in pyramidal, terminal, loose clusters 3–6 cm across, pedicels pubescent, becoming reddish, 2–2.5 cm long; small and strongly convex; creamy white; calyx minute, petals 4, recurved; stamens 4. Insect pollinated.

Fruit. Drupe; July–September; globose, white (rarely bluish) when ripe, 4–5 mm across, on bright red pedicel; pit 2-celled and 2-seeded, seeds 13,000/lb (28,600/kg).

Distribution. Common to locally abundant in lower Michigan, becoming less common to absent northward. County occurrence (%) by ecosystem region (Fig. 19): I, 95; II, 53; III, 43; IV, 12. Entire state 67%. Widely distributed throughout e. North America, QC to MB, s. to FL and TX.

Site-Habitat. Wet-mesic to dry, open sites of diverse landforms and soils; cutover forests of diverse composition, lakeshores, riverbanks, edges of swamps, fens, marshes, wet shrub prairies and thickets, roadsides, fencerows. Occurring in 10 near-natural palustrine (wet)

communities of Michigan (Kost et al. 2007). Apparently colonizing disturbed uplands from open, wet-mesic sites and edges of wetland ecosystems of many kinds. Associates may include nearly all plants of wetland edges, oak-hickory communities, to a lesser extent mesic forests and species that occur in diverse open, disturbed sites.

Notes. Very shade-intolerant; forming dense clones with rounded profiles to 15 m2 in size, especially in full sun. Clones may reach 10 m across in Minnesota (Smith 2008). Easily transplanted. Effective ornamental plant for a wide range of soil-water conditions; dense, massed growth, and red infructescences; sometimes used for hedges. Hybridizes with *Cornus rugosa* to form *C.* × *friedlanderi* W. H. Wagner.

Chromosome No. $2n = 22$, $n = 11$; $x = 11$

Similar Species. Morphologically similar in many respects to silky dogwood, *Cornus amomum,* but not in form or habitat. Michigan plants are subsp. *racemosa* (Lam.) J. S. Wilson, whereas populations of the subsp. *foemina* range south of the western Great Lakes Region.

106

1. Leaves, × 1
2. Flowering shoot, reduced
3. Fruiting shoot, drupe, reduced
4. Winter lateral bud and leaf scar, enlarged
5. Winter twig, reduced

KEY CHARACTERS

- upright, multibranched shrub to 5 m, forming dense multistemmed clones
- leaves opposite, narrowly ovate to lanceolate, veins 3–4 per side, parallel-arcuate
- stems and twigs gray
- fruit a white drupe when ripe, on bright red peduncles

Distinguished from red-osier, *Cornus sericea*, by erect not arching stems, gray twigs and main stems, leaves narrowly ovate to lanceolate, with long-acuminate apex.

Distinguished from silky dogwood, *Cornus amomum*, by gray twigs and main stems; stems more erect, not arching, current shoots lacking pubescence; pith white, becoming light brown, not brown; ripe fruit whitish on bright red peduncles.

107

CORNACEAE
Cornus rugosa Lamarck

Round-leaved Dogwood

Size and Form. Coarse, upright shrub, 2–3 m; loosely branched, spreading; forming clones by root sprouting.

Bark. Yellowish-green to gray, purple streaks, warty.

Leaves. Opposite, simple, deciduous, blades 5–12 cm long, 4–10 cm wide; broadly ovate to suborbicular, short-acuminate; rounded; entire; green with closely appressed hairs above, densely villous beneath; veins 6–9 pairs per side, parallel-arcuate; petioles 1–2 cm long.

Stems-Twigs. Slender, dull, greenish with reddish or purple streaks, becoming pink in fall and winter with sun exposure, undersides of twigs remaining greenish; more or less warty, lenticels purplish, young stems with closely appressed dense pubescence, becoming sparsely pubescent; short shoots occasional; pith white, continuous.

Winter Buds. Terminal bud present, much larger than laterals, 0.6–1.4 cm long, flower buds swollen at base, 2 valvate scales exposed, reddish, apex attenuated and curved; lateral buds oblong, sessile or short-staked, 3–7 mm long, 2 valvate scales exposed.

Flowers. May–June; bisexual; borne in dense, flat-toped terminal cymes, 5–7 cm across, peduncles stout, usually pubescent, 1–4 cm long; small; white; calyx minute, 4-toothed; petals 4, creamy white, 3–4 mm long; stamens 4, exceeding the petals. Insect pollinated.

Fruit. Drupe; August–September; globose, puberulent, pale blue to blue when ripe, rarely greenish-white, 5–6 mm across; pit 2-celled and 2-seeded; seeds 19,000/lb (41,800/kg).

Distribution. Occasional to locally common throughout the state. County occurrence (%) by ecosystem region (Fig. 19): I, 53; II, 63; III, 71; IV, 88. Entire state 61%. Widely distributed in n. North America, QC to MB, s. New England to VA, w. Great Lakes Region, Midwest.

Site-Habitat. Marked diversity of upland landforms and ecosystems, dry-mesic to mesic sites, often in well-drained sandy soils with calcareous substrate within the rooting zone, open to moderately shaded conditions, absent in wetlands; open oak and mixed forests, wooded lakeshores and dunes; stream banks, rock outcrops, roadsides, fencerows. Associates include many species from upland oak forests of southern Michigan and coastal ecosystems of northern Michigan to forests with northern and boreal species.

Notes. Shade-intolerant, persisting in moderate shade. Vigorous sprouts from the root collar exhibit few lateral branches. In southern lower Michigan usually an indicator of sites with calcareous substrates. Attractive for planting because of its white flowers borne in large, dense clusters, but it is rarely used in landscaping.

Chromosome No. $2n = 22$, $n = 22$; $x = 11$

1. Flowering shoot, × 1/2
2. Hairs on upper leaf surface, × 18
3. Fruiting shoot, reduced
4. Winter bud and leaf scar, enlarged
5. Winter twig, reduced

KEY CHARACTERS

- upright, loosely branched clonal shrub, 2–3 m
- leaves opposite, broadly ovate to suborbicular, 6–9 veins per side
- stems greenish, with reddish or purple streaks, becoming pink with exposure, especially fall and winter, more or less warty, lenticels purplish
- fruit a pale blue to blue drupe when ripe

CORNACEAE
Cornus sericea Linnaeus
Red-osier

Size and Form. Erect shrub with branches arching, 1–3 m high; branches loose, broad-spreading, some prostrate, forming clones of relatively few stems by basal sprouting, tip rooting, or layering. Michigan Big Tree: girth 25.4 cm (10.0 in), diameter 3.2 cm (8.1 in), height 5.2 m (17.0 ft), Oakland Co.

Bark. Bright red in full sun, greenish in deep shade; gray, smooth nearly to base of stems, corky lenticels conspicuous, small or in patches with increasing frequency toward the base of the stem.

Leaves. Opposite, simple, deciduous, blades 4–10 cm long and about half as wide; ovate to ovate-lanceolate; acuminate; rounded; entire, glabrous or appressed pubescent and dark green above, glabrous or appressed-pilose and glaucous beneath; usually 5 (4–6) pairs of veins per side, parallel-arcuate; petioles 1–2 cm long.

Stems-Twigs. Slender, smooth, not forming conspicuous fissures; green to reddish and slightly pubescent on current shoots, soon becoming glabrous except for finely appressed hairs at the shoot apex; twigs and branches bright red to the base when exposed, with prominent whitish lenticels that often coalesce, forming patches; leaf scars crescent-shaped, bundle scars 3; pith small, white, continuous.

Winter Buds. Terminal bud often stalked, pubescent, conical-ovoid, 3–5 mm long; lateral buds, oblong, pubescent, appressed, 1–3 mm long, 2 valvate scales exposed.

Flowers. June; bisexual; borne in flat-topped terminal cymes 3–5 cm across; small; dull white; calyx minute, 4-toothed; petals 4, 3–4 mm long; stamens 4. Insect pollinated.

Fruit. Drupe; August–October; globose, whitish when ripe, 6–7 mm across; pit 2-celled and 2-seeded; seeds 18,500/lb (40,700/kg).

Distribution. Common to locally abundant throughout the state. County occurrence (%) by ecosystem region (Fig. 19): I, 100; II, 90; III, 100; IV, 88. Entire state 95%. Widely distributed throughout North America except for the s.e. and Gulf Coast plains.

Site-Habitat. Open, wet to wet-mesic sites; lakeshores, stream and riverbanks, first bottoms of river floodplains, swamps, marshes, sand dunes; usually circumneutral to basic soils. Able to thrive in many different types of wetlands. Associates include American elm, red and silver maples, black and red ashes, yellow birch, tamarack, red raspberry, silky dogwood, common elder, beaked hazelnut, wild black currant, winterberry, mountain holly.

Notes. Very shade-intolerant. Degree of stem redness directly related to amount of exposure to sunlight; shade-grown plants tend to have greenish stems; twigs only with fine appressed hairs near the terminal bud in fall-winter condition. Easily transplanted and propagated. Not densely clone-forming as in *C. foemina*. Valuable as a landscape plant because of its showy red stems during winter and spreading form; various cultivars are in use. Widely used by Native Americans across North America for a variety of purposes, including medicine, technology, smoking, and rituals (Marles et al. 2000, 165–67). In older literature, its name was *C. stolonifera* Michaux.

Chromosome No. $2n = 22$, $n = 11$; $x = 11$

Similar Species. This species is very widely distributed in temperate to boreal North America, Asia, and Europe. In Europe, Siberia, and Northeast China to Korea, the counterpart is *C. alba* L.

110

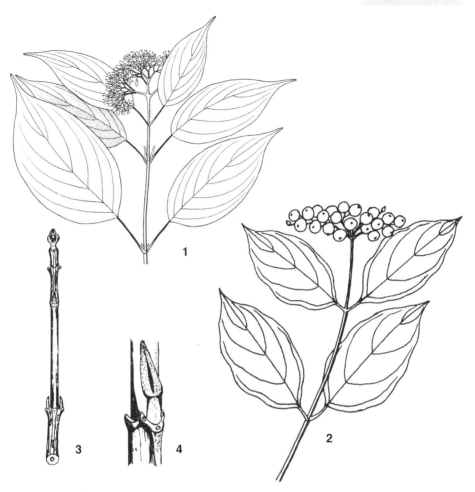

1. Flowering shoot, × 1/2
2. Fruiting shoot, drupe reduced
3. Winter lateral bud and leaf scar, enlarged
4. Winter twig, × 1

KEY CHARACTERS

- erect shrub to 3 m with arching branches
- stems bright red, smooth above, with distinctive whitish lenticels, pith white
- leaves opposite, 4–6 parallel-arcuate veins per side
- fruit a whitish drupe when ripe
- habitat diverse open, disturbed, wet to wet-mesic sites; often edges of wetlands, streams, rivers

Distinguished from silky dogwood, *Cornus amomum*, by bright, smooth, red stems and twigs nearly to the base, not striated, current shoots glabrate except at the apex, pith white, ripe fruit whitish.

Distinguished from gray dogwood, *Cornus foemina*, by bright red stems; stems arching; habitat diverse wet sites generally lacking in uplands; not forming large, dense clones of many erect stems.

BETULACEAE
Corylus americana Walter

American Hazelnut

Size and Form. Erect, bushy shrub, to 3 m; numerous stems arising from the root collar forming a clonal clump, as well as spreading locally by sprouting from rhizomes; spreading branches form a low, rounded crown. Michigan Big Tree: girth 1.0 m (40 in), diameter 32.3 cm (12.7 in), height 10.7 m (35 ft), Washtenaw Co.

Bark. Smooth, grayish-brown, shiny, becoming mottled and slightly fissured near base.

Leaves. Alternate, simple, deciduous, blades 6–15 cm long, 4–12 cm wide; broad-ovate to oval or orbicular; short-acuminate; oblique, rounded, or subcordate; more or less irregularly, doubly-serrate; coriaceous; slightly hairy, dark green above, soft-pubescent and paler beneath; veins 8 or less; petiole with gland-tipped bristles, 0.5–2 cm long with paired scalelike stipules at the base, early deciduous.

Stems-Twigs. Slender to moderate, brown to reddish-brown, current shoots covered with pinkish gland-tipped bristles, becoming stiff and dark brown in autumn; older twigs and branches glabrous, covered with horizontal lenticels; male catkins conspicuous in fall and winter near the twig ends, 0.5–2.5 cm long, often with a characteristic kink or curl; leaf scars half circular to triangular; bundle scars 3–7; stipule scars elongated; pith small, tan to brown.

Winter Buds. Terminal bud absent, lateral vegetative buds sessile, ovoid, obtuse, 2–5 mm long, brown, becoming reddish or purplish with short, white hairs, especially near the tip, 4–6 scales visible; flower buds globose, otherwise similar to vegetative buds.

Flowers. March–April; unisexual, plants monoecious; male flowers in stalked catkins formed in late summer and conspicuous at ends of current year shoots, anthesis in early spring when drooping catkins are 4–8 cm long; female flowers solitary or in small, bud-like clusters in leaf axils near the ends of twigs. Wind pollinated.

Fruit. Nut; August–September; large, 2–4 in a cluster, each nut tightly enclosed by 2 downy, coarsely toothed, leaflike bracts about twice as long as the nut but not forming a beak; nut flattened-globose, 1–1.5 cm long; seeds edible, 491/lb (1,080/kg).

Distribution. Common to abundant in the southern half of the Lower Peninsula, rare elsewhere. County occurrence (%) by ecosystem region (Fig. 19): I, 74; II, 10; III, 14; IV, 38. Entire state 42%. Widely distributed throughout e. North America and n. Great Plains, New England, QC to SK, s. to OK, e. to GA.

Site-Habitat. Open to lightly shaded conditions in dry, moist, or seasonally flooded sites; common in river floodplains, also roadsides, fencerows, edges of woods, old fields. Indicator of disturbed sites and usually circumneutral to nutrient-rich soils. River floodplain associates include American elm, red ash, silver maple, ninebark, *Cornus* spp.; riverbank grape, wild black currant; reported to occur in 11 of Michigan's near-natural communities (Kost et al. 2007) and thus associated with many additional species.

Notes. Shade-intolerant, persisting in light to moderate shade. In Minnesota, rhizome sprouting can form dense clones to 10 m across (Smith 2008). Nuts rapidly removed by animals and rarely found or collected by humans except in early autumn. Nuts about as good as the larger, imported hazelnut or filbert, from the European *C. avellana* L. Not widely used in landscaping, but a native species of potential importance in dry or moist sites. About 15 hazel species are reported in northern temperate regions of North America, Europe, and Asia, 4 of which qualify as trees, for example, the Turkish hazel, *C. colurna* L., which occurs in southeastern Europe and Asia Minor.

Chromosome No. $2n = 22$, $n = 11$; $x = 11$

Similar Species. In eastern North America, the beaked hazelnut, *Corylus cornuta* Marsh., is similar; in central and northern Europe, the

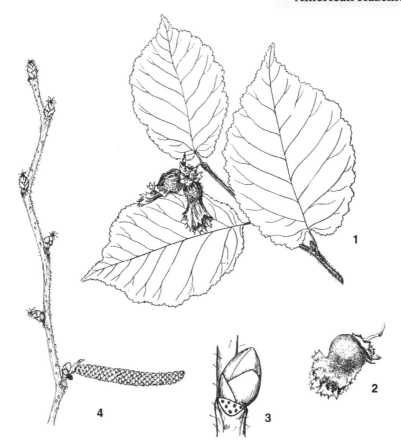

1. Fruiting shoot, nut enclosed by bracts, × 1/2
2. Fruit, nut enclosed by bracts, × 1
3. Winter lateral bud and leaf scar, enlarged
4. Winter twig with male catkin, × 1

KEY CHARACTERS

- erect, bushy, clonal shrub to 3 m high
- current shoots covered with gland-tipped bristles; pith tan to brown; stalked male catkins near twig ends in autumn and winter
- leaves broad, oval, irregularly doubly serrate, coriaceous; petioles bristly
- fruit a large nut enclosed in two leaflike bracts
- habitat open to lightly shaded sites; common to abundant in southern lower Michigan

Distinguished from the beaked hazelnut, *Corylus cornuta*, by current shoots and petioles covered with gland-tipped bristles; pith tan to brown; bracts surrounding the nut not forming a beak; male catkins stalked; rare in northern lower Michigan and upper Michigan.

counterpart is *C. avellana* L; in northeastern Asia, *C. mandshurica* Maxim., *C. yunnanensis* (Franch.) A. Camus, and *C. heterophylla* Fisch. ex Trautv. are similar disjunct species in China; in Japan, *C. heterophylla* var. *thunbergii* Blume.

BETULACEAE
Corylus cornuta Marshall

Beaked Hazelnut

Size and Form. Erect, bushy, clone-forming shrub, to 4 m; numerous stems arising from root collar and sprouting from rhizomes to form a small clone; spreading branches form a medium-high, rounded crown.

Bark. Smooth, gray with scattered lenticels, becoming rough.

Leaves. Alternate, simple, deciduous, blade 4–13 cm long, 3–8 cm wide; ovate or ovate-oblong; acuminate; subcordate; more or less irregularly doubly serrate of slightly lobed; mostly glabrous, dark green above, paler and pubescent on veins beneath; 8 veins or less; turning yellow in autumn; petioles puberulent, not bristly-hairy, 0.5–1.5 cm long, bearing a pair of early-deciduous scalelike stipules.

Stems-Twigs. Slender, reddish-brown; glabrous or with a few pale hairs near nodes or twig tip, zigzagged; leaf scars triangular; bundle scars 7, often obscure; stipule scars elongated; pith brown, continuous.

Winter Buds. Terminal bud absent, laterals small, globose, with acute tip, 2–5 mm long, reddish-brown; 4–6 scales exposed, outer scales somewhat hairy to glabrous.

Flowers. April–May, before the leaves; unisexual, plants monoecious; male flowers borne in nonstalked or very short-stalked catkins, conspicuous near ends of twigs in autumn and winter; female flowers solitary or in small, budlike clusters near the ends of twigs. Wind pollinated.

Fruit. Nut; August–September; 2–4 in a cluster, each nut tightly enclosed in a light green, tubular involucre of united bracts, covered with stiff hairs, much constricted and prolonged above the nut to form a beak, 2.5–4 cm long; nut flattened-globose; seeds brown, edible, 249/lb (548/kg).

Distribution. Common in the Upper Peninsula and the northern half of the Lower Peninsula. County occurrence (%) by ecosystem region (Fig. 19): I, 21; II, 50; III, 86; IV, 100. Entire state 45%. Transcontinental northern range in USA and adjacent Canadian provinces, s. to SC, GA, and AL in the East and CA and CO in the West.

Site-Habitat. Diverse mesic to dry and open to moderately shaded sites of low or moderate nutrient availability; edges and openings in hemlock–northern hardwoods forests, oak and oak-pine barrens. Wide range of associates from north to south.

Notes. Shade-intolerant, persisting in light shade. Edible nuts rapidly removed by animals in early autumn. Nuts nearly as good as the larger, imported European hazelnut or filbert, *C. avellana* L. Native Americans twisted the long, flexible shoots into rope. Not widely used in landscaping, but a native species of potential importance in dry or moist sites.

Chromosome No. $2n = 22$, $n = 11$; $x = 11$

Similar Species. *Corylus americana* Walter, American hazelnut, is closely related, with an overlapping range in lower Michigan and much of the eastern United States. It is rare in the Upper Peninsula and has a more southerly range than *C. cornuta*. The California hazelnut, *C. cornuta* var. *californica* (A. DC.) Sharp, is widely distributed in California and coastal Oregon and Washington. A favorite food of northwest coast people, who often burned bushes to the ground to increase productivity. The counterpart species in Japan is *C. sieboldiana* Blume.

114

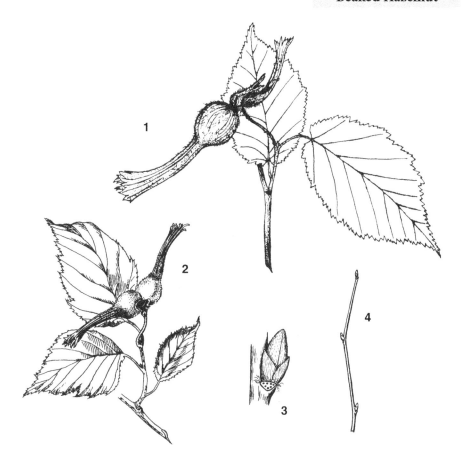

1. Fruiting shoot, nut enclosed by bracts, × 1
2. Fruiting shoot, nut enclosed by bracts, × 1/2
3. Winter lateral bud and leaf scar, enlarged
4. Winter twig, reduced

KEY CHARACTERS

- erect bushy shrub to 4 m high
- current shoots and petioles puberulent, not bristly-hairy
- leaves broad, oval, irregularly doubly serrate, coriaceous
- nut enclosed in a tubular involucre, projecting above the nut to form a beak
- common in the northern Lower Peninsula and Upper Peninsula

Distinguished from American hazelnut, *Corylus americana*, by twigs and petioles not glandular or bristly-hairy; pith brown; leafy bracts surrounding the fruit covered with stiff hairs, much constricted and prolonged above the nut to form a beak, male catkins not stalked or very short-stalked; uncommon in southern lower Michigan.

THE HAWTHORNS—*CRATAEGUS*

Shrubs and small trees of the genus *Crataegus* are relatively easily recognized in all seasons by their widespread occurrence in open, disturbed sites. In spring their showy white to pink clusters of flowers attract our attention and insect pollinators. In summer the green, lustrous leaves of a compact, much branched crown hold our interest. In fall colorful fruits—bright to dull red to orange and yellow (rarely blue-black) attract birds and mammals, including human collectors. The fruits often last into the winter if not taken earlier by birds. In winter the twiggy, often zigzag or crooked, horizontal branching pattern is striking and serves to display the long, sharp, lustrous chestnut-brown thorns for which this genus is renowned. The common name is often shortened from hawthorn to "thorn" as in cockspur thorn.

Now the fun is over. Starting with scarcely two dozen species, varieties, and forms before 1900, a naming binge of putative new species resulted in over 1,000 for North America alone by about 1925 (Voss 1985). Since then taxonomists and systematic botanists have sought to reduce the number drastically and organize them into distinctive groups (Phipps 1983; Voss 1985; Gleason and Cronquist 1991). Phipps (2003) recognizes 140 species worldwide in 40 series; Judd et al. (2008) recognize 265. The most detailed research in Michigan was done by Edward Voss (1985, 386–418), based in part on the research of Ernest Palmer (1925, 1946) and other botanists and collectors in Michigan. Voss and Reznicek (2012) provide a key and descriptions for 29 species. We treat 5 native species with relatively wide distributions in Michigan, the upper Great Lakes Region, and eastern North America. Taxa of *Crataegus* in North America are geographically differentiated but not strongly habitat differentiated. In our area they indicate high light irradiance and disturbance; "moist open places" succinctly describes their home place. They are not absolutely particular about soil, but they show a decided preference for alkaline soils and are most abundant and varied in limestone regions (Braun 1989). Once established in forest gaps, they can tolerate light shade, although their characteristic dense and wide-spreading branch system is lost. Today they are more commonly plants of forest edges, roadsides, fencerows, abandoned fields, and pastures.

The natural history of eastern American hawthorns suggests that before the Pleistocene glaciations and habitation of North America by humans, hawthorns were mostly relegated to diverse riverine and moist habitats. Here adequate but not widespread disturbance was always available for their persistence. Due to the enormous network of rivers, streams, and lakes in eastern North America, hawthorns had easy access via bird and mammal dispersal to a variety of disturbed sites following human activity. These were first created locally and regionally by Native American burning and agriculture and in post-European settlement time by widespread clearing of forests for agriculture, logging, and postlogging fires, followed by many decades of fire exclusion, which contributed to their establishment and spread (Voss and Reznicek 2012). Phipps (2003) gives a comprehensive treatment of general knowledge of hawthorn ecology, reproductive biology, folklore, uses (food, medicine, ornament), cultivation, and the most important species for ornamental horticulture.

The hawthorns are typically small, much-branched, wide-spreading shrubs or small trees from 3 to 10 m. The trunk and major branches may have formidable branched thorns. The numerous stout, flexuous branches are usually armed with stiff, sharp, unbranched thorns. Hawthorns bear vegetative long shoots from overwintering terminal buds. Lateral buds of previous years produce shorter floral shoots with leaves and flowers. Leaves of floral shoots are consistently smaller, longer than wide, and less lobed or incised compared to those of vegetative long shoots. In the species descriptions and key characters lists, leaves of vegetative shoots are described unless otherwise noted as being characters of floral-shoot leaves.

The bark varies in color from dark red to gray and is scaly or shallowly fissured, with narrow, shredded vertical plates, which sometimes become loose at both ends. The twigs are usually slender and zigzagged; leaf scars are somewhat raised, narrow, narrowly crescent-shaped; bundle scars 3; stipule scars present, small. Winter buds are small, round or oblong-ovoid, and covered with 4–8 reddish or brownish scales; terminal bud is present. Two buds often occur side by side on the current shoot, one of which develops into a new shoot (floral or vegetative) and the other into a thorn. The wood is heavy, hard, tough, close-grained, and reddish brown, with thick, pale sapwood.

The leaves are alternate, simple, and deciduous. Leaf blades are doubly serrate to serrate and often lobed or incised in the "hawthorn way," quite in contrast to that of oaks or most temperate maple species. The leaves are highly variable even on the same plant depending on whether they are on determinate or indeterminate vegetative shoots or on shorter, flower-bearing shoots. Blades of leaves of strictly vegetative shoots generally exhibit larger and more incised lobes than those on floral shoots, which are sharply doubly-serrate but unlobed. Leaves provide food for many butterflies, moths, and other insects (Eastman 1992), as well as fungal pathogens, including including cedar-hawthorn rust (*Gymnosporangium globosum).*

The bisexual flowers, whitish or pinkish, appear in April, May, or June, with or after the leaves, in simple or compound corymbs. They are not very fragrant, and some people find they have a distinctive but unpleasant pungent smell. They are insect pollinated by bees, flies, and beetles. The fruit is typically a reddish pome, subglobose to pear-shaped, with 1–5 seed-containing nutlets. In describing species of *Crataegus,* botanists use the term *nutlet* to characterize the bony or hard-walled structure, or "pit," which encloses tiny seeds. The flesh is usually dry and mealy but may be sweet and succulent. The seeds of *Crataegus* are widely dispersed by birds and mammals. Birds occasionally become intoxicated on fermented fruits when feeding in autumn. Some species of hawthorn form small clones by means of root sprouts occurring away from the main trunk.

The genus name is from the Greek *kratos*, "strength," because of its extremely hard wood and fine grain. The common name "hawthorn" is said to come from the Anglo-Saxon *haguthorn*, "a fence with thorns." Another interpretation is that the fruit is often called a "haw," and species of this group are sometimes called "thorns," and voilà, hawthorn(s). Hawthorns for centuries have been favorite park and hedge plants in England and Europe, and their flowering branches are traditionally used in May Day celebrations. Native peoples used the thorns for lancing blisters and boils, extracting splinters, ear piercing, and fishhooks. The hard wood was made into tool handles and weapons. Their medicinal history, especially in China and Europe, is very long, with more recent accounts in North America (Phipps 2003). In the Pacific Northwest, the bark of the native *Crataegus douglasii* (a disjunct in eight counties in upper Michigan) was used to treat diarrhea, dysentery, stomach pains, and venereal disease and to thin the blood and strengthen the heart (Pojar and Mackinnon 1994). In China *C. laevigata* is reported to provide a leaf-infusion tea to reduce blood pressure, a coffee substitute from the seeds, and a tobacco substitute from the leaves (Hora 1980). The fruits of many species are made into jellies and preserves.

Of the twenty-eight native species in Michigan (Voss and Reznicek 2012), twenty-one occur in less than 25 percent of the counties. We describe five of the more commonly occurring species, ranging from 27 to 57 percent of county occurrence, and their key is presented below. Although they are not treated in detail, it is important to mention three other species, two of which are common and conspicuous in southern Michigan, *Crataegus mollis* and *C. coccinea*. Both differ from all the species we describe except *C. macrosperma* in having the leaf blades widest clearly below the middle and the bases truncate or even subcordate. *Crataegus mollis* is one of the larger species, with conspicuously large flowers and leaves. The inflorescence, calyx, and lower leaf blades (at least along main veins) are villous-tomentose, and even young fruits are quite hairy. *Crataegus coccinea* is reasonably similar, a little smaller but more widespread into northern lower Michigan; it differs in being essentially glabrous. The third species, *Crataegus calpodendron,* conspicuous because of its late flowering (into mid-June), is a lowland and upland forest edge and understory species. It resembles *C. succulenta* but is more pubescent and lacks glossy leaves.

Two other species are worth mentioning for special reasons. Whereas all species have red pomes when ripe (rarely yellow), the northern *Crataegus douglasii*, black hawthorn, is unique in being black fruited. Perhaps the most commonly cultivated species (and also occasionally escaping) is the Washington thorn, *C. phaenopyrum*. It has unique, 3-lobed leaves, looking rather like those of a small red maple, and very persistent large clusters of small, bright red fruits. Many hawthorns become small trees; the most characteristic in our region are *C. punctata* and *C. mollis.*

1. Leaf blades primarily broadest above the middle, obovate; base mostly narrowly cuneate; usually a small tree to 10 m.

   2. Leaves glossy above, glabrous, not impressed-veined above; nutlets 1–3; trunks not bearing branched thorns; occurring occasionally in southern lower Michigan, absent or rare in northern lower Michigan and upper Michigan . . . *C. crus-galli*, p. 122

   2. Leaves dull above, pubescent or persistently villous beneath; impressed-veined above; nutlets 3–5; trunks bearing branched thorns; occurring frequently in southern and northern lower Michigan . . . *C. punctata*, p. 126

1. Leaf blades broadest at or below the middle; broadly elliptic, oblong-ovate to broadly ovate or sometimes nearly orbicular; base cuneate, acute to truncate or rounded; tall shrub or small tree, <5–8 m.

   3. Leaf blades mostly broadest at or slightly below the middle; the base cuneate, narrowly tapered or acute; petiole winged or not; thorns straight or slightly curved; trunk and larger branches not fluted or buttressed.

      4. Leaf margins shallowly lobed, mostly above the middle; base strongly cuneate to acute; teeth not gland-tipped; petioles stout, winged and grooved near the leaf blade; glandless; stamens 10 or 20; nutlets 2–3, prominently ridged on the back, deeply pitted on the inner surfaces . . . *C. succulenta*, p. 128

      4. Leaf margins with conspicuous lobes below the middle; base narrowly tapered or cuneate; teeth gland-tipped; petioles not winged, slender with scattered glands near the blade base; stamens 10; nutlets 3–5, smooth on inner surfaces . . . *C. chrysocarpa*, p. 120

   3. Leaf blades mostly broadest well below the middle or even near the broadly rounded or truncate base; petiole slender and somewhat winged near the blade base; thorns stout and strongly curved; trunk and larger branches fluted or buttressed . . . *C. macrosperma*, p. 124

ROSACEAE
Crataegus chrysocarpa Ashe

Fireberry Hawthorn, Round-leaf Hawthorn

Size and Form. Much-branched, round-topped shrub or rarely a small tree to 6 m.

Bark. Dark reddish-brown, becoming gray, flat-topped narrow vertical scales on old stems.

Leaves. Leaves of vegetative shoots are alternate simple, deciduous; vegetative long-shoot blades 2–9 cm long, 1.5–6 cm wide; broadly elliptic, ovate, or nearly orbicular, broadest at or below the middle; acute; narrowly tapered or cuneate; dark yellowish-green and somewhat lustrous above, glabrous or slightly pubescent, veins slightly impressed; paler below, glabrous or pubescent along the veins; margin irregularly doubly with 4–6 distinct triangular, gland-tipped lobes per side above the middle. Blades of flowering-shoot leaves smaller, narrower, and lack such deeply incised lobes compared to blades of vegetative long shoots; petioles slender, sparsely pubescent with small red glands near the blade, 1.5–2.5 cm long.

Stems-Twigs. Stout, young shoots greenish, villous-pubescent, becoming reddish-brown and glabrate and marked by numerous nearly white lenticels by autumn of first year, older twigs ashy-gray or grayish-brown; thorns numerous, deep chestnut-brown, spotted with gray, straight or slightly curved, 3–6 cm long; leaf scars somewhat raised, narrowly crescent-shaped, bundle scars 3; stipule scars present, small; pith rather small, round, continuous.

Winter Buds. Terminal bud present; buds small, globose, about 6 scales exposed, reddish-brown.

Flowers. May–June, with the leaves; bisexual; white, 1.5–2 cm wide, numerous in glabrous or slightly villous corymbs; sepals 5, lobes narrowly lanceolate, serrate, with stalked glands along the margins, persistent with reflexed lobes; petals 5; stamens 10, anthers pale yellow. Insect pollinated.

Fruit. Pome; late August to October; short oblong to subglobose, 0.6–1.5 cm wide; sepals persistent, lobes reflexed; dark or bright red or orange, rarely golden yellow; nutlets 3–5, smooth on inner surfaces; flesh soft, becoming mellow or succulent; seeds 10,750/lb (23,650/kg).

Distribution. Common in northern lower Michigan and upper Michigan; rare in southern lower Michigan. County occurrence (%) by ecosystem region (Fig. 1): I, 8; II, 50; III, 86; IV, 75. Entire state 36%. Second most wide-ranging species in North America, NL and New England to BC, s. to OR, UT, CO, e. to VA.

Site-Habitat. Open, dry-mesic to wet-mesic sites, usually with circumneutral to basic soils or substrate; edges of forests, along shores (especially of the Great Lakes); sandy bluffs, dunes, gravelly to rocky sites, fence-rows, roadsides, abandoned fields. Associates highly diverse, often shade-intolerant pioneer species.

Notes. Shade-intolerant, persisting in light shade. Seeds dispersed widely by birds and mammals. Leaves polymorphic in shape; leaf blades on flowering shoots are smaller and ca. 1.3 times longer than wide compared to those on vegetative long shoots. The ovate to nearly orbicular leaves on vegetative shoots are characterized by doubly-serrate margins and short, acute lobes, typical of many hawthorns (i.e., unlike lobes of oaks or maples). It has value for landscaping in northern regions where few species are hardy.

Chromosome No. $2n = 68$, $n = 34$; $x = 17$

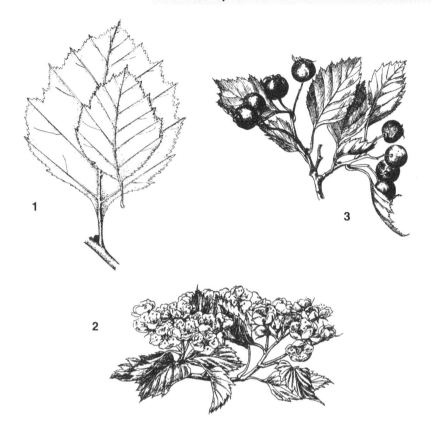

1. Long-shoot leaf on twig and floral leaf, × 1
2. Flowering shoots, × 1/2
3. Fruiting shoots, pome, × 1/2

KEY CHARACTERS
- much-branched shrub or rarely a small tree to 6 m
- vegetative long-shoot leaf blades broadly elliptic, ovate, or nearly orbicular, broadest at or below the middle; margin doubly serrate and typically with short, acute lobes having gland-tipped teeth; base cuneate
- thorns numerous, deep chestnut-brown spotted with gray, straight or slightly curved, 3–6 cm long
- fruit a pome, sepals persistent; dark or bright red or orange; nutlets 3–5, smooth on inner surfaces

ROSACEAE
Crataegus crus-galli Linnaeus
Cockspur Thorn

Size and Form. Tall shrub or usually a small tree to 10 m; low broad or depressed crown, distinctly layered horizontally, with intricate, wide-spreading stiff branches, branchlets flexuous with many long thorns. Michigan Big Tree: girth 96.5 cm (38 in), diameter 30.7 cm (12.1 in), height 8.8 m (8.8 ft), Wayne Co.

Bark. Dark grayish-brown, scaly on old boles.

Leaves. Alternate, simple, deciduous; vegetative long-shoot blades 2.5–9 cm long, 1–3 cm wide; firm, mostly obovate; acute to rounded; cuneate; finely serrate above the middle usually with gland-tipped teeth, rarely incised or lobed; veins inconspicuous not impressed above; dark glossy green above, paler and dull below; glabrous; floral-shoot blades are smaller and narrowly obovate to elliptic; petioles often winged at junction with the blade base, 0.3–1.5 cm long.

Stems-Twigs. Slender and flexible, with many long, straight, or curved thorns, 5–9 (11.5) cm long; flowering shoots short; glabrous, yellowish-brown or reddish becoming glossy; leaf scars somewhat raised, narrowly crescent-shaped, bundle scars 3; stipule scars present, small; pith rather small, round, continuous.

Winter Buds. Terminal bud present; buds small, globose, 6 scales exposed, shiny, reddish-brown.

Flowers. May–June, with the leaves; bisexual; white, 1.5–2 cm wide, numerous in glabrous corymbs; sepals 5, narrowly lanceolate, entire, glandular-serrate; petals 5; stamens 10 or 20, anthers pink or yellow. Insect pollinated.

Fruit. Pome; September–October; ellipsoid-ovoid to subglobose; green to dull red or bright red, 1–1.4 cm wide; nutlets 1–3, rounded at ends, inner surface flat, ridged on back, 0.6–1.5 cm wide; flesh hard, dry, inedible.

Distribution. Frequent in southern lower Michigan. County occurrence (%) by ecosystem region (Fig. 19): I, 58; II, 0; III, 0; IV, 0. Entire state 27%. Widely distributed throughout e. North America, from QC s. New England to FL and from ON s. MN IA, KA, OK to TX.

Site-Habitat. Open to lightly shaded, dry-mesic to wet-mesic and seasonally wet sites, usually with circumneutral to basic soils or substrates; river and stream floodplains; fencerows, edges of forests, open woodlands, fields, and meadows. Associates highly diverse, often shade-intolerant pioneer species.

Notes. Shade-intolerant, persisting in light shade. In addition to vegetative long shoots and floral shoots, lateral buds occasionally give rise to vegetative short-shoots, which may produce floral shoots the following year. Seeds dispersed widely by birds and mammals. One of the best hawthorns for ornamental use, with many attractive variants.

Chromosome No. $2n = 51, 68$; $n = 17, 34$; $x = 17$

122

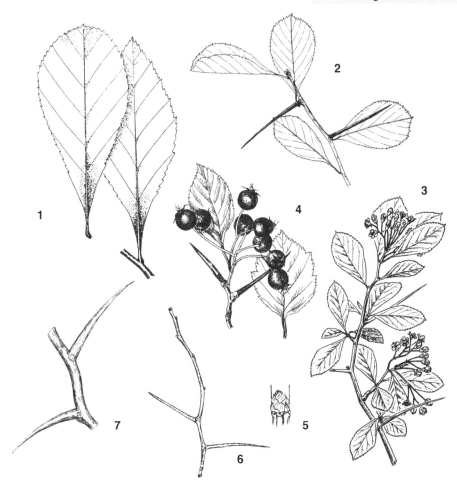

1. Long-shoot leaf (*left*) and floral shoot leaf (*right*), × 1
2. Long shoot with leaves and thorns, reduced
3. Branch with long-shoot leaves, floral shoots, and shorter vegetative shoots, reduced
4. Fruiting shoot, pome, × 1/2
5. Winter terminal bud and leaf scar, enlarged
6. Winter twig, reduced
7. Winter twig with thorns in axil of lateral buds, enlarged

KEY CHARACTERS

- tall shrub or small tree to 10 m with distinctly layered horizontal branches
- thorns long, slender, straight or slightly curved, gray, 5–9 cm long
- vegetative long-shoot leaf blades mostly obovate, finely serrate above the middle, usually with gland-tipped teeth, rarely incised; veins inconspicuous, not impressed above; base cuneate
- fruit a pome, green to dull red or bright red, often dark dotted, nutlets 1–3, with smooth inner surfaces; flesh hard and dry, inedible
- frequent in southern lower Michigan, absent in northern lower and upper Michigan

ROSACEAE
Crataegus macrosperma Ashe

Fan-leaf Hawthorn, Large-seed Hawthorn

Size and Form. Much-branched shrub or small tree with stout ascending branches 5–7 m; trunk and larger branches fluted or buttressed, not round in cross section.

Bark. Pale gray or brownish-gray, brown, larger stems with narrow platelike scales.

Leaves. Alternate, simple, deciduous; vegetative long-shoot blades thin, 4–9 cm long, 3–8 cm wide; thin, broadly ovate to nearly orbicular, broadest below the middle; acute to short acuminate; broadly rounded or truncate; margins sharply doubly serrate with 4–6 short, deeply incised lobes per side, often with reflexed and gland-tipped, acuminate tips; young leaves bronze-green and slightly pubescent above, becoming dark yellowish-green and scabrous, much paler and glabrous beneath; more deeply lobed, with 4–6 short, broad lobes per side, often with reflexed and gland-tipped, acuminate tips; leaf blades of floral shoots are smaller and markedly longer than wide compared to those of vegetative long shoots; petioles slender, 1–3 cm long, somewhat winged and grooved near the blade base.

Stems-Twigs. Young shoots bronze-green, glabrous, becoming dark reddish-brown by end of the first growing season, with moderately pronounced lenticels, very soon turning gray or grayish-brown; thorns numerous, stout and strongly curved, dark chestnut brown, becoming gray, 2–5 cm long; vegetative short-shoots occasional; leaf scars somewhat raised, narrowly crescent-shaped, bundle scars 3; stipule scars present, small; pith rather small, round, continuous.

Winter Buds. Terminal bud present; buds small, globose, 4–8 scales exposed, shiny, reddish-brown.

Flowers. May–June, with the leaves; bisexual; white, 1.3–2 cm wide, numerous in very loose, glabrous corymbs; sepals 5, lobes linear, usually entire or with 1 or 2 gland-tipped teeth; petals 5; stamens usually 5–10, anthers pink. Insect pollinated.

Fruit. Pome; September–October; oblong to subglobose, scarlet to crimson, 1–2 cm wide, calyx lobes persistent, erect or spreading; nutlets 3–5, thick, ridged on back, not pitted on inner surfaces, 7–8 mm long; flesh thin, becoming soft and succulent when ripe.

Distribution. Common in lower Michigan, locally frequent in western upper Michigan. County occurrence (%) by ecosystem region (Fig. xx): I, 50; II, 57; III, 14; IV, 62. Entire state 52%. Widely distributed in e. North America except FL.

Site-Habitat. Open to lightly shaded, dry to mesic and seasonally wet sites usually with circumneutral to basic soils or substrates; river floodplains and stream banks, oak and pine barrens, edges of or openings in hardwood and conifer forests, rocky ridges, cutover lands, fields and pastures, roadsides. Associates highly diverse, often shade-intolerant pioneer species.

Notes. Shade-intolerant, persisting in light shade. In addition to vegetative long shoots and floral shoots, lateral buds occasionally give rise to vegetative short-shoots, which may produce floral shoots the following year. Seeds dispersed widely by birds and mammals.

Chromosome No. $2n = 68$, $n = 34$; $x = 17$

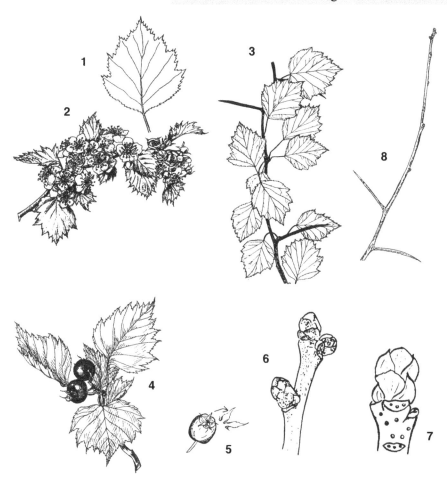

1. Leaf, × 1/2
2. Flowering shoots, × 1/2
3. Twig with vegetative shoots and thorns, × 1/2
4. Fruiting shoot, pome, × 1/2
5. Fruit, pome, reduced
6. Winter lateral and terminal buds, enlarged
7. Winter terminal bud and leaf scar, enlarged
8. Winter twig with thorn, reduced

KEY CHARACTERS

- much-branched shrub or small tree with ascending branches to 7 m
- leaf blades broadly elliptic-ovate, broadest below the middle, with 4–6 short, broad lobes per side; bases broadly rounded, truncate, or subcordate
- thorns numerous, stout and strongly curved, dark chestnut brown, becoming gray, 2–5 mm long
- fruit a scarlet to crimson pome, oblong to subglobose, calyx lobes persistent; nutlets 3–5, thick, not pitted on inner surfaces; flesh soft when ripe, succulent

ROSACEAE
Crataegus punctata Jacquin
Dotted Hawthorn, White Thorn

Size and Form. Tall shrub or more often a small, flat-topped tree to 10 m with open crown and stiff, crooked, horizontal or slightly ascending branches and stout, thorny branchlets; trunks bearing compound thorns. Michigan Big Tree: girth 127 cm (50 in), diameter 40.4 cm (15.9 in), height 11.9 m (39 ft), Oakland Co.

Bark. Dark gray or brown, fissured and ridged on the trunk.

Leaves. Alternate, simple, deciduous; vegetative long-shoot blades longer than wide, 4–8 cm long, 1.5–3.5 cm wide; obovate to elliptic-oblong, broadest above the middle; acute; narrowly cuneate to rounded; dull yellowish-green to gray-green above, paler below and persistently villous, especially along the veins; serrate to doubly serrate above the middle, rarely shallowly lobed near the apex; veins impressed on upper surface; blades of flowering shoots are smaller and markedly longer than wide compared to blades of vegetative long-shoots; petioles winged, nearly to the base, 0.6–2 cm long.

Stems-Twigs. Moderate to stout, young twigs brownish with gray pubescence, becoming glabrate and ashy- or silvery-gray; thorns usually few, slender, straight or slightly curved, gray, 3–10 cm long; vegetative short-shoots infrequent; leaf scars somewhat raised, narrowly crescent-shaped, bundle scars 3; stipule scars present, small; pith rather small, round, continuous.

Winter Buds. Terminal bud present; buds small, globose, 4–8 scales exposed, shiny, reddish-brown.

Flowers. May–June with the leaves; bisexual; white, 1.5–2 cm wide, numerous in villous corymbs; sepals 5, lobes lanceolate, mostly entire; petals 5; stamens 20, anthers pink or yellow. Insect pollinated.

Fruit. Pome; August–September; at first pear-shaped, becoming elongate to subglobose; yellow to dull or bright red, conspicuously brown-dotted, 1–1.5 (2) cm wide; nutlets 3–5, smooth on inner surfaces, 5–6 mm long; mellow to scarcely succulent; seeds 4,700/lb (10,340/kg).

Distribution. Frequent in lower Michigan. County occurrence (%) by ecosystem region (Fig. 1): I, 79; II, 50; III, 29; IV, 0. Entire state 57%. Widely distributed in e. North America, NS, QC to MN, s. from New England to GA and MN s. to OK, AR, MS, AL.

Site-Habitat. Open to lightly shaded, dry to wet-mesic and seasonally wet sites usually with circumneutral to basic soils or substrates; river and stream floodplains; lake plain oak openings, edges of forests, open woodlands, gravelly or rocky fields, meadows, pastures, ravines, ditches, fencerows. Associates highly diverse, often shade-intolerant pioneer species.

Notes. Shade-intolerant, persisting in light shade. Slow-growing, moderately long-lived. Seeds dispersed widely by birds and mammals.

Chromosome No. $2n = 34$, $n = 17$; $x = 17$

126

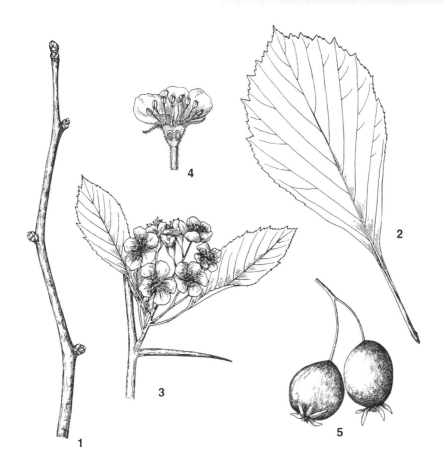

1. Winter twig, × 1
2. Leaf, × 1
3. Flowering shoot, × 1/2
4. Vertical section of flower, enlarged
5. Fruit, pome, × 1

KEY CHARACTERS

- tall shrub or more often a small tree to 10 m
- thorns slender, straight or slightly curved, gray, 3–10 cm long; trunks bearing compound thorns
- vegetative long-shoot leaf blades obovate to elliptic-oblong, broadest above the middle, lacking distinctive marginal lobes; veins impressed above; base mostly narrowly cuneate
- fruit a pome, dull to bright red, conspicuously brown-dotted, nutlets 3–5, with smooth inner surfaces
- frequent in lower Michigan, rare in upper Michigan

ROSACEAE
Crataegus succulenta Link

Fleshy Hawthorn

Size and Form. Much-branched shrub, usually becoming a small tree, 6–8 m; branches mostly ascending, forming a narrow crown; thorns numerous, long, and stout.

Bark. Reddish-brown, scaly; thorns often present on lower part of the trunk.

Leaves. Alternate, simple, deciduous; vegetative long-shoot blades 4–8 cm long and less than half as wide; leathery, especially at margins, broadly elliptic to ovate or rhombic, broadest at the middle; acute; strongly cuneate; margins coarsely and doubly serrate with 4–5 short, acute, often shallow, glandless lobes per side mostly above the middle; leathery, dark glossy green, glossy above, with strongly impressed veins; much paler beneath and pubescent, especially along the veins, the pubescence forming a fringe of short, stiff hairs on each side of principal veins; veins usually strongly impressed above; leaf blades of floral shoots smaller and markedly longer than wide compared to those of vegetative long-shoots; petioles stout, glandless, 1–2 cm long, winged and grooved near the blade.

Stems-Twigs. Moderate, young twigs glabrous, yellowish-green, becoming orange-brown and lustrous by autumn, later becoming dark gray; thorns very long, numerous, stout, straight or slightly curved, bright chestnut-brown, lustrous, 3–9 cm long; vegetative short-shoots occasional; leaf scars somewhat raised, narrowly crescent-shaped, bundle scars 3; pith rather small, round, continuous.

Winter Buds. Terminal bud present; buds small, globose, about 6 scales exposed, reddish-brown.

Flowers. Mid-May to June, with the leaves; bisexual; white, 1.6–2 cm wide, numerous in villous corymbs; sepals 5, lobes lanceolate, glandular-serrate, becoming reflexed; petals 5; stamens 10–20, pink or sometimes yellow. Insect pollinated.

Fruit. Pome; September–October; subglobose; dark to bright red, pale dotted, 0.7–1.5 cm wide, sometimes remaining hard and dry into winter; nutlets 2–3, prominently ridged on the back, deeply pitted on the inner surfaces, 6–8 mm long; pulp soft, succulent when ripe; seeds 20,600/lb (45,320/kg).

Distribution. Occasional to rare throughout the state. County occurrence (%) by ecosystem region (Fig. 19): I, 37; II, 30; III, 57; IV, 50, Entire state 37%. Widely distributed throughout North America except for CA, NV, and the deep South (TX to FL).

Site-Habitat. Open to lightly shaded, dry to wet-mesic and seasonally wet sites, usually with circumneutral to basic soils or substrates; river and stream floodplains; lake plain oak openings, fencerows, edges of forests, open woodlands, gravelly or rocky fields, meadows, ravines, ditches. Associates include those of the riverine sites and many others of the habitats noted above.

Notes. Shade-intolerant, persisting in light shade. In addition to vegetative long-shoots and floral shoots, lateral buds occasionally give rise to vegetative short-shoots, which may produce floral shoots the following year. Seeds dispersed widely by birds and mammals.

Chromosome No. $2n = 34, 51$; $n = 17$; $x = 17$

Similar Species. Two varieties are often recognized: var. *succulenta,* only found in lower Michigan; and var. *macracantha* (Loud.) Eggl., widespread from lower to upper Michigan (Voss and Reznicek 2012).

128

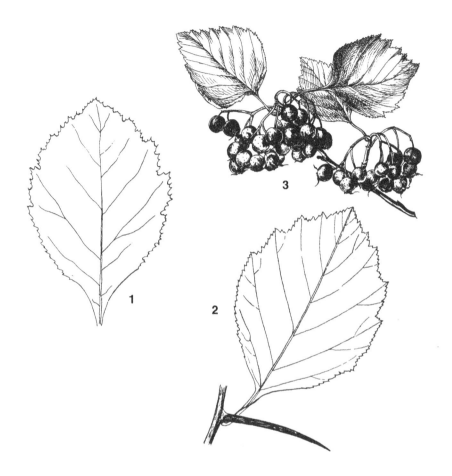

1. Leaf, × 1
2. Leaf with thorn in leaf axil, × 1
3. Fruiting shoots, pome, × 1/2

KEY CHARACTERS

- much-branched shrub, more often a small tree with ascending branches to 8 m
- thorns very long, numerous, stout, straight or slightly curved, bright chestnut-brown
- vegetative long-shoot leaf blades broadly elliptic to ovate or rhombic; coarsely and doubly serrate, the larger blades with 4–5 short, acute, often shallow, glandless lobes; petioles glandless, winged and grooved near the blade
- flowers late May or June, white, calyx lobes glandular-serrate
- fruit a pome; dark to bright red, pale dotted; nutlets 2–3, deeply pitted on the inner surfaces; pulp soft, succulent when ripe

Shrubby Cinquefoil, Shrubby Five-Fingers

Size and Form. Low, erect shrub, 0.2–1 m; much-branched compact form; clone-forming by sprouting from root collar and layering of prostrate branches.

Bark. Light reddish-brown to grayish-black on old stems, shreddy, readily exfoliating.

Leaves. Alternate, pinnately compound, deciduous; 1–3 cm long with very short (3–5 mm) rachis (3–6 mm); leaflets 5–7, crowded, narrow, 1–1.5 cm long, 0.2–0.6 cm wide, terminal leaflets often united at base; elliptic to oblong-lanceolate, pointed both ends; entire, revolute; dark green and sparsely hairy above, paler and silky-pubescent below; petioles yellow to white, pubescent, short, 0.5–1 cm long; stipules conspicuous, lanceolate, papery, about as long as the petiole.

Stems-Twigs. Very slender, covered with long, silky hairs; bark of old stems conspicuously exfoliating; leaf scars very small, bundle scar 1; stipules persistent; pith small, brown.

Winter Buds. Flower buds 3–6 mm long and 2–4 mm across, sessile, oblong to globose, with 4 densely hairy, striated scales exposed, reddish; lateral buds small.

Flowers. June–September; bisexual, borne solitary or a few on terminal shoots, 2–3 cm across; usually bright yellow; sepals 5, triangular-ovate; petals 5, roundish, spreading, 6–11 mm; ovaries hairy; stamens numerous. Insect pollinated.

Fruit. Achene, borne in a small, compact head of many achenes, about 1.5 cm long, each achene covered with long hairs and surrounded by 5 dry, persistent sepals; fruit persistent through winter.

Distribution. Common to locally abundant throughout the state, especially counties bordering lakes Huron, Michigan, and Superior; exceptions are west-central lower Michigan and interior western upper Michigan. County occurrence (%) by ecosystem region (Fig. 19): I, 66; II, 70; III, 100; IV, 38. Entire state 67%. Circumboreal and widely distributed throughout North America from NL to AL, s. to Mexico, Rocky Mountains, n. Great Plains, w. Great Lakes region, and New England, s. to NC.

Site-Habitat. Open, wet to wet-mesic sites typically with circumneutral to basic soil or substrate; wet meadows; prairies; fens; swamps; marsh edges; river, stream, and lake margins; rocky sites in the north such as alvar and limestone bedrock lakeshores; reported from 16 natural communities (Kost et al. 2007). Associates include species in open sites of the ecosystems and communities noted above. As many as 28 shrub species occur in rich tamarack swamps. Although associates include predominantly species of basic sites, ericaceous shrubs such as bog-rosemary, leatherleaf, Labrador-tea, and cranberries also rarely may be associates.

Notes. Very shade-intolerant. Tolerant of low oxygen availability. Although all ecosystems that support it have basic soil or a calcareous substrate, the poor fen has a strongly acidic mat (1–3 m deep) supporting many ericaceous and acid-tolerating species. However, roots of this and other shrub species extend below the mat into the calcareous substrate. One of the few species of shrubs that flowers more or less continuously throughout the summer. Highly variable in habit, flower color, leaf dimensions, and pubescence. Formerly recognized as *Potentilla fruticosa* L. Also known as *Pentaphylloides floribunda* Á. Löve in the West (Kershaw, MacKinnon, and Pojar 1998) to distinguish the shrubby species from herbaceous species of the genus *Potentilla*. One of the most widely tolerant of all shrubs to seashore and deicing salts. A great many cultivars have been developed.

Chromosome No. $2n = 14, 28$; $n = 7, 14$; $x = 7$

Similar Species. *Dasiphora fruticosa* is widely distributed worldwide in the northern hemisphere. Similar East Asian species include *Dasiphora glabra* Soják and *D. parvifolia* (Fisch. ex Lehm.) Juz.

1. Leaf, × 2
2. Flowering shoot, × 1/2
3. Fruiting shoot with head of achenes, achene, × 1
4. Winter lateral bud and leaf scar, × 2

KEY CHARACTERS

- low, bushy shrub to 1 m
- leaves small, pinnately compound; leaflets 5–7, crowded; elliptic to oblong-lanceolate, silky-pubescent below
- twigs covered with long, silky hairs; bark exfoliating
- flowers usually bright yellow
- fruit an achene born in a small head of many hairy achenes, persistent in winter

DIERVILLACEAE
Diervilla lonicera Miller

Bush-honeysuckle,
Northern Bush-honeysuckle

Size and Form. Low, upright shrub with arching stems, 0.4–1 m; growing singly or forming clones by rhizomes.

Bark. Grayish-brown, shredding on older stems.

Leaves. Opposite, simple, stalked, blades 6–12 cm long, 2–5 cm wide; ovate to lanceolate, long acuminate; rounded to slightly cuneate, occasionally inequilateral; serrate, ciliate; glabrous; green above, paler and sometimes pubescent on main veins beneath; petioles short, 3–8 mm long, ciliate.

Stems-Twigs. Slender, nearly round, grayish to yellowish-brown, with 2 hairy-lined ridges decurrent from each node, otherwise glabrous; pith pale.

Winter Buds. Terminal bud present; laterals solitary or superposed, sessile, oblong, sharply pointed, appressed, V-shaped to crescent-shaped; bud scales 4 or more, keeled and pointed, not persistent; leaf scars opposite or occasionally whorled, opposing scars meeting around stem or connected by a transverse ridge, bundle scars 3.

Flowers. June; bisexual; usually 3-flowered on peduncled cymes; small; calyx 5-lobed; corolla tubular, 2-lipped, yellow, turning orange to reddish; stamens 5. Insect pollinated.

Fruit. Capsule; September–October, persisting in winter; greenish-brown, glabrous, 2-celled, 0.7–1.5 cm long, beaked, with a shriveled calyx; seeds many, minute, not winged.

Distribution. Common throughout the state, the frequency increasing northward. County occurrence (%) by ecosystem region (Fig. 19): I, 82; II, 90; III, 100; IV, 100. Entire state 88%. Widely distributed in e. North America, NL to SK, New England and upper Great Lakes Region s. to AL, and GA.

Site-Habitat. Open, disturbed, dry to mesic sites of marked diversity of landform and soil; outwash and lake plains, moraines, lakeshores, bedrock shores and cliffs, dune areas, old fields, stream banks. There are many associates of diverse communities of these ecosystems and sites. Reported in 14 of Michigan's natural communities (Kost et al. 2007), 12 of them with rocky substrates.

Notes. Moderately shade-tolerant, persisting and flowering in deep shade, although more abundant and clone-forming with moderate to high light irradiance. Flowers are less conspicuous than those of true honeysuckle species; rarely used in ornamental planting. The 3 shrub species of this genus occur only in eastern North America.

Chromosome No. $2n = 36$, $n = 18$; $x = 18$

Similar Species. The southern bush-honeysuckle, *Diervilla sessilifolia* Buckl., ranges from western Virginia and Tennessee south to Georgia and northern Alabama. In the deep South, the range of mountain bush-honeysuckle, *D. rivularis* Gatt., overlaps that of the southern bush-honeysuckle. The taxa of bush-honeysuckles in the southeastern United States are described by Hardin (1968).

1. Leaves, × 1/2
2. Flowering shoot, × 1/2
3. Fruiting shoot, reduced
4. Fruit, capsule, × 1
5. Winter lateral bud and leaf scar, enlarged
6. Winter twig, reduced

KEY CHARACTERS

- low shrub to 1 m
- leaves opposite, blade margins serrate, apex long-acuminate
- twigs with 2 hairy-lined ridges decurrent from the nodes
- flowers yellow, turning orange to reddish, in terminal clusters, forming persistent capsules

Distinguished from *Lonicera* spp. by serrate leaves; bud scales not persistent at base of current year's shoots; flowers inconspicuous; fruit a capsule, not a fleshy berry.

THYMELAEACEAE
Dirca palustris Linnaeus
Leatherwood

Size and Form. Erect shrub to 2 m; single main stem, soon dividing into ascending and spreading crooked branches that form a dense, rounded crown; lacking root-collar sprouts.

Bark. Smooth, gray to brownish, moderately thick, tough, leathery, exceedingly pliable.

Leaves. Alternate, simple, deciduous; blades 4–9 cm long and about half as wide; early leaves broadly elliptical, oblong to obovate; late leaves less broad and more narrowly elliptical; obtuse; rounded to cuneate; entire; light green above, somewhat glaucous beneath; pubescent when young, becoming glabrous; turning yellowish, with characteristic pale circular patches; petioles very short, 3–5 mm, enlarged at the base, covering the winter buds.

Stems-Twigs. Slender, rounded, with silky hairs, becoming glabrous, prominently enlarged at nodes, appearing jointed, light brown to olive green, with conspicuous white lenticels, leaf scars raised, deeply U-shaped, and nearly encircling the buds; bundle scars 5; wood soft and weak.

Winter Buds. Terminal bud absent; buds solitary, short-conical, bud scales densely brown-hairy, covered by the petiole.

Flowers. April–early May, before the leaves; bisexual; borne in pendent clusters of 2–4, pale yellow, 6–10 mm long; floral tube lacking lobes; stamens 8, strongly exserted; subtending bud scales persistent for several weeks. Insect pollinated.

Fruit. Drupe; June; shed early; ellipsoid, pale green to yellowish-green when ripe, oval, 1–1.5 cm long, containing a single, shiny, dark brown to black seed.

Distribution. Occasional to locally frequent throughout the state. County occurrence (%) by ecosystem region (Fig. 19): I, 58; II, 47, III, 71; IV, 100. Entire state 59%. Distributed throughout e. North America, NS to ON, s. to FL and LA; ND, KA, OK in the Great Plains.

Site-Habitat. Mesic, moderately to deeply shaded, usually basic, nutrient-rich sites; sandy, loamy, rocky soils; understory or ground-cover layer of mesic to wet-mesic beech–sugar maple forests, wooded dunes. Associates include sugar maple, beech, basswood, northern red oak, bitternut hickory, tulip tree, hop-hornbeam, blue-beech, striped maple, alternate-leaved dogwood, running strawberry-bush, spicebush, maple-leaved viburnum, elderberries.

Notes. Shade-tolerant; slow-growing. Unlike typical shrub species, *Dirca* has a single main stem, soon dividing into a dense, rounded crown of many crooked branches. One of few shrubs able to reproduce in forest understories, due in large part to early flowering and leaf flushing before leaves of overstory trees leaf out. Excellent indicator of moist, fresh, relatively fertile sites that are usually associated with beech–sugar maple or hemlock–northern hardwoods forests and excellent displays of wildflowers in spring. Common name from the leathery, pliable bark of stems, not the weak wood; bark very difficult to break; used by Native Americans in basket weaving and for bow strings, thongs, and fishlines. One of the earliest shrubs to flower in spring. Potentially valuable in landscaping for moist, nutrient-rich, shady situations.

Similar Species. The closely related western leatherwood, *Dirca occidentalis* A. Gray, is an uncommon plant of moist slopes in California. The rare *D. mexicana* G. L. Nesom & M. H. Mayfield occurs in northeastern Mexico.

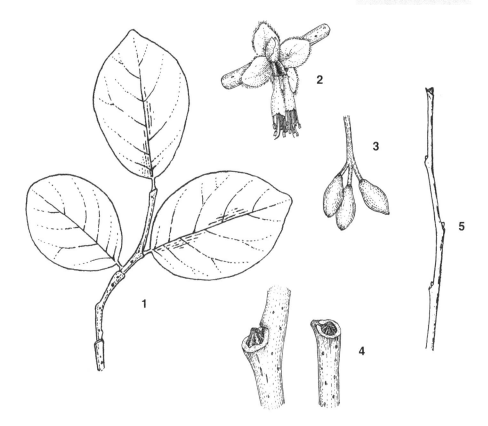

1. Leaf, × 1
2. Flowering shoot, reduced
3. Fruiting shoot, drupe, × 1
4. Winter lateral bud (left) and end bud (right), enlarged
5. Winter twig, reduced

KEY CHARACTERS

- erect shrub to 2 m, single main stem soon developing spreading crooked branches
- bark gray, tough, leathery, extremely pliable
- stems prominently enlarged at nodes, appearing jointed; conspicuous white lenticels, leaf scars nearly encircling the buds
- leaf blades broadly elliptical, tip obtuse, margin entire, petioles very short and enlarged at the base, covering the winter buds
- drupe, ellipsoid, pale green when ripe
- habitat mesic, shaded, relatively nutrient-rich sites

ELAEAGNACEAE
Elaeagnus umbellata Thunberg
Autumn-olive

Size and Form. Tall, much-branched shrub or small tree to 7 m, commonly with arching, pendulous branches, usually with thorny branches; forming clonal clumps of many ramets.

Bark. Trunks 10–16 cm in diameter with relatively long, light grayish, flat-surfaced plates, interspersed with crevices to 1–2 cm wide, gray and revealing dark reddish-brown inner bark.

Leaves. Alternate, simple, deciduous, remaining green long into winter, shed when green; blade 3–8 cm long, 2–4 cm wide; elliptic to ovate-oblong; obtuse to short-acuminate; rounded to broad-cuneate; entire, wavy or crinkled, green above, silvery beneath, with scattered rust-brown, peltate scales, midrib prominent, with many brown scales; petioles 3–8 mm long.

Stems-Twigs. Round, slender to stout; both determinate and indeterminate shoots present, vigorous indeterminate shoots with sylleptic branches that bear either leaves or thorns; color varying with exposure from golden brown or glistening brown (exposed) to silvery (in shade), small twigs to 1 cm completely covered with circular tan to silvery scales with a rusty-brown central dot, larger branches becoming dark brown to gray with fissures and distinct brown lenticels; larger branches often bearing lateral shoots with thorns; small lateral shoots may bear leaves and either a terminal bud or a thorn; basal sprouts to 2 m with axillary thorns; thorns persistent; leaf scars raised, half round, minute, bundle-scar 1; pith medium, round, brown, continuous.

Winter Buds. Terminal bud narrow, angled, blunt, brown or silvery, scaly; lateral buds small, broadly triangular or globose, acute and distinctly 2-lobed, solitary or occasionally collateral, tiny scales silvery.

Flowers. June; 1–2 cm long, bisexual, showy, fragrant, solitary or borne in clusters of 3–10 in the axils of leaves or at nodes of twigs of the preceding season; apetalous, calyx yellowish-white; stamens 4. Insect pollinated.

Fruit. Achene, enclosed in a fleshy perianth (drupelike), the enlarged pulpy base of the calyx tube; August–September; subglobose to ovoid, 4–8 mm across, at first silvery or silvery brown, becoming red, dotted with pale scales, on pedicels 4–12 mm long; juicy, edible; seeds 28,200/lb (62,040/kg).

Distribution. Native in southern Europe, western Asia, the Himalaya mountain system, China, and Japan. Hardy throughout the state. County occurrence (%) by ecosystem region (Fig. 19): I, 42; II, 40; III, 0; IV, 12. Entire state 35%. Widely distributed in e. North America, ON s. to FL and LA, also NE, KA, MT, WA, OR.

Site-Habitat. Full sunlight and light, dry to mesic sites; tolerant of drought and calcareous soils. Naturalizing in urban environments, roadsides, fencerows, stream banks, forest edges, and generally in open sites wherever birds and mammals disperse its seeds; absent or rare in heavily shaded forest understories. Associates include all species in oak-hickory forests and shade-intolerant and midtolerant forests of Michigan. Lacking in wetlands except on open, mesic to dry-mesic microsites.

Notes. Shade-intolerant, persistent in light shade, fast-growing. Native in Eurasia. Planted for wildlife cover and food and spreading aggressively by seeds dispersed widely by birds and animals. Highly attractive to birds for food, cover, and nesting. Clone-forming by multiple sprouts from the root collar and occasional root sprouts located near the parent plant. Larger arching branches give rise to indeterminate shoots to 2.1 m with sylleptic branches. Both determinate and indeterminate shoots abundant. Extremely aggressive in open areas, along roads, in fencerows, and in relatively open forest stands. One of the most harmful invasive shrubs to forest regeneration by shading and killing seedlings of forest trees, especially in oak and oak-hickory forests. An invasive species, avoid planting in urban or natural landscapes.

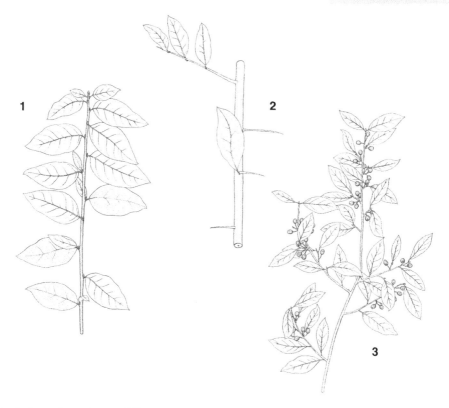

1. Indeterminate shoot, × 1/4
2. Stem with spines and determinate shoot, × 1/4
3. Branch with fruiting shoots, berry, reduced

KEY CHARACTERS

- tall arching shrub or small tree of Eurasian origin, spreading aggressively in open forests and elsewhere in open areas
- leaves alternate, elliptic to ovate-oblong, green above, silvery below, with scattered rusty-brown glands
- stems from glistening golden brown to silvery, branches becoming dark gray, branches often with shoot-bearing terminal thorns
- fruit an achene enclosed in a fleshy, juicy envelope, nearly round, becoming red and coated with silvery scales, edible

Distinguished from Russian-olive, *Elaeagnus angustifolia*, by bushy clonal clumps with widely spreading branches; larger and broader leaves, not willowlike; green above; fruit red and coated with silvery scales; aggressively colonizing almost any open area and open-canopy forests.

Chromosome No. $2n = 28$, $n = 14$; $x = 14$

Similar Species. The silverberry, *Elaeagnus commutata* Rydb., is an upright, much-branched, unarmed clonal shrub or small tree, 0.5–3 m, of northern and western distribution. Twigs with rusty-brown scales and stellate pubescence; leaves silvery on both sides; flowers yellow, very fragrant; mealy fruit covered with silver scales. Very hardy; sometimes used as hedge plant. The Russian-olive, *Elaeagnus angustifolia* L., a native of Asia, is a tall shrub or small tree >10 m high (Smith 2008; Voss and Reznicek 2012).

ERICACEAE
Epigaea repens Linnaeus

Trailing Arbutus

Size and Form. Prostrate, evergreen, vinelike shrub with creeping stems and few ascending branches, 10–15 cm high and to 60 cm long; stems rooting at the nodes.

Leaves. Alternate, simple, evergreen; blades stiff and leathery, often wrinkled, 2–8 cm long and 1.5–4 cm wide; ovate to oblong; rounded and mucronate; rarely acute; rounded or cordate; margin entire, ciliate; green and more or less bristly-hairy above and below; petioles pubescent, 0.5–3 cm long.

Stems-Twigs. Slender, dark reddish-brown to black, exfoliating, covered with coarse reddish or reddish-brown hairs; leaf-scars linear or absent, the scars being shed with exfoliating bark; pith moderate, rounded, brown, continuous.

Winter Buds. Terminal bud where present short-stalked, 5–7 mm long, conical, brown, with several hairy scales exposed; lateral buds minute.

Flowers. Early spring, mid-April to May; plants functionally unisexual, dioecious; flowers borne singly or in clusters of 2–6 on short racemes terminal or from the axil of the upper leaf; *either* female with vestigial stamens *or* apparently bisexual with functional stamens but unexpanded stigmas; white to pink, very fragrant; each flower subtended by 2 ovate bracts; sepals 5, lobes ovate to acuminate, 5–8 mm long; petals united to form a slender salverform corolla tube ca. 1.5 cm long, petals 5; stamens 10; ovary hairy. Insect pollinated.

Fruit. Capsule, 5-celled, small, 8–9 mm wide, densely hairy, fleshy, globose, red.

Distribution. Infrequent to locally common throughout the state except in parts of southern Michigan; increasing in abundance northward. County occurrence (%) by ecosystem region (Fig. 19): I, 68; II, 23; III, 100; IV, 100. Entire state 77%. Widely distributed throughout e. North America.

Site-Habitat. Sandy-gravelly, dry to dry-mesic, acidic soils and fire-prone sites of outwash plains and ice-contact terrain. Typical of pine barrens with jack pine and northern pin oak; also in dry and dry-mesic northern forests. Associates include jack pine, red pine, northern pin oak, black cherry, paper birch, bigtooth and trembling aspens, red maple, wintergreen, bearberry, sand cherry, sweetfern, and blueberries.

Notes. Shade-intolerant, persisting in light shade. *Epigaea* from Greek *eip*, "on," and *gaea*, "Earth"; because it grows along the ground it is easily covered with leaves and overlooked. Ants are attracted to the fleshy part of the capsule when it opens; they disperse the small oval seeds (Voss 1996; Smith 2008).

Chromosome No. $2n = 24$, $n = 12$; $x = 12$.

Similar Species. The only counterpart species, *Epigaea asiatica* Maxim., occurs in Japan.

1. Branch with vegetative and flowering shoots, × 1/2
2. Flower, enlarged
3. Fruiting shoots, capsule × 1/2
4. Capsule, enlarged

KEY CHARACTERS

- prostrate evergreen shrub with creeping stems and few ascending branches
- leaves alternate, evergreen, blades leathery, ovate-oblong, margin entire, tip rounded-mucronate, bristly-hairy both sides
- stems covered with coarse reddish-brown hairs
- dry sand barrens and dry to dry-mesic northern forests

CELASTRACEAE
Euonymus atropurpureus Jacquin
Burning-bush, Wahoo

Size and Form. Tall, erect, sometimes straggling shrub with spreading branches, rarely a small tree to 4 m; stems often single or forming small, open clones by rhizomes (Musselman 1968).

Bark. Gray, smooth, becoming roughened at the base.

Leaves. Opposite, simple, deciduous, blades 4–12 cm long, 1–4 cm wide; oval-elliptic, acuminate; cuneate or rounded; finely serrate; thin; dull green, glabrous above, finely pubescent beneath, turning scarlet in autumn; petioles 1–2 cm long.

Stems-Twigs. Slender, stiff, green at first, becoming dark purplish-brown, glabrous, rounded but often 4-sided with 4 corky lines or ridges below the nodes; leaf scars half elliptic, small; bundle scar 1; pith greenish, spongy.

Winter Buds. Terminal bud conical, 3–5 mm long, purplish, 3–5 scales exposed; lateral buds small, oblong, green tinged with red.

Flowers. June; bisexual; borne in small axillary cymes; 4-parted; purplish-brown, 6–8 mm across; sepals semicircular, about 1 mm long; petals dark purple, broadly ovate, 2–2.5 mm long. Insect pollinated.

Fruit. Capsule; September; fleshy, rough-surfaced, glabrous, about 4 mm across, rounded and deeply 4-lobed, pinkish-red when ripe; seed light brown covered with a scarlet aril in drooping clusters, usually persistent until midwinter; seeds 16,900/lb (37,180/kg).

Distribution. Occasional and exclusively in the southern half of lower Michigan. County occurrence (%) by ecosystem region (Fig. 19): I, 37; II, 0; III, 0; IV, 0. Entire state 14%. Widely distributed throughout e. North America and the Great Plains, QC, ON, ND, s. to TX and all Gulf coastal plain states.

Site-Habitat. Characteristic of riverbanks, moist bottomlands of stream and river floodplains not inundated during the growing season; mesic, circumneutral-basic, nutrient-rich sites. Associates include shagbark hickory, American basswood, northern hackberry, slippery elm, blue-beech, hop hornbeam, common hazel, running strawberry-bush.

Notes. Shade-intolerant, persistent in light shade. Sometimes cultivated as an ornamental for its showy scarlet foliage in autumn, for which it is named. Often planted is the closely related European strawberry-bush or spindle tree, *Euonymus europaeus* L., with smaller leaves and pale flowers. More highly prized as an ornamental is the winged wahoo, *E. alatus* (Thunb.) Siebold, native from northeastern Asia to central China. It is a tall shrub to 7 m high, with leaves obovate-elliptic, turning brilliant red in autumn. It is noted for its stiff, spreading twigs with 4 conspicuous, flat, corky wings.

Similar Species. The western counterpart is the western burning-bush, *Euonymus occidentalis* Nutt. ex Torr., which ranges from eastern Washington and Oregon to coastal coniferous forests of northern California and the peninsular ranges of Southern California below 2,100 m. Species of Europe (*E. europaeus*) and Asia (*E. alatus*) are identified in "Notes."

140

1. Vegetative shoot, × 1/2
2. Long shoot with leaves and flowering shoots, reduced
3. Fruiting shoot and single capsule, × 1/2
4. Winter lateral bud and leaf scar, enlarged
5. Winter twig, reduced

KEY CHARACTERS

- tall, erect shrub to 4 m or rarely a small tree
- leaves opposite, simple, oval-elliptic, turning scarlet in autumn
- twigs slender, stiff, often 4-sided, with 4 corky ridges below the nodes
- fruit a capsule, in drooping clusters, rounded and deeply 4-lobed, splitting to reveal a scarlet aril covering the seeds

Distinguished from running strawberry-bush, *Euonymus obovatus,* by tall, erect shrub habit, oval-elliptic leaves, capsule 4-lobed; and stems with corky ridges below the nodes.

CELASTRACEAE
Euonymus obovatus Nuttall

Running Strawberry-bush

Size and Form. Small creeping shrub, with a few ascending flowering branches to 30–40 cm; stems taking root at the nodes and forming clones in moist soil.

Leaves. Opposite, simple, deciduous, 2–5 pairs on each erect shoot, 3–8 cm long and 1.5–4 cm wide; terminal leaves obovate; abruptly tapering to a blunt apex; cuneate; finely crenate-serrate; thin; dull light green above, paler beneath; glabrous; petioles short, grooved, 2–5 mm long.

Stems-Twigs. Trailing, green, slender, shoots ascending, 4-sided, sometimes slightly winged; leaf scars opposite, half elliptical, bundle scar 1; pith greenish.

Winter Buds. Terminal bud narrowly ovoid, greenish, with 3–4 pairs of outer scales; lateral buds small, solitary, sessile, fusiform, often not evident.

Flowers. April–May; bisexual; borne 1–3 flowers borne on long pediceled cymes, 5-parted; very flat, 5–6 mm across; greenish-yellow, with shiny nectary; petals 5, stamens 5. Insect pollinated.

Fruit. Capsule; September; shallowly 3-lobed, rough-warty, pink to crimson; about 1.5 cm wide; each seed covered by a scarlet or orange aril; persistent in winter; seeds 25,500/lb (56,100/kg).

Distribution. Common to locally abundant in southern lower Michigan; rare in northern lower Michigan, absent in upper Michigan. County occurrence (%) by ecosystem region (Fig. 19): I, 100; II, 10; III, 0; IV, 0. Entire state 49%. Distributed throughout the upper Great Lakes Region and adjacent ON, s. throughout the Midwest to AR, GA, SC.

Site-Habitat. Fertile topsoil or organic matter of mesic and seasonally flooded sites, tolerating moderate to deep shade. Typical in mesic beech–sugar maple forests, the levee and second bottom of floodplains, intolerant of inundation during the growing season; northward in moist, circumneutral, northern white-cedar swamps. Occasionally present in dry-mesic forests where organic matter is deep, circumneutral-basic, and nutrient-rich enough for its establishment and growth. Diverse tree associates of beech–sugar maple and floodplain forests, northern white-cedar, eastern hemlock, leatherwood, Canadian fly honeysuckle, common elder, red elderberry, maple-leaved viburnum, alternate-leaved dogwood, burning-bush.

Notes. Shade-tolerant. Useful as a ground cover under tree plantings. This large genus includes 170 to 200 species in northern temperate regions, most of which are located in China, Japan, Korea, and the Himalayan region; 3 occur in eastern North America, and one, the western burning-bush, *Euonymus occidentalis* Torrey, occurs in California.

Similar Species. The strawberry-bush or hearts-a-bursting-with-love, *Euonymus americanus* L. ($2n = 64$), is a southern relative with an erect, somewhat straggling shrub form to 2 m. A native of China and Japan, *Euonymus fortunei* (Turcz.) Handel.-Mazz., wintercreeper, is a creeping or climbing evergreen vine. Its many varieties and hardiness make it popular in cultivation, especially in English gardens (Coats 1992). Dirr (1990) describes 38 cultivars.

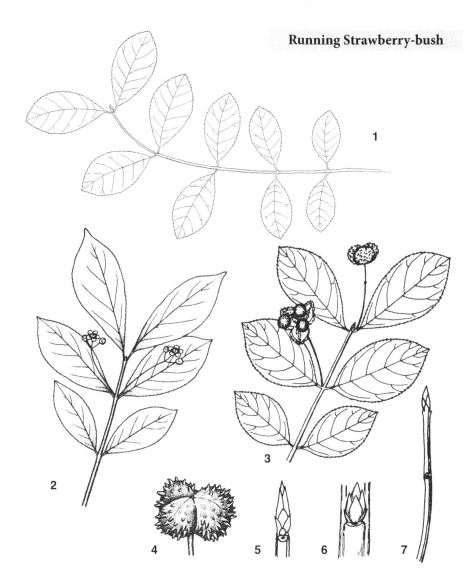

1. Leaves on creeping stem, reduced
2. Shoot with flower buds, reduced
3. Fruiting shoot, capsule, reduced
4. Capsule, × 1 1/2
5. Winter terminal bud, enlarged
6. Winter lateral bud and leaf scar, enlarged
7. Winter twig, reduced

KEY CHARACTERS

- trailing vine or small creeping shrub
- leaves opposite, simple, terminal leaves obovate, margin finely crenate-serrate
- twigs green, trailing or erect, 4-sided, rooting at nodes
- fruit a capsule, 3-lobed, pink to crimson, each seed covered with a scarlet aril
- occurs naturally almost exclusively in the southern half of lower Michigan

Distinguished from burning-bush, *Euonymus atropurpurea*, by its low, trailing habit; terminal leaves obovate; capsule 3-lobed; and stems lacking corky ridges below the nodes.

RHAMNACEAE
Frangula alnus Miller
Glossy Buckthorn

Size and Form. Tall shrub or small tree to 6 m with an open crown, single individuals or dense, multistemmed clones sprouting from the root crown. Michigan Big Tree: girth 58.4 cm (23 in), diameter 18.6 cm (7.3 in), height 10.7 m (35.0 ft), Oakland Co.

Bark. Thin, brownish to reddish-brown, becoming dark gray, pale white lenticels conspicuous.

Leaves. Alternate, occasionally subopposite on long indeterminate shoots (see diagram on facing page); simple, deciduous; blades 3–7 cm long, 2–5 cm wide; oblong to obovate; acute; broadly cuneate or rounded; entire to irregularly crenulate or wavy; glabrous, dark lustrous glossy-green above, lighter and pubescent beneath, at least along midrib and larger veins; 8–9 pairs of veins, pinnate but curving upward near the margin; petioles 0.6–2 cm long; stipules small, slender, early deciduous.

Stems-Twigs. Stout, unarmed, at first pubescent and grayish-brown, becoming glabrous and reddish-brown to gray; dotted with numerous pale white, vertically elongated lenticels; leaf scars half elliptical, small, somewhat raised; bundle scars 3; stipule scars small; pith small, tan, continuous.

Winter Buds. Terminal buds 3–6 mm long, naked, 4–6 scales exposed, dark yellow-brown, densely woolly, with reddish-brown or orange-brown hairs; lateral buds 1–4 mm, obliquely sessile, appressed, otherwise similar to the terminal.

Flowers. May–July, after the leaves, bisexual, pale greenish-yellow, inconspicuous, 2–3 mm across; borne singly or in few-flowered, sessile umbels in the axils of lower leaves, pedicels glabrous or nearly so, 5–10 mm long; sepals 5, narrowly ovate, acute; petals 5, broadly obovate. Insect pollinated.

Fruit. Drupe; July–August, globose, 6–8 mm wide, conspicuous, red turning dark purple or black, enclosing 2–3 nutlike, non-gooved pits containing seeds.

Distribution. Native of Europe, North Africa, western Asia; long cultivated and naturalized in North America. Frequent in the southern three tiers of counties in southern lower Michigan, occasional in northern lower Michigan and upper Michigan. Hardy throughout the state. County occurrence (%) by ecosystem region (Fig. 19): I, 39; II, 20; III, 57; IV, 62. Entire state 36%.

Site-Habitat. Relatively open and disturbed wetlands and adjacent forests and power lines; grows on almost any disturbed soil site; fens, bogs, swamps, wet shrub thickets, floodplains; typically sites with mildly basic or calcareous substrates. Most common in the vicinity of urban areas where it has been cultivated.

Notes. Moderately shade-tolerant. Seeds of the weedy, invasive European alder buckthorn are widely dispersed by birds. Root-collar sprouts vigorous, extending vertically, with few lateral branches. An aggressive and serious pest, invasive, with a history of invading swamps and fens and threatening and replacing native plants. First collected in Michigan in 1934, Delta Co. (Voss and Reznicek 2012). Avoid use in landscaping. Formerly it was widely known as *Rhamnus frangula* L.

Chromosome No. $2n = 20$, $n = 10$; $x = 10$

Similar Species. The Carolina buckthorn or Indian cherry, *Frangula caroliniana* (Walter) A. Gray, is similar in its tall shrub or small tree habit and many morphological characteristics; Midwest, central and southern Appalachian Mountains south to Florida and Texas. The western North American counterpart is, cascara buckthorn, *F. purshiana* (DC.) Cooper. Native peoples steeped or boiled its bark in water to make a strong laxative tea (or syrup) whose properties have been medically substantiated (Pojar and MacKinnon 1994; Parish, Coupe, and Lloyd et al. 1996; Kershaw, MacKinnon, and Pojar 1998). A widespread and similar low shrub of wetlands in Michigan is *Rhamnus alnifolia* L'Hér.

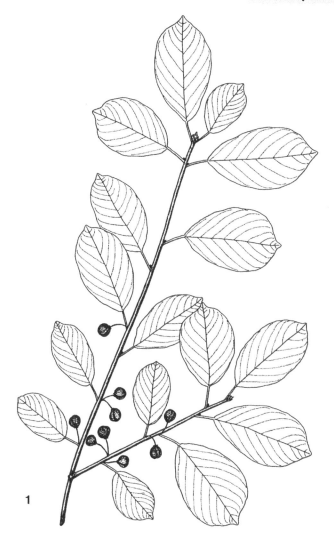

1. Individual with vegetative long shoots and fruits, drupe, reduced

KEY CHARACTERS

- tall shrub or small tree to 6 m, clone-forming, stems unarmed
- leaves alternate, rarely subopposite, blades oblong to obovate, entire; glabrous, dark lustrous glossy-green above; veins pinnate, upturned at margins
- terminal bud 3–5 mm long, naked, dark yellow-brown, densely woolly with reddish-brown hairs
- fruit a globose drupe, conspicuous, red turning dark purple or black
- Eurasian origin and aggressively invading wetlands and adjacent disturbed sites, especially near urban areas

Distinguished from alder-leaved buckthorn, *Rhamnus alnifolia* L'Hér., by its tall shrub or small tree form, leaf blades entire, glossy green upper leaf surface, naked terminal bud with densely woolly reddish-brown hairs; a nonnative invasive species of wetlands and other disturbed sites, most abundant in southern lower Michigan.

ERICACEAE
Gaultheria hispidula (L.) Bigelow
Creeping-snowberry

Size and Form. Prostrate, creeping, very small evergreen shrub, aromatic; 10–50 cm long; rooting at nodes and forming clonal mats.

Leaves. Alternate, simple, evergreen, very small and numerous, wintergreen odor when bruised or chewed; blades 4–10 mm long and about half as wide; broadly elliptic; acute to abruptly bristle-tipped, rounded or broadly tapered; entire and revolute; smooth and lustrous green above, paler with closely appressed, sparse, dark brown hairs beneath; petioles very short.

Stems-Twigs. Very slender to threadlike, creeping and branching, with appressed brownish hairs.

Winter Buds. Solitary, minute, sessile, ovoid.

Flowers. May–June; bisexual; few; most commonly solitary in the axils of the leaves on short recurved pedicels 1 mm long; small; white to pinkish; calyx 4-parted; corolla bell-shaped, 2–5 mm, with rounded lobes one-third as long as the tube, stamens 4. Insect pollinated.

Fruit. Capsule, surrounded by a persistent, pulpy, expanded calyx forming a fleshy, berrylike fruit; July–September; white, 0.5–1 cm across, usually with some bristly brownish hairs; aromatic when crushed or chewed; borne on a recurved pedicel 1 mm long; edible, mealy. Seeds 3,09,300/lb (680,460/kg).

Distribution. Infrequent in southern lower Michigan, common to abundant in northern lower Michigan and upper Michigan. County occurrence (%) by ecosystem region (Fig. 19): I, 32; II, 77; III, 100; IV, 100. Entire state 35%. Widely distributed in the n. part of e. North America, NL to BC, s. New England to WV, OH, western Great Lakes Region, ID, WA.

Site-Habitat. Wet, poorly drained, conifer and mixed hardwood–conifer swamps of acidic or basic substrates. Avoiding the wettest microsites, it is common on rotting tree trunks (i.e., coarse woody debris), moss-covered rocks, and organic soil. Associates include tamarack, black and white spruces, eastern hemlock, northern white-cedar, balsam poplar, trembling aspen, leatherleaf, Labrador-tea, bog-rosemary, blueberries, and wild-raisin.

Notes. Moderately shade-tolerant. All parts contain oil of wintergreen, formerly produced commercially from *Gaultheria* species. A major constituent is methyl salicylate, closely related to the main ingredient in aspirin.

Chromosome No. $2n = 44, 88$; $n = 22, 44$; $x = 11$

Similar Species. A common associate in the western Great Lakes Region is the related wintergreen, *Gaultheria procumbens*. The counterpart species in Japan is *G. japonica* (A. Gray) Sleumer. In alpine rhododendron forests and grassland sites of Yunnan, China, a counterpart species is *Gaultheria cardiosepala* Handel-Mazzetti. The cespitose, low shrub native in Chile, *G. minima* (Phil.) Sandw., is likely a closely related alpine species.

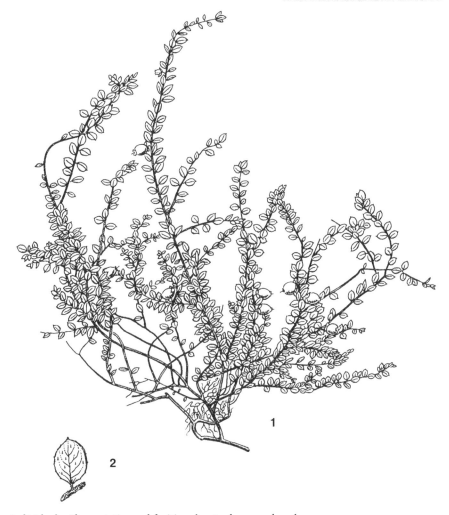

1. Individual with vegetative and fruiting shoots, drupe, reduced
2. Stem with leaf, × 1 1/2

KEY CHARACTERS

- prostrate, creeping, evergreen shrub, all parts aromatic
- leaves alternate, very small and numerous, less than 1 cm long, wintergreen odor when bruised or chewed, petiole very short
- fruit a many-seeded capsule surrounded by a persistent, fleshy white calyx, <1 cm across, edible
- habitat wetland sites, including conifer and hardwood-conifer swamps

Distinguished from wintergreen, *Gaultheria procumbens*, by its mostly procumbent stems and white fruits instead of bright red fruits on upright stems, many markedly smaller leaves, and confinement to wetlands.

Distinguished from small cranberry, *Vaccinium oxycoccos*, and large cranberry, *V. macrocarpon*, by its tolerance of shade and growth in forested wetlands, all parts with wintergreen odor and flavor, stems with dark brown, strongly appressed hairs, leaves with green lower surfaces with more acute tips, and white fruit borne on short pedicels.

Size and Form. Low, small, creeping evergreen shrub, 5–15 cm; stems trailing on the ground or slightly below the surface, fruiting stems erect to 15 cm; clone-forming by sprouting from rhizomes.

Leaves. Alternate, simple, evergreen, wintergreen odor when crushed or chewed, blades 2–5 cm long, 1–3 cm wide; ovate, elliptic to narrowly obovate, rounded and apiculate; cuneate; shallowly crenate-serrate; thick and leathery; lustrous, dark green above, paler but not whitened beneath, sometimes turning red with cold weather; glabrous; petioles more or less pubescent, 2–5 mm long.

Stems-Twigs. Slender, glabrous or nearly so, greenish or brownish; fruiting branches erect, arising from main stems that are usually below ground, aromatic when crushed.

Winter Buds. Solitary, sessile, minute, 1–2 mm long, flattened ovoid, several scales exposed.

Flowers. July–August; bisexual; most commonly solitary in the axils of the leaves, on pedicels 4–8 mm long; 5-parted; small; white to pinkish; calyx 5-parted; corolla urn-shaped, 3 rounded lobes, 6–7 mm long; stamens 10. Insect pollinated.

Fruit. Capsule, surrounded by a persistent, pulpy, expanded calyx forming a fleshy, berrylike fruit; October, persistent through winter into spring; bright red, 0.8–1 cm across, aromatic when crushed, edible, mealy, spicy. Seeds 3,885,000/lb, 8,547,000/kg

Distribution. Common to abundant throughout the state. County occurrence (%) by ecosystem region (Fig. 19): I, 82; II, 97; III, 100; IV, 88. Entire state 89%. Wide ranging in e. North America, NL to MB, s. to GA, AL, and MN.

Site-Habitat. Open to moderately shaded, dry and dry-mesic, acidic, nutrient-poor sites, typically sandy soils; less common in wetland sites. Associates include diverse species of muskeg and nutrient-poor conifer swamps (e.g., tamarack, black spruce, leatherleaf, Labrador-tea, bog-rosemary, and creeping-snowberry) and many species of dune and swale, barrens, and dry to dry-mesic forests oak and pine forests (e.g., jack pine, northern pin oak, bearberry, blueberries, black huckleberry, trailing arbutus, and wintergreen).

Notes. Shade-intolerant, persistent in light to moderate shade. Fruit production is low, so maintenance and spread of populations is likely primarily asexual (Smith 2008). All parts contain oil of wintergreen, formerly produced commercially from the plant. A major constituent is methyl salicylate, closely related to the main ingredient in aspirin. Of the 100–150 species of *Gaultheria* occurring mostly in Asia, Australia, and South America, only 6 occur in North America north of Mexico. Turkey, grouse, bobwhite, pheasant, black bear, white-tailed deer, and other animals feed on fruits and leaves.

Chromosome No. $2n = 44, 88$; $n = 22, 44$; $x = 11$

Similar Species. The creeping-snowberry, *Gaultheria hispidula* (L.) Bigelow, is a creeping plant of conifer and conifer-hardwood swamps, especially in northern lower Michigan and upper Michigan. Related dwarf species, western teaberry, *G. ovatifolia* A. Gray (up to 5 cm tall), and alpine wintergreen, *G. humifusa* (Graham) Rydb., (1–3 cm tall), occur in montane to alpine sites of bogs and moist to dry forests in mountains of the West (British Columbia, Washington, Oregon, California, Idaho, and Montana). Common to the point of invasiveness is the West Coast relative salal, *Gaultheria shallon* Pursh, an erect woody shrub to 3 m. It ranges from Alabama to the Pacific coast to and inland into the Cascades and Sierra Nevada. In Japan the similar red-fruited species is *G. adenothrix* Maxim. Thirty-two species, prostrate to small trees to 4 m, are reported from China; one prostrate, red-fruited species of western China is *G. suborbicularis* W. W. Smith.

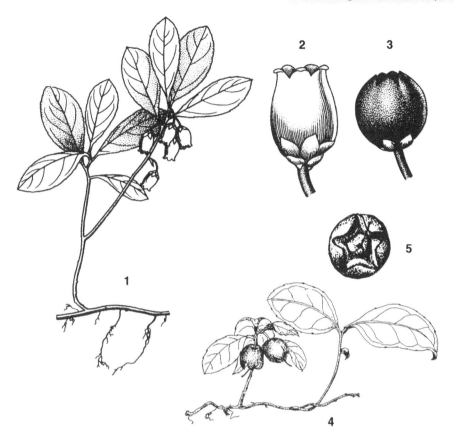

1. Flowering shoot, × 1/2
2. Flower, × 2
3. Fruit, capsule, side view, × 2
4. Fruiting and vegetative shoots, × 1/2
5. Fruit, top view, × 1 1/2

KEY CHARACTERS

- low, creeping, small evergreen shrub with erect flowering and fruiting stems
- all parts emitting a wintergreen fragrance when crushed or chewed
- leaves alternate, ovate, evergreen, thick and leathery, shiny dark green above
- fruit a many-seeded capsule surrounded by a persistent, pulpy, bright red calyx, edible
- habitat usually open to moderately shaded, dry to dry-mesic, acidic sites

Distinguished from creeping-snowberry, *Gaultheria hispidula*, by its more erect form, much larger and fewer shiny dark green leaves, bright red fruits, and typically dry and dry-mesic sites.

ERICACEAE
Gaylussacia baccata (Wangenh.)
K. Koch

Black Huckleberry

Size and Form. Low, erect, much-branched shrub, 0.3–1.5 m; clone-forming by sprouting from rhizomes.

Bark. Thin, brown, becoming dark gray to blackish, outer layer peeling with age.

Leaves. Alternate, simple, deciduous 2–5 cm long and about half as wide; oval or oblong-ovate; acute, obtuse, or rounded; cuneate; entire, ciliate; moderately thick and leathery; yellowish-green above, paler beneath; pubescent on both sides at first, becoming glabrous; lower surface more densely covered with shiny, yellow, resinous dots than on upper surface, sticky when young; petioles pubescent, short, 2 mm long.

Stems-Twigs. Slender, stiff, 3-sided to roundish, more or less pubescent; current shoots greenish, previous year's peach or pink above and green beneath, earlier years brownish with white stripes—giving a "rainbow" effect; leaf scars small, crescent-shaped; bundle scar 1; pith small, continuous.

Winter Buds. Terminal bud absent; flower buds 3 mm long with yellow resin dots near the apex, ovoid, 5–7 scales exposed; vegetative buds 1–2 mm long, pointed, 2–3 scales visible.

Flowers. May–June; bisexual; 1–7 flowers borne on short, dense, drooping, 1-sided racemes; glandular-dotted, calyx 5-lobed; corolla tubular-campanulate, 5-lobed, reddish; stamens 10. Insect pollinated.

Fruit. Drupe; August–September, globose, 6–8 mm across, black, lustrous, with 10 1-seeded pits, about 2 mm long; edible but seedy.

Distribution. Locally common throughout Michigan, especially northern lower Michigan. County occurrence (%) by ecosystem region (Fig. 19): I, 71; II, 87; III, 100; IV, 62. Entire state 78%. Widely distributed in e. North America, NL to ON, s. New England to GA, AL, and MN s. to AR, MS.

Site-Habitat. Open to lightly shaded, dry and dry-mesic, acidic, sandy-gravelly sites; also frequent in wetland bogs, swales, and nutrient poor conifer swamps. Reported from 10 diverse native communities of Michigan (Kost et al. 2007), it has many tree and shrub associates of the above sites, especially early successional and ericaceous species.

Notes. Shade-intolerant, persistent in light shade; sprouting vigorously following fire and spreading clonally. Rubbing the undersurface of a leaf blade slowly and vigorously across a white sheet of paper reveals a faint but distinctive yellowish streak—a confirming identification test. Fruit ripens markedly later than that of blueberries, with which it often grows.

Chromosome No. $2n = 24$, $n = 12$; $x = 12$

Similar Species. Huckleberries are relatively common and thrive on the acidic soils of the southeastern and southern United States in the southern Appalachian mountains, piedmont, and coastal plain. Besides black huckleberry, three deciduous species are reported, including erect shrubs to 1.5–2 m: dangleberry (*G. frondosa* [L.] T. & A. Gray); bear-huckleberry (*G. ursina* [M. A. Curtis] Torr. & A. Gray); and Mosier's huckleberry (*G. mosieri* Small). Huckleberries are noted for their spreading clones; a clone of the box huckleberry (*G. brachycera* [Michx.] A. Gray), a low evergreen shrub, is reported to cover 100 acres (40.5 ha) on a mountain slope in Pennsylvania; its estimated age is 13,000 years (Willaman 1965).

150

1. Branch with flowering shoots, × 1/2
2. Flowering shoot, × 1
3. Lower leaf surface with resin glands, × 10
4. Fruiting shoot, drupe, reduced
5. Winter lateral bud and leaf scar, enlarged
6. Winter twig, reduced

KEY CHARACTERS

- low, erect, much-branched, clone-forming shrub
- leaves alternate, simple, entire, ciliate; lower surface densely covered with shiny, yellow, resinous dots
- twigs brownish with white stripes
- fruit a drupe; globose, black, lustrous, edible
- habitat open to lightly shaded dry and dry-mesic acidic sites and also some acidic, nutrient-poor wetlands

HAMAMELIDACEAE
Hamamelis virginiana Linnaeus
Witch-hazel

Size and Form. Tall understory shrub or small tree to 8 m and 12 cm in diameter. Several large, crooked stems usually growing together in a clump (clone), forming an irregular, open crown; stems confined to a single root plate. Michigan Big Tree: girth 43.2 cm (17 in), diameter 13.7 cm (5.4 in), height 13.1 m (43 ft), Muskegon Co.

Bark. Thin, smooth, becoming scaly, light brown to grayish-brown, with conspicuous lenticels; inner bark reddish-purple.

Leaves. Alternate, simple, deciduous, blades 6–15 cm long, 3–10 cm wide; irregularly oval-obovate; acute to obtuse; base strongly asymmetrical; crenate; thin; dark green to yellowish-green, lighter and nearly glabrous beneath; veins straight, parallel, and widely spaced, 5–7 per side; turning yellow in autumn; glabrous; petioles short, 0.5–1.5 cm long.

Stems-Twigs. Slender, zigzag, scurfy stellate-pubescent, becoming glabrous, reddish-brown to light orange; leaf scars 2-ranked, bundle scars 3; pith small, tan, continuous.

Winter Buds. Terminal bud naked, 7–14 mm long, stalked, elongate, flattened, sickle- or crescent-shaped, scurfy, yellowish-brown; lateral buds smaller, appressed, commonly superposed.

Flowers. October–November, during and after leaf fall; bisexual; showy, in clusters of 3; peduncles finely pubescent; calyx 4-parted; petals 4, very narrow and straplike, bright yellow, twisted; 1.5–2 cm long, ca. 1 mm wide. Insect pollinated in autumn.

Fruit. Capsule; October–November of the year after flowering; woody, broad-obovoid, 1–1.5 cm long, tan, pubescent; splitting open at the top when dry and discharging 2 lustrous black seeds, which are forcibly ejected up to 10 m from the tree (Meyer 1997), empty capsules remaining on the twigs another year or more; seeds ca. 9,800/lb (21,560/kg).

Distribution. Common in southern lower Michigan, occasional in northern lower Michigan, rare in upper Michigan. County occurrence (%) by ecosystem region (Fig. 19): I, 97; II, 90, III, 14; IV, 62. Entire state 84%. Wide ranging throughout e. North America, NS to ON, s to FL, w. to TX.

Site-Habitat. Primarily dry to dry-mesic, often shaded sites, sandy, acidic to basic soils, often with calcareous substrate; oak-hickory and oak forests, to a lesser extent mesic mixed hardwood forests and swamps; sandy ridges of southeastern Michigan lake plains. Associates include a great variety of tree and shrub species of very diverse sites and communities (Kost et al. 2007). Typically occurs with species of oak, oak-hickory, and mixed hardwood forests.

Notes. Moderately shade-tolerant; slow-growing; short-lived. Forked twigs were used by the water diviners, or "well witchers," to seek underground sources of water. Witch-hazel astringent is obtained from the leaves, twigs, and bark and has long been used in lotions and medicinal extracts. Occasionally used ornamentally as a shrub border for its abundant flowering in late autumn, the flowers remaining after the leaves are shed. Other species, native in both North America and Asia, flower in the spring. The common name is said to be applied due to the hazellike straight veins of the leaves.

Chromosome No. $2n = 24$, $n = 12$; $x = 12$

Similar Species. The closely related and smaller *Hamamelis vernalis* Sarg. flowers in the late winter to early spring and occurs in the mid- to deep South. In East Asia, *H. mollis* Oliv. occurs in central China, and *H. japonica* Sieb. & Zucc. is native in Japan.

1. Winter twig, × 1
2. Portion of twig, side view, × 1
3. Leaf, × 1/2
4. Flowering shoots, × 1
5. Fruit, capsule open, with seeds disseminated, × 2

KEY CHARACTERS

- tall shrub with several large, upright trunks, typically growing together in a clump (clone), or small tree with multistemmed trunks
- leaf blades oval, margin crenate, base strongly asymmetrical
- terminal bud naked, stalked, yellowish-brown, stellate-pubescent
- flowers appearing in late autumn, petals narrow, straplike, bright yellow
- fruit a woody capsule, persisting on branches over winter and discharging 2 seeds in autumn of the year after flowering

HYPERICACEAE
Hypericum kalmianum Linnaeus

Kalm's St. John's-wort

Size and Form. Low, erect, shrub, much-branched, 0.2–0.8 m.

Leaves. Opposite, simple, deciduous, sessile, often with tufts of smaller leaves in axils of larger leaves; blades 2–5 cm long, 3–8 mm wide, often revolute; linear-oblong to oblanceolate; obtuse to acute, slightly narrowed; margin revolute, entire, occasionally slightly wavy near the tip; dark green above, dotted with translucent glands, glaucous beneath, with only midrib showing; petiole short, 1–4 mm long.

Stems-Twigs. Slender, with exfoliating, pale brown, papery bark, ascending 4-angled branches, young twigs 2-edged, with winglike ridges below the nodes, leaf scars somewhat raised, triangular, opposing leaf scars connected by a transverse ridge, bundle scar 1; pith relatively large, relatively porous, large stems hollow.

Winter Buds. Small, solitary, sessile, green, glabrous.

Flowers. July-August; bisexual; in small, compound, terminal cymes, 2–3.5 cm wide; petals 5, bright yellow, 2–3.5 cm across, showy; ovary 5-celled, styles 5, united below; stamens very many, distinct, united below. Insect pollinated.

Fruit. Capsule; August–September; ovoid, 6–10 mm across, dark brown, beaked at first; splitting into 5 cells to release many tiny seeds.

Distribution. Common in coastal ecosystems of lakes Michigan and Huron; rare in the western Upper Peninsula. County occurrence (%) by ecosystem region (Fig. 19): I, 37; II, 67; III, 100; IV, 12. Entire state 51%. A species of northern distribution, QC, ON, s. through the central Great Lakes Region to OH, IN, IL.

Site-Habitat. Open, cool, wet to wet-mesic, oxygen-deficient sites having circumneutral to basic soils or calcareous substrates; fens, interdunal wetlands, wet-mesic prairies, rocky limestone shores, calcareous stream banks; common along lakes Michigan and Huron. Associates include a diverse array of species that root in acidic topsoil (leatherleaf, Labrador-tea, bog-laurel, small and large cranberries), as well as those whose roots extend below the acidic peaty surface to a calcareous substrate (northern white-cedar, willows, shrubby cinquefoil, meadowsweet, wild-raisin).

Notes. Shade-intolerant. Common St. John's-wort, *Hypericum perforatum* L., a native of Europe, is widely naturalized and distributed throughout North America. It is widely recognized nationally and by many states as a noxious weed. In Michigan it has been reported from at least 93% of the 83 counties.

Hypericum is a large genus with about 450 species, primarily annual or perennial herbs; it occurs in temperate regions and mountains of the tropics. The group is most easily distinguished by opposite leaves whose blades are dotted with many translucent glands. The flowers are typically large, yellow, and showy. The species are generally intolerant of shade. The introduced herb, common St. John's-wort, *H. punctatum*, Lam., has become a noxious weed that has spread across low-elevation landscapes of the West. Leaves and flowers of *Hypericum* contain a complex substance that, if prepared just right, mixed with vodka and aged for a week, "is a perfect remedy for the 'blues,' sadness, irritability, insomnia, and the general grouchies" (Moore 1979).

The term *wort* is Anglo-Saxon and was used to distinguish low-growing or herbaceous plants from trees. Many of the species are herbs. The name St. John's apparently arose from the custom among country people in England of gathering plants of this genus on St. John's Day to protect them from evil spirits (Deam 1924).

Chromosome No. $2n = 18$, $n = 9$; $x = 9$

Similar Species. In the Great Lakes region, shrubby St. John's-wort, *Hypericum prolificum* L.

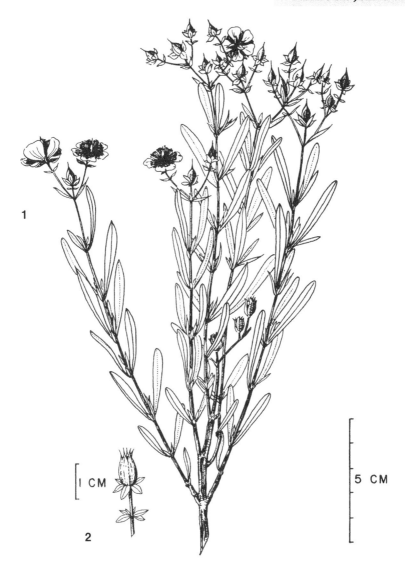

1 CM

5 CM

1. Individual with vegetative and flowering shoots
2. Fruiting shoot, capsule

KEY CHARACTERS

- low, erect, much-branched shrub, 0.2–0.8 m.
- stems angled, with exfoliating bark
- leaves opposite, entire, sessile, often with tufts of smaller leaves in axils of larger leaves; dark green above, dotted with translucent glands
- inflorescences terminal, flowers bright yellow, showy
- habitat open, cool wet sites having circumneutral to basic soils or calcareous substrates

Distinguished from shrubby St. John's-wort, *Hypericum prolificum*, by its low shrubby form, leaves sessile with narrower blades, 5-celled ovary, and inflorescences all terminal.

HYPERICACEAE
Hypericum prolificum Linnaeus
Shrubby St. John's-wort

Size and Form. Medium, upright shrub, 1–2 m; usually branched at or near the ground, larger individuals often widely branching.

Leaves. Opposite, simple, deciduous, often with tufts of smaller leaves in axils of larger leaves; 2–5 cm long, 3–18 mm wide; linear to oblong; obtuse and generally mucronate, gradually tapering; margin entire, often revolute; glabrous, green above, dotted with translucent glands, glaucous beneath, with only the main vein showing; petioles 1–5 mm long.

Stems-Twigs. Slender, bark exfoliating, light brown; stout, ascending and sharply 2-angled, with winglike ridges below the nodes, leaf scars somewhat raised, triangular; opposing leaf scars connected by a transverse ridge, bundle scar 1; pith relatively large, brown, large stems hollow.

Winter Buds. Small, solitary, sessile, green, glabrous.

Flowers. July–August; bisexual; in small, compound cymes, pairs in small terminal clusters and in axils of the uppermost leaves; 2–3.5 cm wide, sepals 5, petals 5, bright yellow, showy; styles 3, united below; ovary 3-celled; stamens many, >100, distinct. Insect pollinated.

Fruit. Capsule; August–September; cylindrical to narrowly ovoid, 6–13 mm across, dark brown, splitting when ripe into 3-celled clusters, each cell tipped with a conspicuous, persistent style; seeds many.

Distribution. Rare to locally common in southern lower Michigan. County occurrence (%) by ecosystem region (Fig. 19): I, 68; II, 17; III, 0; IV, 0. Entire state 37%. Widely distributed in e. North America.

Site-Habitat. Open, wet to wet-mesic, oxygen-deficient sites having circumneutral to basic soils or calcareous substrates; fens, interdunal, wet-mesic prairies, calcareous stream banks, open oak forests with calcareous substrate, meadows, roadsides. Associates include many early successional species of the diverse sites noted above.

Notes. Shade-intolerant. Common St. John's-wort, *Hypericum perforatum* L., an herbaceous native of Europe, is widely naturalized and distributed throughout North America. It is widely recognized nationally and by many states as a noxious weed. In Michigan it has been reported from at least 93% of the 83 counties.

Chromosome No. $2n = 18$, $n = 9$; $x = 9$

Similar Species. Kalm's St. John's-wort, *Hypericum kalmianum* L. in the Great Lakes Region. Multistemmed shrubs reaching 2–2.5 m include the bushy St. Johns-wort, *H. densiflorum* Pursh, of the Atlantic coastal plain west to Texas, and the naked St. Johns-wort, *H. nudiflorum* Michx. ex Willd., of the mid-South and South. Widely distributed in China is the bushy shrub *H. patulum* Thunberg, which is widely planted in gardens of Japan, India, South Africa, and elsewhere.

5 CM

1 CM

1
2

1. Individual with vegetative and flowering shoots
2. Fruiting shoot, capsule

KEY CHARACTERS
- medium, much-branched shrub to 2 m with stout angled stems branching at or near the base
- leaves opposite, entire, often with tufts of smaller leaves in axils of larger leaves; green above, dotted with translucent glands; petioles short
- inflorescences axillary as well as terminal, flowers bright yellow, showy
- habitat open, wet to wet-mesic sites with basic soils or calcareous substrates, occurrence primarily in southern lower Michigan

Distinguished from Kalm's St. John's-wort, *H. kalmianum*, by its larger, shrubby form, leaves with larger blades and short petioles, 3-celled ovary, inflorescences axillary as well as terminal, and occurrence primarily in southern lower Michigan.

AQUIFOLIACEAE
Ilex mucronata (L.) M. Powell,
V. Savolainen & S. Andrews

Mountain Holly

Size and Form. Tall, slender shrub or rarely a small tree to 4 m; single stems or few-stemmed clonal clumps. Michigan Big Tree: girth 33.0 cm (13 in), diameter 10.5 cm (4.1 in), height 6.1 m (20 ft), Oakland Co.

Bark. Thin, smooth, light gray.

Leaves. Alternate, simple, deciduous, blades of long-shoot leaves 2–6 cm long, 2–3 cm wide (long-shoot, No. 1 and upper part of No. 4); elliptic-oblong, often with nearly parallel sides; acute or abruptly blunt, with mucronate tip; rounded to slightly acute; entire or rarely with a few sharp teeth; dark green above, paler, grayish-green beneath; glabrous; falling early in autumn; blades of short shoots smaller, narrower, otherwise similar; petioles very slender, purple, 0.5–2 cm long.

Stems-Twigs. Long-shoots slender, smooth, glaucous-purplish, becoming maroon or gray; short-shoots similar but with very short internodes; leaf scars raised, crescent-shaped, bundle scar 1; pith small, greenish, continuous.

Winter Buds. Small, ovoid, sessile; pointed at the apex, 2 ciliate scales exposed.

Flowers. May; before the leaves fully expand; unisexual. plants dioecious; small; solitary or up to 4 in a cluster on threadlike pedicels 1–3 cm long; calyx lacking; petals yellowish, linear, 1.5–2 mm long and about 4 mm wide. Insect pollinated.

Fruit. Drupe; July–August; subglobose, bright pink to purplish-red, 5–10 mm long; 4 seeds, persistent in winter.

Distribution. Frequent throughout the state, becoming common northward; rare or absent in the lake plain of southeastern Michigan, absent in Monroe, Wayne, St. Clair, Macomb Co. County occurrence (%) by ecosystem region (Fig. 19): I, 58; II, 80; III, 100; IV, 100. Entire state 73%. Northern range: NL and QC to MN, s. to IN and WV.

Site-Habitat. Cool, open to moderately shaded, acidic to circumneutral wet-mesic sites, especially along the margins of lakes, marshes, bogs, poor fens, swamps, and swales. Associated with tamarack, yellow birch, speckled alder, winterberry, leatherleaf, highbush blueberry, poison-sumac, American highbush-cranberry, *Kalmia* spp., Labrador-tea, and creeping-snowberry.

Notes. Moderately shade-tolerant; forming small, multistemmed clones by root-collar sprouting. Formerly, and in Voss 1985, its scientific name was *Nemopanthus mucronatus* (L.) Loes.

Chromosome No. $2n = 40$, $n = 20$; $x = 20$

Similar Species. The related Appalachian mountain holly, *Ilex collina* Alexander, is a rare species of swamps, bogs, and mountain streams in Virginia, West Virginia, and North Carolina.

1. Fruiting shoot, drupe, × 1/2
2. Female flower, × 5
3. Male flower, × 5
4. Branch with long vegetative shoots and short fruiting shoots, drupe, reduced
5. Winter lateral bud and leaf scar, enlarged
6. Winter twig, reduced

KEY CHARACTERS

- tall, often clonal shrub to 4 m
- leaves elliptic-oblong, margin entire or nearly so with a mucronate tip, petioles purplish
- current shoots and small twigs purplish
- fruit a bright pink to purplish-red drupe when ripe
- habitat wetlands and wet-mesic sites

Distinguished from Michigan holly, *Ilex verticillata*, by its single stem or few-stemmed clones; leaves entire and lacking bristle tips, apex mucronate, veins not impressed above; young shoots purplish; fruit a bright pink to purplish-red drupe.

AQUIFOLIACEAE
Ilex verticillata (Linnaeus) A. Gray
Winterberry, Michigan Holly

Size and Form. Tall shrub or rarely a small tree, to 5 m; slender, densely branched stems forming a round or oval crown; roots shallow. Clone-forming in clumps by sprouting from root collar and locally by root sprouting. Michigan Big Tree: girth 17.8 cm (7 in), diameter 5.7 cm (3.2 inches), height 10.1 m (33 ft), Van Buren Co.

Bark. Thin, smooth, olive green with warty lenticels.

Leaves. Alternate, simple, deciduous, blades 3–10 cm long and nearly half as wide; obovate, oblanceolate, or oblong; acute or acuminate; cuneate; sharply serrate or sometimes doubly serrate, teeth shallow with short bristle-tips that point up (perpendicular to blade surface); thin- to firm-coriaceous; glabrous and dull, dark green above, paler and sometimes pubescent on veins beneath, dark green long into the fall, turning black after frost; veins conspicuously impressed or sunken above, strongly protruding beneath; stipules tiny, black; petioles purplish, more or less pubescent, about 1 cm long.

Stems-Twigs. Very slender, thin, young shoots greenish, becoming gray to dark olive, smooth, becoming roughened by conspicuous lenticels; epidermis of small stems exfoliating and looking whitish where epidermis has separated from the stem; short-shoots common on older twigs; stipule scars minute; leaf scars crescent-shaped, bundle scar 1; pith small, white, continuous.

Winter Buds. Terminal bud present, at least on some shoots, 1–2 mm long, conical, blunt, chestnut brown, finely ciliate, scales keeled; lateral buds smaller, ca. 1 cm long, sessile, appressed, blunt, often superposed with a smaller bud at the base of a larger bud, pointing outward at an angle to the stem, brown with scattered white hairs along upper part of scales.

Flowers. May–July, after leaves are fully developed; functionally unisexual; plants dioecious; 1–3 crowded together on short pedicels 1–2 mm long; small; greenish or yellowish-white; 4- to 6-parted; calyx toothed; petals greenish or yellowish-white, oblong-elliptic, <1 mm long; corolla shortly connate at base. Insect pollinated.

Fruit. Drupe; September–October; bright red, persisting into midwinter, 6–8 mm across,

so crowded as to appear whorled, 4–6 bony pits; seeds 92,000/lb (202,400/kg).

Distribution. Common to locally abundant throughout the state. County occurrence (%) by ecosystem region (Fig. 19): I, 89; II, 97; III, 86; IV, 100. Entire state 93%. Widely distributed in e. North America, NL to ON s. to FL, w. to TX.

Site-Habitat. Low, cool, wet-mesic, basic or acidic sites with high water tables; deciduous swamps, wet-mesic woods, lake, marsh, and pond margins; tolerant of wet soil but not seasonal flooding. Associates include American elm, red maple, blue-beech, yellow birch, black ash, tamarack, mountain holly, highbush blueberry, common elder, poison-sumac.

Notes. Moderately shade-tolerant. The species name *verticillata* refers to the axillary clusters of flowers. The showy red fruits make it a potentially important native landscape species, especially in wet soils. Often forming clones of many ramets but not wide-spreading. *Ilex* is in the holly family, and the majority of some 300 holly species belong to this genus. About 20 species are native to eastern North America, and most are highly valued for ornamental plantings. All are good food sources for wildlife. Evergreen hollies long have been associated with the Christmas and winter solstice season. Of the evergreen species, the English holly, *Ilex aquifolium* L., Japanese holly, *I. crenata* Thunb., Chinese holly, *I. cornuta* Lindl. & Paxon, and American holly, *I. opaca* Ait., are the most highly prized. Not surprisingly, there are many more cultivars and hybrids than species in the genus.

Chromosome No. $2n = 36$, $n = 18$; $x = 18$

Similar Species. Closely related is the mountain winterberry, *Ilex montana* Torr.

1. Vegetative shoot, × 1/2
2. Fruiting shoot, drupe, reduced
3. Winter twig with fruit, drupe × 1/2
4. Winter lateral bud and leaf scar, enlarged
5. Side view of superposed buds, vegetative (*lower*) and flower (*upper*), enlarged
6. Winter twig, × 1

KEY CHARACTERS

- Tall clonal shrub or small tree to 5 m
- leaves obovate with short bristle tips that point up, perpendicular to blade surface, sharply serrate; stipules tiny, black
- twigs slender, olive-colored, with exfoliating epidermis
- fruit of female clones a bright red drupe, persisting into winter
- habitat wet-mesic sites, edges of wetlands

Distinguished from mountain holly, *Ilex mucronata*, by its multistemmed clonal clumps; twigs with exfoliating epidermis; leaves sharply serrate, teeth with short bristle tips that point upward (perpendicular to blade surface), veins impressed above and protruding below; young shoots greenish; fruit a bright red drupe, persisting into winter.

& A. Gray, of uplands in eastern and southeastern states west to Tennessee and south to Alabama. Farther south and west, the possumhaw, *I. decidua* Walter, also is similar. Many deciduous species of *Ilex* are found in wetlands and uplands of Japan and China. In Japan the counterpart species is *Ilex macropoda* Miq.

CUPRESSACEAE
Juniperus communis Linnaeus
var. *depressa* Pursh

Ground Juniper

Size and Form. Low, straggly, evergreen shrub, rarely over 1.5 m; numerous stiff branches erect or ascending from a prostrate base; forming broad, flat-topped, saucer-shaped mats or patches to 4 m or more across. Michigan Big Tree: girth 58.4 cm (23.0 in), diameter 18.6 cm (7.3 in), height 2.7 m (9.0 ft), Leelanau Co.

Bark. Thin, gray, becoming dark reddish-brown, scaly or exfoliating in papery strips.

Leaves. In whorls of 3, jointed at base, crowded on 3-angled stems, awl-shaped, loose or spreading, rounded at base; tips acute to mucronate; thin, straight or somewhat curved inward, usually shallowly keeled, 12–20 mm long, 1–1.5 mm wide; grayish or green, concave above, with a broad pale blue or white stomatal band one-third as wide as leaf; pungent.

Stems-Twigs. Young shoots green, becoming ridged to 3-angled and dark reddish-brown and scaly with age.

Strobili. Unisexual; plants dioecious. Males and females borne in leaf axils of 2nd-year twigs; May–June; male strobilus oblong, small, 2–3 mm long, erect, catkinlike structure of 6–12 scales on a central axis, each bearing several pollen sacs; female strobilus a tight pointed cluster of 3–8 small scales, each of which may bear 1–3 ovules; after fertilization the upper scales become fleshy and enclose the ovules, maturing after 2–3 years but remaining closed. Wind pollinated.

Seed cones. Subglobose, 5–10 mm in diameter, axillary; dark blue or black, covered with a bloom; fleshy, scales imbricate; pungent; seeds usually 3, seeds 36,500/lb (80,300/kg).

Distribution. Common and locally abundant on dunes and sandy areas in counties bordering the Great Lakes and also in old fields, pastures, and cutover forests in lower Michigan. County occurrence (%) by ecosystem region (Fig. 19): I, 63; II, 67; III, 100; IV, 62. Entire state 67%. All Canadian provinces, AB, s. in the Rocky Mountains to AZ, NM; New England and western Great Lakes Region.

Site-Habitat. Originally in open, rocky, or other non-fire-prone sites but now found widely on dry to dry-mesic upland soils where open conditions prevail and fire is excluded; upland sites, soil substrate usually calcareous (pH >7.2). Often found as a small individual in moderately open wetlands of various kinds on small rises and coarse woody debris. Associated in one place or another with most trees and shrubs of the dry and dry-mesic upland forest communities, as well as at edges of mesic and wet-mesic forests in Michigan and the upper Great Lakes Region.

Notes. Shade-intolerant, once established persisting in light to moderate shade. Very slow-growing, only a few cm per year. Birds spread seeds very widely; plants often found under trees where birds roost.

Juniperus communis has the widest distribution worldwide of any conifer; native in Europe, Asia Minor, the Caucasus, Iran, Afghanistan, North Africa, the Himalaya, China, Siberia, and throughout North America (Vidaković 1991). Overall, it tolerates a great range of climatic and soil conditions. It is exclusively a low shrub, var. *depressa*, in North America (Adams 2008). Highly susceptible to damage and death by fire, a main reason for its absence in fire-prone forests of our region prior to European settlement. Wood with yellowish sapwood and reddish-brown heartwood, aromatic; durable. Cones provide the flavor for gin. Its low-spreading, evergreen form and complex branching act to catch and retain deciduous leaves, which upon decomposition provide the nursery for deciduous species, including many dry-site oaks and associated species.

Collections of the common or dwarf juniper were made on the Lewis and Clark expedition, probably by Lewis (e.g., October 17, 1804, mouth of Cannonball River in present-day North Dakota, and July 7, 1806, near

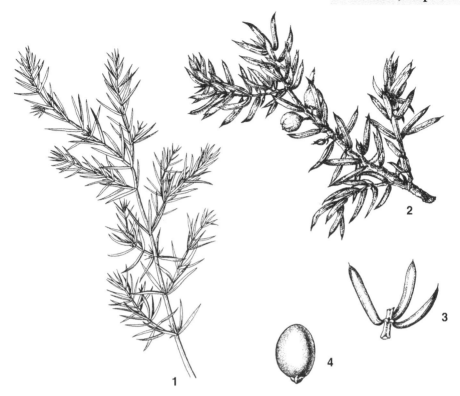

1. Vegetative shoot, × 1/2
2. Twig bearing cones, × 1
3. Whorl of awl-shaped leaves, × 2
4. Female cone, × 2

KEY CHARACTERS

- low, straggly, evergreen shrub, branches ascending and spreading from a prostrate base
- needles awl-shaped, in whorls of 3, one distinct band of stomata on upper leaf surface
- cones, fleshy, dark blue or black, glaucous, berrylike
- habitat dry and dry-mesic, open sites with calcareous substrate, often sandy

Distinguished from eastern red-cedar, *J. virginiana,* by its low shrub habit to 1.5 m, absence of scalelike leaves, and needles with a single white band on the upper surface.

Distinguished from creeping juniper, *J. horizontalis,* by shrub habit to 1.5 m, absence of scale-like leaves, and widespread occurrence in cutover forests and open sites throughout the state.

present-day Lincoln, Montana (Phillips 2003). The variety name *depressa* was assigned by Frederick Pursh, who studied the plant collections of Lewis and in 1814 published *Flora of North America.* Many cultivars of *J. communis* are used in horticulture (Dirr 1990).

Chromosome No. $2n = 22$, $n = 11$; $x = 11$

Similar Species. Four related prostrate or small shrubs are described by Adams (2008).

In California the variety *saxatilis* Pall., mountain juniper, is a rare prostrate shrub found at high elevations (Lanner 2002, 229). The East Asian counterpart is var. *nipponica* (Maxim.) Silba. It is closely related to the European *Juniperus communis* var. *communis* L., German common name Wacholder, a tall shrub or tree to 4–5 m of diverse open sites of northern and central Europe to central Russia.

CUPRESSACEAE
Juniperus horizontalis Moench
Creeping Juniper

Size and Form. Decumbent shrub with long, gnarled, trailing primary stems to 5 m; many short, erect, ascending branches, 10–30 cm high; often spreading and forming matlike ground-cover clones by rooting of stems lying on the surface.

Bark. Reddish-brown to blackish, scaly or exfoliating in plates.

Leaves. Opposite, evergreen, of two kinds: (1) sessile, awl-shaped, loosely ranged in whorls of 2–3, 0.6–1.3 mm long; found on juvenile shoots; and (2) mainly sessile, scale-like, closely appressed, overlapping, 4-ranked, 2 mm long; ovate, acute, cuticle toughened; found on adult shoots; green, becoming reddish-purple in winter; persistent 4–5 years; pungent.

Stems-Twigs. Greenish-brown, covered with overlapping scalelike leaves for several years, becoming reddish-brown, flaking, and peeling.

Winter Buds. Tiny vegetative buds enclosed and concealed between the upper pair of leaves.

Strobili. Unisexual; plants dioecious. Male and female strobili borne at the tip of previous year's twigs; May–June; male strobilus small, 2–3 mm long, with 6–10 scales, each bearing several pollen sacs; female strobilus similar to that of *J. communis* var. *depressa*. Wind pollinated.

Seed cones. Subglobose, 5–8 mm in diameter on short, recurved peduncles; dark blue to black, covered with bloom; fleshy, scales imbricate; pungent; seeds 3–5, not pitted.

Distribution. Occasional to locally common along shores of the Great Lakes in northern lower and upper Michigan; nearly transcontinental in Canada. County occurrence (%) by ecosystem region (Fig. 19): I, 3; II, 37; III, 100; IV, 50. Entire state 28%. All Canadian provinces, AB, s. to northern Great Plains, e. to western Great Lakes Region and New England.

Site-Habitat. Open to lightly shaded conditions, dry to dry-mesic, rocky, seasonally wet, often circumneutral or basic sites; sand dunes, dolomite alvars, bedrock glades and lakeshores, fens with marl substrate. Associates include northern white-cedar, tamarack, balsam fir, trembling aspen, balsam poplar, alders, ground juniper, *Salix* spp., *Vaccinium* spp., Labrador-tea, bearberry, shrubby cinque-foil, sweet gale, trailing arbutus, sand cherry, bog-rosemary, bog-laurel.

Notes. Very shade-intolerant; slow-growing. Highly susceptible to injury and death by fire, thus rocky and sandy, fuel-poor sites are favored. Forms ground cover in dune woodlands of northern white-cedar and balsam fir. Birds disperse seeds widely. Cultivars used in horticulture; Dirr (1990) describes 55. On dunes and sandy beach shores, the habit of beach-heath, *Hudsonia tomentosa* Nutt., an evergreen angiosperm, might be mistaken for *Juniperus horizontalis*. However, beach-heath is a bushy shrub with bright yellow flowers and tiny leaves obscured by dense whitish pubescence.

Chromosome No. $2n = 22$, $n = 11$; $x = 11$

Similar Species. Related genetically to *Juniperus virginiana* L., eastern red-cedar, and *J. scopulorum* Sarg., Rocky Mountain juniper, and hybridizing with both (Adams 2008).

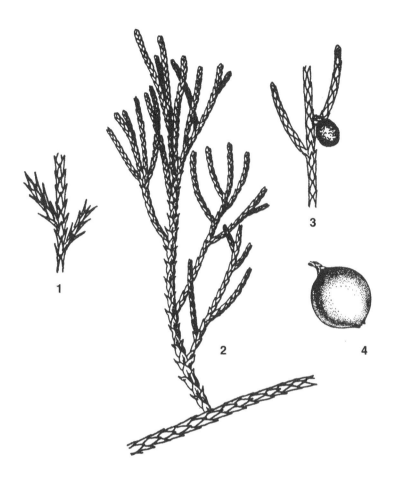

1. Juvenile shoot with awl-shaped leaves, × 1
2. Vegetative branch, adult shoots with scale-like leaves
3. Female cone on adult shoot, enlarged
4. Female cone, × 3

KEY CHARACTERS

- decumbent shrub with long, trailing primary stems and ascending branches
- needles awl-shaped on juvenile stems and scalelike on adult stems
- cones small, fleshy, dark blue to black and covered with bloom, berrylike
- habitat open, dry to seasonally wet sites

Distinguished from eastern red-cedar, *J. virginiana*, and ground juniper, *J. communis* var. *depressa*, by its trailing, decumbent habit. Juvenile and adult foliage similar to that of *J. virginiana*.

ERICACEAE
Kalmia angustifolia Linnaeus
Sheep-laurel, Lambkill

Size and Form. Low, slender, erect shrub to about 1.7 m; forming clonal clumps by root-collar sprouting.

Leaves. In whorls of 3 or rarely opposite, simple, evergreen; blades flat, slightly leathery, narrowly oblong to elliptic; 2–5 cm long, 0.5–2.5 cm wide; obtuse to rounded; cuneate to rounded; entire, revolute; bright green and obscurely hairy above, paler green and glabrous or young leaves with scattered stalked glands beneath; petioles short, 3–10 mm long.

Stems-Twigs. Moderate to slender, terete, strongly ascending, glabrous, light brown to grayish-brown; leaf scars half round to shield-shaped, bundle scar a transverse line; pith small, continuous.

Winter Buds. Terminal bud present, small, 2–4 mm long, 2 green outer scales; lateral buds solitary, sessile, 2–3 mm long, 2 green outer scales; flower buds large, with many green overlapping scales coated with gland-tipped hairs.

Flowers. Late June to early July; bisexual; many-flowered axillary, drooping clusters with glandular pedicels; calyx glandular, lobes ovate; corolla 5-parted, deep pink to purple, about 1 cm across, saucer-shaped, showy; stamens 5, spring-loaded. Insect pollinated.

Fruit. Capsule, 3–5 mm across, flattened globe shape with persistent style nearly as long as the capsule; borne in axillary clusters; pedicels long; many-seeded; capsules persistent over winter and may persist 3 years or more.

Distribution. Locally common in northern lower Michigan and adjacent counties, rare to absent in upper Michigan. County occurrence (%) by ecosystem region (Fig. 19): I, 11; II, 53; III, 14; IV, 0. Entire state 25%. A northern and boreal species, ranging from NL to ON, s. to WV and the western Great Lakes Region.

Site-Habitat. Wet, open, acidic, nutrient-poor sites; bog, muskeg, and conifer swamp edges, open patches of jack pine and spruce-dominated woodlands. Associates include tamarack, black spruce, jack pine, American mountain-ash, leatherleaf, Labrador-tea, blueberries, willows, bog-laurel, bog-rosemary, mountain holly, wintergreen, creeping-snowberry, black chokeberry.

Notes. Shade-intolerant, persisting in moderate shade. The stamens are tucked into pouches in the corolla and are spring-loaded.

As one of the common names indicates, the leaves, stems, and roots contain a white alkaloid that is toxic to livestock, especially sheep.

Chromosome No. $2n = 24$, $n = 12$; $x = 12$

Similar Species. The low shrub bog-laurel, *Kalmia polifolia* Wangenheim, is similar and a common associate in northern bogs. Smaller, related, and sometimes considered distinct are the Carolina sheep-laurel, *Kalmia angustifolia* var. *carolina* (Small) Fern., and the whitewicky, *K. cuneata* Michx. They occur infrequently on acidic sites of the coastal plains of the Carolinas, Georgia, and Florida.

166

1. Flowering shoot, × 1/2
2. Fruiting shoot, capsule, × 1/2
3. Fruit, capsule, × 3

KEY CHARACTERS

- low, erect shrub to 1.7 m, clump-forming
- leaves opposite, evergreen to blunt apex, entire or somewhat revolute
- flowers late spring to early summer in axillary clusters, showy, deep pink
- fruit a small, many-seeded capsule, persistent for up to 3 years
- habitat bogs, muskegs, acidic swamps, open woodlands; almost exclusively in northern lower Michigan

Distinguished from bog-laurel, *Kalmia polifolia*, by its larger size, leaves with short petioles, blades larger and narrowly oblong to elliptic, margins only slightly revolute, lower surface not white-pubescent; late flowering, flowers in drooping axillary clusters; occurring almost exclusively in northern lower Michigan.

ERICACEAE
Kalmia polifolia Wangenheim
Pale-laurel, Bog-laurel

Size and Form. Low, straggling shrub to about 0.7 m; stems single or clump-forming.

Bark. Dark brown to blackish, with vertical fissures, becoming roughened or scaly.

Leaves. Opposite, simple, evergreen, 1–4 cm long, 0.6–1.2 cm wide; pointing up, sessile or nearly so, leathery; blades linear to lanceolate; obtuse; cuneate; revolute; glossy dark green, glabrous above, conspicuously whitened beneath by minute pubescence; petioles absent or about 1 mm.

Stems-Twigs. Young stems, slender, 2-edged, with flattened sides and conspicuous nodes, glabrous, light brown at first, becoming dark brown to blackish with vertical lines or fissures; leaf scars half round to shield-shaped, bundle scar 1; pith small, continuous.

Winter Buds. Terminal buds present, small, 2–4 mm long, 2 green outer scales; lateral buds solitary, sessile, small, 2 outer scales; terminal flower buds large, with many green overlapping scales coated with gland-tipped hairs.

Flowers. Late May to early June; bisexual; many-flowered terminal umbels on slender, nonglandular pedicels 1–3 cm long; calyx lobes ovate-oblong; corolla 5-parted, 1–2 cm across, saucer-shaped, pink to rose-purple, showy; stamens 10, spring-loaded. Insect pollinated.

Fruit. Capsule, 6–7 mm across, 5-celled, brown, subglobose, with persistent style nearly as long as the capsule; many-seeded; capsules persistent for several years.

Distribution. Infrequent in southern lower Michigan, common to locally abundant in northern lower Michigan and upper Michigan. County occurrence (%) by ecosystem region (Fig. 19): I, 28; II, 87; III, 86; IV, 100. Entire state 61%. A northern and boreal species ranging from NL to AB, n. to AK, s. to PA and the w. Great Lakes Region.

Site-Habitat. Wet, open to lightly shaded, acidic, nutrient-poor sites; bogs, muskeg, acidic peat mounds and islands in fens, conifer swamps, interdunal swales. Associates include diverse tree and shrub species of acidic sites and microsites of the above ecosystems and sites.

Notes. Shade-intolerant, persisting in light shade. The stamens are tucked into pouches in the corolla and are spring-loaded; they pop up when an insect lands in the flower (Smith 2008, 256). In many respects, similar morphologically and in its habitat to bog-rosemary, *Andromeda glaucophylla*, but leaves opposite. Leaves and stems contain a white alkaloid, which is toxic to livestock, especially sheep. A tetraploid species, whereas *Kalmia angustifolia* L. and *K. latifolia* L., the major representatives of this genus in the southern Appalachian Mountains, are diploids.

Chromosome No. $2n = 48$, $n = 24$; $x = 12$

Similar Species. In Michigan the most similar species is sheep-laurel, *Kalmia angustifolia* L. Closely related to the western bog-laurel, *Kalmia microphylla* (Hook.) Heller, a widely distributed species in western North America in subalpine, alpine, and coastal bogs and wet meadows (Hayes 1960).

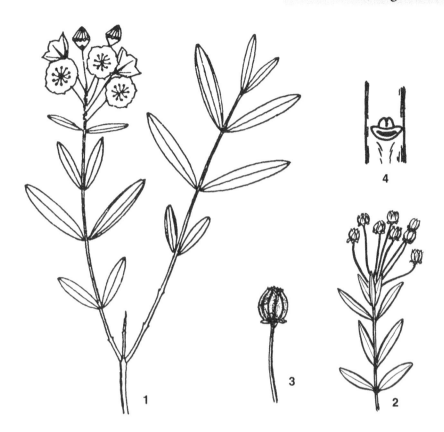

1. Flowering shoot, reduced
2. Fruiting shoot, capsule, reduced
3. Fruit, capsule, × 1
4. Winter lateral bud and leaf scar, enlarged

KEY CHARACTERS

- low, straggling shrub to <1 m, clump-forming
- leaves opposite, evergreen, leathery, sessile or nearly so; blades linear to lanceolate; lower surface conspicuously white with dense pubescence
- twigs two-edged with conspicuous nodes
- flowers May–June in terminal clusters, showy, rose-pink, anthers spring-loaded
- fruit a many-seeded capsule, persistent for several years
- habitat wet, acidic bogs and open swamps in northern lower and upper Michigan

Distinguished from sheep-laurel, *Kalmia angustifolia,* by its smaller and more straggling form; leaves sessile, blades linear to lanceolate, lower surface conspicuously whitened by pubescence; early flowering, flowers in terminal clusters; more common in northern lower and upper Michigan.

Distinguished from bog-rosemary, *Andromeda glaucophylla,* by its habit of typically erect, straggling stems, not trailing; leaves opposite, glossy above; flowers less than 8 mm across.

OLEACEAE
Ligustrum vulgare Linnaeus
Common Privet

Size and Form. Tall shrub, to 5 m; stout, much branched; sprouting from the base to form multistemmed clones. Clone formation by basal sprouting and layering.

Bark. Thin, gray, nonexfoliating.

Leaves. Opposite, simple, deciduous, long persistent but shed before spring; blades moderately thick, 2–4 cm long, 0.7–2 cm wide with 4–5 pairs of indistinct veins; oblong-ovate to lanceolate, leaves of indeterminate shoots lanceolate; obtuse to acute; base cuneate on determinate shoots, rounded on indeterminate shoots; entire; dark green above, paler beneath; glabrous both sides; petioles 0.3–1 cm long.

Stems-Twigs. Slender, tan, minutely pubescent when young, becoming glabrous, older twigs grayish to grayish-brown with narrow anastomosing ridges, lenticels round, light brown; leaf scars raised, elliptical or crescent-shaped, opposing scars not meeting and not connected by a distinct traverse line; bundle scar 1; pith light brown, continuous.

Winter Buds. Terminal bud present, vegetative buds ovoid at base, with short acuminate apex, 3–6 mm long, 2 pairs of scales exposed, margin ciliate, brown, the outer 2 scales elongated, 7–10 mm long and nearly enclosing the bud; flower buds swollen at base, apex acute, 3–5 mm long; lateral buds small, ovoid, blunt, 2 pairs of scales exposed, brown, margin ciliate.

Flowers. June; bisexual; borne in rather dense terminal panicles, 3–6 cm long; white; unpleasant odor; calyx short-tubular, 4-toothed; corolla tubelike, 2.5–3 mm long and abruptly expanded into 4 spreading lobes, anthers exerted. Insect pollinated.

Fruit. Drupe; September, persisting through winter until spring; in terminal clusters, lustrous, blackish, ovoid to subglobose, 6–8 mm long, thin fleshy covering of the pit; 1–4 seeded.

Distribution. Native in central and southern Europe and northern Africa. Infrequent in southern lower Michigan; rare in northern lower Michigan; absent in upper Michigan. County occurrence (%) by ecosystem region (Fig. 19): I, 53; II, 7; III, 0; IV, 0. Entire state 19%.

Site-Habitat. Wide tolerance of soil-site conditions in uplands, especially open edges of roads and fields and cutover forests.

Notes. Shade-intolerant, persisting in light to moderate shade. Seeds widely disseminated by birds. Becoming occasionally naturalized in urban environments and disturbed forests. Widely planted as a low or tall hedge; stands pruning and shaping well. Several cultivars are in use, as well as several related species. An invasive species, not recommended for use of any kind.

Chromosome No. $2n = 46$, $n = 23$; $x = 23$

Similar Species. About 50 species in the genus, mainly in East Asia and Malaysia to Australia; many are evergreen. The one species from Europe and North Africa is the common privet, *L. vulgare*. Similar species cultivated in Michigan and rarely escaping into fields and woods are *L. obtusifolium* Sieb. & Zucc., native in Japan and China, and the California privet, *L. ovalifolium* Hassk. (Voss and Reznicek 2012). Closely related in China is the naturally wide ranging Chinese privet, *L. sinense* Lour.

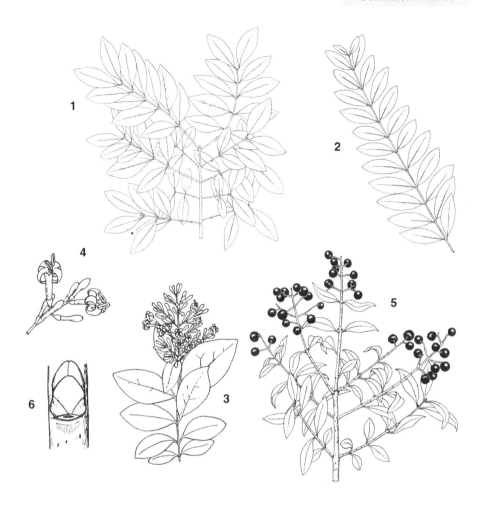

1. Branch with determinate shoots, early leaves, reduced
2. Indeterminate shoot, distal part, reduced
3. Flowering shoot, reduced
4. Flowers, × 1
5. Fruiting shoot, × 1/4
6. Winter lateral bud and leaf scar, enlarged

KEY CHARACTERS

- tall, much-branched shrub to 5 m
- twigs round, leaf scars opposite, opposing scars not meeting, pith white
- leaves deciduous but persisting into winter, blades oblong-ovate to lanceolate, glabrous, entire
- fruit a small, blackish drupe, ovoid to subglobose, persisting through winter

LAURACEAE
Lindera benzoin (Linnaeus) Blume
Spicebush

Size and Form. Tall, clonal, much-branched, aromatic shrub to 5 m; several stems to 3 cm in diameter arise from a common base; arching branches form a wide-spreading, open, rounded, or flat-topped crown; forming clonal clumps by sprouting from the root collar; roots also give rise to ramets that initiate a new clump. Michigan Big Tree: girth 25.4 cm (10 in), diameter 8.1 cm (3.2 in), height 7.0 m (23 ft), Wayne Co.

Bark. Thin, dark gray, slightly roughened by corky lenticels.

Leaves. Alternate, simple, deciduous; blades 5–15 cm long and about half as wide, oblong-ovate to elliptical, widest at or above the middle; indeterminate shoots bearing leaves markedly different in size, early leaves ovate to nearly orbicular, late leaves oblong-ovate to oval and markedly larger than early leaves; apex of early leaves blunt or broadly rounded, late leaves with acute or short acuminate apex; narrowed or rounded at base; entire; thin; glabrous, light green above, paler and sometimes pubescent on veins beneath, turning yellow in autumn; veins prominent; pleasantly aromatic, with citronellalike odor when crushed; petioles 0.5–2 cm long.

Stems-Twigs. Slender, green or olive brown, with conspicuous, vertical, pale corky lenticels; glabrous at maturity; in winter with clusters of spherical flower buds; aromatic, spicy to taste, leaf scars slightly raised, half round or crescent-shaped, bundle scars 3; pith moderate, white, continuous.

Winter Buds. Terminal bud absent; lateral vegetative buds small, elongated, appressed, brown; flower buds globose, greenish, paired collateral to the vegetative bud, stalked, conspicuous in clusters in autumn and winter.

Flowers. March–May, mostly unisexual, plants dioecious; before the leaves; borne in clusters of 3–6 at the nodes on nearly sessile small umbels; 0.5–1 cm across; yellow; fragrant. Insect pollinated.

Fruit. Drupe, single-seeded; July–September; subglobose-oblong, bright red, spicy-aromatic, 6–12 mm long; seeds light violet brown, speckled darker brown.

Distribution. Common in southern lower Michigan; infrequent in northern lower Michigan; absent in upper Michigan. County occurrence (%) by ecosystem region (Fig.

19): I, 84; II, 27; III, 0; IV, 0. Entire state 48%. Widely distributed in e. North America, also KS s. to TX.

Site-Habitat. Wet-mesic to mesic, well-drained to poorly drained, fertile, circumneutral to basic soils of equable, moderately shaded sites; typical of hardwood (i.e., deciduous) and hardwood-conifer swamps, beech-sugar-maple-dominated ecosystems (especially wet-mesic microsites), and floodplain ecosystems (levees, second bottoms) with low-frequency, short-duration flooding. Associates include species of communities in the above-mentioned nutrient-rich, wet-mesic to mesic sites (Kost et al. 2007).

Notes. Moderately shade-tolerant; moderately slow-growing; tolerant of seasonally high water tables but not inundation during the growing season. Sprouts regularly from the root collar of a common base to form a many-stemmed clone. Killed back severely by extreme winter temperatures but sprouting back rapidly. All parts of the plant aromatic. The species name *benzoin* refers to the similarity in odor to the resin of the Asiatic tree *Styrax benzoin.* The aromatic oil was used by pioneers in making a medicinal tea. The bark is reported to be pleasant to chew, a tonic, astringent, and stimulant.

The genus name is after John Linder, a Swedish physician (1676–1723). Excellent native ornamental for its early, conspicuous, yellow flowers, yellow foliage in autumn, and bright red fruits, remaining after the leaves are shed. Caterpillars of the lovely Promethea moth and the green-clouded swallowtail feed on spicebush and the related sassafras. Grows well in moist or dry-mesic situations.

Chromosome No. $2n = 24$, $n = 12$; $x = 12$

Similar Species. The southern spicebush,

1. Indeterminate vegetative shoot, × 1/2
2. Male flower, enlarged
3. Female flower, enlarged
4. Fruiting shoot, drupe, reduced
5. Winter vegetative bud and leaf scar (*lower*), female flower bud (*upper*), enlarged
6. Winter vegetative bud and leaf scar (*lower*), male flower buds (*upper*), enlarged
7. Winter twig, × 1

KEY CHARACTERS

- tall shrub forming clonal clumps of several to many stems; all parts aromatic
- leaves usually obovate, entire, apex blunt, veins prominent
- stems olive green with dense sessile clusters of globose flower buds
- female clones with bright red drupes in autumn
- habitat, wet-mesic to mesic fertile sites in lower Michigan

Lindera melissifolia (Walter) Blume, with leaves and young twigs pubescent-hairy, occurs chiefly on the Atlantic coastal plain from North Carolina to Florida and west to Texas and in the Mississippi Valley lowlands north to southern Missouri. The bog spicebush, *L. subcoriacea* Wofford, occurs in acid bogs and swamps on the coastal plain from North Carolina south to Florida, Mississippi, and Louisiana. Several East Asian species are similar, including the most notable, *L. erythrocarpa* Makino.

CAPRIFOLIACEAE
Lonicera canadensis Marshall
Canadian Fly Honeysuckle

Size and Form. Low, erect, or semierect straggling shrub to 1.5 m; branches flexible, spreading, occasionally becoming prostrate; clonal by basal sprouting and layering.

Bark. Light brown, peeling or shredding in thin strips.

Leaves. Opposite, simple, deciduous, blades 3–9 cm long and about half as wide; ovate to ovate-oblong; acute or blunt; rounded to subcordate; entire, ciliate; thin, bright green above, paler beneath; villous-pubescent when young, becoming more or less glabrous both sides; veins pinnate; petioles ciliate, slender, short, about 5 mm long.

Stems-Twigs. Slender, rounded, flexible, glabrous, ashy gray, ridged downward from nodes; bud scales persistent at base shoot; leaf scars opposite and usually connected by a definite ridge or line, crescent-shaped; bundle scars 3; pith white, continuous between nodes.

Winter Buds. Terminal bud present; laterals divergent, 3–6 mm long, short-ovoid or nearly globose, glabrate; bud scales persistent at base of current shoot.

Flowers. May–June; bisexual; borne in sessile pairs on long axillary peduncles, 2–2.5 cm long; pale yellow to creamy white, sometimes tinged with red or purple; calyx 5-toothed; corolla tubular-funnelform, 1.5–2 cm long, 2-lipped; stamens 5. Insect and hummingbird pollinated.

Fruit. Berry; July–September; cylindric-elongate, 1–1.5 cm long, red to purplish, usually in pairs on a long glabrous peduncle in axils of leaves, the berries diverging in opposite directions at the apex of the peduncle; seeds 3–4.

Distribution. Common throughout the state except in the lower tier of counties in southern Michigan, becoming increasingly frequent in northern lower Michigan and upper Michigan. County occurrence (%) by ecosystem region (Fig. 19): I, 71; II, 93; III, 100; IV, 100. Entire state 84%. Ranges from NL, QC, ON, w. Great Lakes Region s. to TN, NC, GA.

Site-Habitat. Open to moderately shaded, wet to dry-mesic habitats; associated especially with mesic to wet-mesic forest communities, including uplands and wetlands; common in northern white-cedar and hardwood-conifer swamps, sandstone bedrock lakeshores.

Indicates a mesic to wet-mesic site, but not necessarily a nutrient-rich one. Associated with a many tree and shrub species over this wide range of ecosystems and sites.

Notes. Moderately shade-tolerant; once established in open to lightly shaded sites, it persists vegetatively in increasingly shaded conditions. The berries of western North American species are reportedly inedible and may be poisonous (like, e.g., *Lonicera ciliosa* DC, western trumpet honeysuckle); nausea and vomiting have been reported (Kershaw, MacKinnon, and Pojar 1998).

Chromosome No. $2n = 18$, $n = 9$; $x = 9$

Similar Species. Closely related is the swamp fly honeysuckle, *Lonicera oblongifolia* Hooker, a common species of basic fens, swamps, and marshy shores. A widely distributed species in North America, bracted honeysuckle, *L. involucrata* (Richards) Banks ex Spreng., ranges from Quebec, the Keweenaw Peninsula, and Isle Royale to Alberta and British Columbia, south to California and northern Mexico. In western North America, the counterpart is the Utah honeysuckle, *L. utahensis* S. Watson.

174

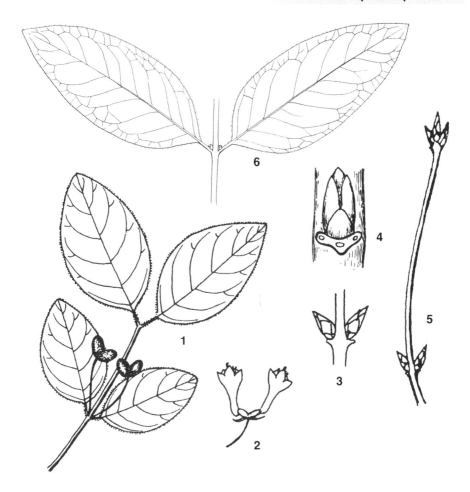

1. Fruiting shoot, berry, reduced
2. Flowers, reduced
3. Winter pair of lateral buds, ×1
4. Winter lateral bud and leaf scar, enlarged
5. Winter twig, reduced
6. Vegetative shoot, leaves, × 1

KEY CHARACTERS

- upright, straggling shrub to 1.5 m
- leaves opposite, entire, ovate-oblong, with ciliate margin
- twigs with white pith, continuous between nodes
- paired tubular flowers on long peduncles in axils of leaves
- berries paired, red, divergent, on long peduncles in axils of leaves

Distinguished from red honeysuckle, *L. dioica* L., by its more erect form, leaves borne on short petioles, and all leaves separate, not united around the stem.

Distinguished from Tartarian honeysuckle, *L. tatarica* L., by low to medium form, lack of stout ascending branches, early deciduous leaves, and white, continuous pith.

CAPRIFOLIACEAE
Lonicera dioica Linnaeus

Glaucous Honeysuckle, Red Honeysuckle

Size and Form. A trailing or twining, climbing vine or semierect, straggling shrub; climbing stems, 1–3 m long.

Bark. Gray to brown, peeling on old stems.

Leaves. Simple, opposite, deciduous, sessile or petiolate, blades of early leaves 4–14 cm long, 1–6 cm wide; late leaves smaller; leaves below inflorescences united around the stem; apex obtuse, rounded, and mucronate; entire, lacking cilia, often wavy, dark green, and glabrous above, glaucous, smooth, or hairy beneath; petioles of lower leaves subsessile to very short.

Stems-Twigs. Very slender, young shoots green to purplish, glabrous; winter twigs gray or straw-colored, glabrous, glaucous; leaf scars opposite and usually connected by a definite ridge or line, crescent-shaped; bundle scars 3; pith lacking, stem hollow between the nodes.

Winter Buds. Terminal bud present, laterals small, ovoid, scales ovate, sharp-pointed, pointing upward or slightly outward, the lowest nearly as long as the bud; bud scales persistent at base of current shoot.

Flowers. May–June; bisexual; paired, sessile or short-stalked, typically in stalked clusters from the center of two fused leaves in a terminal position; greenish-yellow, often tinged with purple or becoming deep maroon; corolla tubular-funnelform shaped, 1.5–2 cm long with two spreading lips, glabrous outside, hairy inside; stamens 5, strongly exserted. Insect and hummingbird pollinated.

Fruit. Berry; July–August; orange-red, round to ovate, 4–6 mm across; crowded together, forming dense terminal clusters, subtended by leafy disks, seeds usually 3.

Distribution. Common to locally abundant. County occurrence (%) by ecosystem region (Fig. 19): I, 92; II, 87; III, 71; IV, 75. Entire state 87%. Transcontinental in Canada, NL to BC and YT, widely distributed in e. North America, New England, w. Great Lakes Region, Great Plains s. to OK, AR, GA.

Site-Habitat. Open, disturbed wet to dry-mesic sites; especially circumneutral to basic habitats such as conifer and hardwood-conifer swamps; also mesic to dry forest edges, and rock outcrops. A generalist associated with many tree and shrub species of this wide range of ecosystems and sites.

Notes. Shade-intolerant, establishing and persisting in light shade. Although named *dioica*, evidently because of paired flowers and fruits, the term should not be confused with *dioecy* (i.e., the dioecious condition of different sexes on different plants).

Chromosome No. $2n = 18$, $n = 9$; $x = 9$

Similar Species. The hairy honeysuckle, *Lonicera hirsuta* Eaton, is frequent in northern lower and upper Michigan; occurs in similar upland and wetland habitats. Distinguished by leaves with ciliate margins, upper surface with short, appressed hairs; blade base densely hairy. Similar in western North America are Utah honeysuckle, *L. utahensis* S. Watson, and twining honeysuckle, *L. ciliosa* DC.

Also a vine, the Japanese honeysuckle, *L. japonica* Thunb., is a native of East Asia and an invasive of increasing occurrence in southern lower Michigan (14 counties, 37%) It is prone to forming dense tangles, climbing trees and shrubs, and smothering other vegetation, especially in the southerly parts of its naturalized occurrence in North America. Most similar to our *L. hirsuta*, from which it differs in leaf and flower characters (Voss and Reznicek 2012).

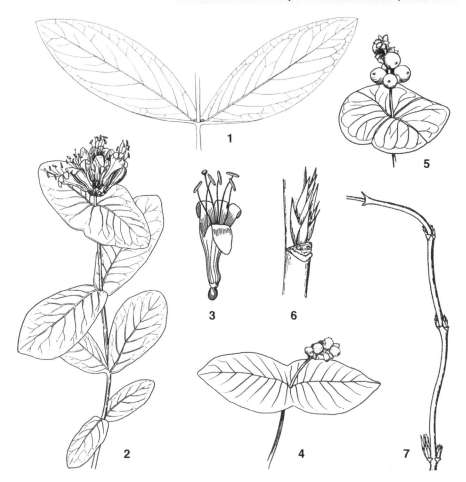

1. Leaves, reduced
2. Flowering shoot, reduced × 1/2
3. Flower, × 1
4. Fruiting shoot, berry, reduced
5. Individual fruits, berry, × 1
6. Winter bud and leaf scar, enlarged
7. Winter twig, reduced

KEY CHARACTERS

- usually a climbing vine by stems 1–3 m long or a low, semierect, straggling shrub
- leaves glabrous above, glaucous beneath, lacking marginal cilia, leaves below the inflorescences united around the stem
- flowers paired, borne on terminal shoots, yellowish-green, becoming deep maroon
- fruit a berry, orange-red, paired in dense clusters

Distinguished from Canadian fly honeysuckle, *L. canadensis,* by its more twining and trailing form and leaves, which below inflorescences and infructescences are united around the stem.

Distinguished from Japanese honeysuckle, *L. japonica,* a native species of common occurrence throughout the state, by leaves all deciduous, glabrous above, glaucous beneath, leaves below inflorescences united around the stem, and berries orange-red.

CAPRIFOLIACEAE
Lonicera maackii (Rupr.) Maxim.

Amur Honeysuckle,
Maack's Honeysuckle

Size and Form. Upright, much-branched, tall shrub to 5 m; large twigs ascending and then arching downward; stems may exceed 6 cm at 1.3 m, diameter at breast height.

Bark. Pale grayish-brown, outer bark longitudinally shallow-fissured and peeling off in thin flakes but soon becoming deeply furrowed with age (Figure 1).

Leaves. Simple, deciduous to semiper-sistent, opposite; blades 4–9 cm long, 2–4 cm wide; ovate to lanceolate, early leaves with acute apex and rounded or occasionally sub-cordate base; late leaves long acuminate near the apex, forming a "drip tip," base rounded; entire, pubescent only on veins both sides, upper side dark green, paler below; petioles glandular-pubescent, 3–5 mm long.

Stems-Twigs. Slender to stout, erect or arching; sylleptic branching on indetermi-nate shoots in the open or along forest edges; pubescent, tan, becoming dark brown to blackish, with longitudinal fissures; stout and arching; leaf scars opposite and usually connected by a definite ridge or line; leaf scars raised, crescent-shaped; bundle scars 3; pith brown at nodes, twigs hollow between nodes.

Winter Buds. Terminal bud present, small, pubescent; lateral buds conical, 2–6 mm long, 1–2 mm across, often larger than terminal; at ca. 45° to stem; whitish pubescent, margins ciliate, 4–6 scales exposed; bud scales per-sistent at base of current shoot.

Flowers. June; bisexual, numerous, in pairs, peduncles 2–3 mm long, distinctly shorter than the petioles of the subtending leaves; white, turning yellow, corolla to 2 cm, glabrous outside, tube short and thin, with a rich, fragrant aroma. Insect and hummingbird pollinated.

Fruit. Berry; September; orange, bright red to purplish-red; pairs in groups of 4, persistent bractlets at base; seeds many, elliptic, com-pressed; seeds 148,000/lb (325,600/kg).

Distribution. Locally common in southern lower Michigan, north to Newaygo and Isabel-la Co. County occurrence (%) by ecosystem region (Fig. 19): I, 34; II, 3; III, 0; IV, 0. Entire state 17%.

Site-Habitat. Open to lightly shaded disturbed, dry-mesic, and mesic sites; cutover forests in uplands and floodplains, old fields, roadsides. Associates include many trees and shrubs of diverse forest communities and early successional species of disturbed areas.

Notes. Moderately shade-tolerant; clone-forming in dense clumps, periodically sprouting from the root collar to form new stems. Whorled leaves occasionally occur on vigorous indeterminate shoots at nodes. Seeds dispersed by birds and most abundantly near a landscaped source in urban centers. Native in northern and central Honshu in Japan, China, and eastern Siberia. Introduced in 1855 or 1860 and widely naturalized, now an increas-ingly serious invasive species, which prevents the regeneration of native trees, shrubs, and herbs, especially in landscapes in and around human habitations. Should not be planted for any reason.

Chromosome No. $2n = 18$, $n = 9$; $x = 9$

1. Summer branch, multiple determinate shoots, developing fruit, berry, × 1/4
2. Fall branch, multiple determinate shoots, fruit ripe, × 1/4

KEY CHARACTERS

- tall, spreading shrub to 5 m with stout branches and ridged bark
- leaves opposite, persisting into late fall and winter, blades ovate-lanceolate, late leaves acuminate at apex, forming a "drip tip"
- twigs hollow between nodes, pith brown at nodes
- berries red, paired in groups of 4 on short peduncles
- nonnative species, common in southern Michigan, especially in disturbed sites near metropolitan areas

Distinguished from Canadian fly honeysuckle, *L. canadensis,* by its tall spreading form, with stout branches and ridged bark, clone-forming in dense clumps, leaves late deciduous, late leaves with blades long-acuminate at the apex, forming a distinctive "drip tip," stems hollow between nodes, and distribution limited to disturbed sites in southern Michigan.

Distinguished from Tartarian honeysuckle, *L. tatarica,* by its thick, ridged bark, leaf blades long-acuminate at the apex, pubescent leaf veins, and berries paired in groups of 4 on short peduncles.

CAPRIFOLIACEAE
Lonicera tatarica Linnaeus
Tartarian Honeysuckle

Size and Form. Medium to tall, erect much-branched shrub to 3 m; periodic sprouting from the root collar forms a moderately dense clonal clump of many stems.

Bark. Brownish-gray to light gray, peeling or shredding in small, vertical strips with age.

Leaves. Opposite, simple, deciduous to semipersistent; blades 3–6 cm long, 1–4 cm wide; ovate to ovate-lanceolate, acute to obtuse; rounded to subcordate; margin entire, smooth or with few sparse hairs; dark green and glabrous above, paler and finely hairy beneath, especially along the veins, rather thin; veins impressed on upper surface; turning yellowish-green in autumn; petioles 2–6 mm long.

Stems-Twigs. Slender to stout, young stems pubescent, becoming glabrous; light brown to mottled tan, splitting in long vertical ridges and peeling on large branches; leaf scars opposite and more or less connected by lines, crescent-shaped, bundle scars 3; pith brown at nodes, stems hollow between nodes.

Winter Buds. Terminal bud oblong, tan, 6–8 scales exposed, pubescent; lateral buds on determinate shoots small, 1–2 mm long, oblong or ovoid, tan, scales sharp-pointed; lateral buds on vigorous indeterminate shoots 2–3 mm long, superposed, the larger lower bud pointing upward, pubescent, oblong, 8-scales exposed, sharp-pointed; bud scales persistent at base of current shoot.

Flowers. May–June; bisexual, numerous, in sessile pairs on slender peduncles, 1.5–2 cm long, from the axils of upper leaves, much longer than petioles of the subtending leaves; pink or whitish, strongly 2-lipped, large, 3 cm or more long, showy, with a rich fragrance, especially in the evening. Insect and hummingbird pollinated.

Fruit. Berry, slightly united at the base, round, red, lustrous, paired on long peduncles; seeds 142,000/lb (312,400/kg).

Distribution. Common in the southern half of the Lower Peninsula, somewhat less frequent farther north and surviving there usually in lake-moderated counties. County occurrence (%) by ecosystem region (Fig. 19): I, 47; II, 37; III, 43; IV, 38. Entire state 42%. Widely distributed from NL and New England w. to SK, s. to NE, MO, e. to VA, NJ.

Site-Habitat. Virtually any open to lightly shaded or disturbed site; old fields in uplands and floodplains; forests of dry to mesic sites. Associates include a great many tree and shrub species throughout its range. Usually associated with Amur honeysuckle, *L. maackii*, in southern lower Michigan.

Notes. Midtolerant. Seeds dispersed by birds and most abundantly found near landscaped sources in towns and metropolitan areas. Hybridizing frequently with relatives (Voss and Reznicek 2012). Native in Eurasia, from southern Russia to the Altai Mountains and central Asia (i.e., formerly in parts of Turkistan). An aggressive invader of oak and other forests that are lightly or moderately shaded in the understory; casts shade that prevents seedling survival of native trees and shrubs. Difficult to eradicate due to vigorous basal sprouting. Should not be planted for any reason. The Japanese species, *L. morrowii* A. Gray, frequently escapes from cultivation and hybridizes with *L. tatarica,* forming the hybrid *L. × bella.* The hybrid is at least as common as *L. tatarica*, occurs in 40 counties (59%), and is recognized by its pubescent leaves and young stems (Voss and Reznicek 2012). Considered an invasive plant, it should not be planted under any circumstances.

Chromosome No. $2n = 18$, $n = 9$; $x = 9$

1. Vegetative shoot, reduced
2. Flowering shoot, × 1/2
3. Flower, × 1
4. Fruiting shoot, berry, reduced
5. Winter lateral bud and leaf scar, enlarged
6. Winter twig, reduced

KEY CHARACTERS

- erect, much-branched medium or large shrub to 3 m
- leaves opposite, entire, with ciliate margins
- twig with brown pith at nodes, stems hollow between nodes
- fruit a red berry, paired on long peduncles
- nonnative, often abundant in disturbed woodlands and roadsides surrounding metropolitan areas

Distinguished from Amur honeysuckle, *L. maackii*, by its lower stature, bark shredding in strips but not furrowed, leaf blades usually ovate with an acute tip, and fruits borne in pairs on long peduncles.

Distinguished from Canadian fly honeysuckle, *L. canadensis*, by its larger stature and erect arching form, semievergreen leaves, pith brown at nodes, and stems hollow between nodes, and as a non-native species naturalizing in disturbed areas.

MENISPERMACEAE
Menispermum canadense Linnaeus
Moonseed

Size and Form. A twining vine, climbing or sprawling, 2–4 m long and to 8 m high; in diameter to 3 cm; forming clones by sprouting from long, horizontal rhizomes held fast in the soil by abundant yellow roots.

Leaves. Alternate, simple, deciduous; blade 6–25 cm long and wide; shallowly 3–7 lobed or angled; preformed leaves tend to be 3-lobed, whereas neoformed leaves have more shallow lobes; orbicular or broadly ovate, palmately veined; pointed or rounded; cordate to subcordate; lobe margins entire; green, tomentose when young, becoming slightly pubescent to glabrous above, paler beneath, and slightly pubescent along the veins; leaves peltate, with the petiole attached to lower blade surface near blade base; petioles 5–20 cm long.

Stems-Twigs. Slender, flexible, current shoots pubescent, becoming glabrous, rounded, with fine longitudinal ridges; stems grayish-green to grayish-brown; leaf scars half circular, often deeply depressed in center; bundle scars 3 or more, indistinct; pith large, white, continuous.

Winter Buds. Very small, partly buried and concealed by dense woolly pubescence; several at each node, the flower bud in a hairy cavity above the leaf scar, 1 or 2 vegetative buds sunken beneath the leaf scar.

Flowers. June; unisexual, plants dioecious; small, borne on loose clusters in axils of leaf-bearing shoots; creamy to greenish-white; sepals 6–10, longer than the 6–9 petals; stamens 12–24. Insect pollinated.

Fruit. Drupe; August–September; single-seeded, oblong to globose, flattened, 8–12 mm wide, bluish-black with bloom, persistent; resembling wild grapes but bitter and reportedly poisonous (Kingsbury 1964); pit and seed crescent-shaped; seeds 7,600/lb (16,720/kg).

Distribution. Common in southern lower Michigan; infrequent in northern lower Michigan; only Menominee Co. in upper Michigan. County occurrence (%) by ecosystem region (Fig. 19): I, 82; II, 43; III, 14; IV, 0. Entire state 54%. Widely distributed throughout e. North America, Great Plains from ND s. to TX.

Site-Habitat. Open to moderately shaded, relatively nutrient-rich sites with circumneutral to basic soils; primarily river floodplains and stream banks, also woodlands, disturbed and cutover sites, fencerows. Associates include species of the floodplain levee and second bottom such as blue-beech, redbud, prickly-ash, hawthorns, spicebush, nannyberry, common elderberry, bladdernut.

Notes. Shade-intolerant, persisting in light to moderate shade. Vigorously climbing over shrubs, small trees, fences, and fencerows. The long yellow roots were reportedly used as ritual medicine by Native Americans of the Eastern Dakota group of the Great Sioux Nation, and the plant was known as "yellow medicine" (Rosendahl 1963; Smith 2008).

Chromosome No. $2n = 52$, $n = 26$; $x = 13$

Similar Species. A closely related species, *Menispermum dauricum* DC., is found in Japan and China.

1. Vine with cluster of fruit, drupe, × 1/3
2. Flowering terminal portion of vine, reduced
3. Fruit, drupe, reduced
4. Leaves, 3 different forms on a single shoot (early leaf at right, late leaves at left), × 1/3
5. Winter lateral bud and leaf scar, enlarged
6. Winter twig, reduced

KEY CHARACTERS

- twining vine, climbing or straggling
- leaves peltate; leaf blades relatively large, orbicular to broadly ovate, palmately veined, shallowly 3- to 7-lobed, lobes entire
- fruit a blue-black drupe, reported poisonous to humans
- habitat river floodplains, stream banks, and adjacent open, circumneutral to basic sites

MYRICACEAE
Myrica gale Linnaeus
Sweet Gale

Size and Form. Low, erect, bushy shrub, 0.3–1.5 m; branches strongly ascending; clonal by basal sprouting to form small clumps and by layering and root suckering.

Leaves. Alternate, simple, deciduous; blades firm, 3–6 cm long, 0.8–1.8 cm wide; oblanceolate; obtuse or rounded; cuneate; slightly toothed toward the apex; glabrous, dark green above, more or less pubescent and grayish beneath, both surfaces dotted with lustrous yellow glands; aromatic when crushed; petioles short, 1–2 mm long.

Stems-Twigs. Slender, dark gray to reddish-brown, downy, resin-dotted when young, lenticels pale; fragrant when crushed; leaf scars half elliptical, more or less raised, bundle scars 3; pith small, continuous, green.

Winter Buds. Terminal bud absent, lateral buds conical-ovoid, 2–4 exposed scales; flower buds at tips of shoots.

Flowers. April–May; unisexual, plants dioecious; male flowers before the leaves in dense cylindrical catkins, 0.7–1.5 cm long, scales brown, lustrous, glabrous; female flowers borne on short, ovoid-oblong catkins, 0.5–1.5 cm long, individual flowers inconspicuous, about 5 mm long. Wind pollinated.

Fruit. Nutlet; July–August; small, obtusely triangular, 0.8–1 cm long and 1.3–2 mm wide; beaked, resinous wax-coated, enclosed by two thick, persistent, ovate, winglike bracts, 2–3 mm long.

Distribution. Locally frequent to common in the northern half of lower Michigan and upper Michigan. County occurrence (%) by ecosystem region (Fig. 19): I, 13; II, 80; III, 100; IV, 100. Entire state 53%. Widely distributed locally in boreal, northern, and mountain sites, NL to AK, s. to OR; New England, w. Great Lakes Region, northern Europe, and northeastern Asia.

Site-Habitat. Open, wet to wet-mesic sites, surficial acidic soils, especially those with circumneutral to basic soils or substrates; fens, interdunal wetlands and swales, northern shrub thickets, lake and stream margins, rocky shores along Lake Superior. Associates include a great many species primarily of moist to wet basic sites (e.g., northern white-cedar, balsam poplar, willows, dogwoods, bog birch, shrubby cinquefoil, wild black currant, Kalm's St. John's-wort), as well as many ericaceous species that invade acidic surfaces of communities such as the northern fen.

Notes. Shade-intolerant, persisting in light shade. All parts pleasantly fragrant when crushed. Root nodules contain nitrogen-fixing bacteria. Nutlets with corky bracts float, which may assist in dispersal via streams and lakes. The only polyploid among North American species of *Myrica*. The essential oil is in demand for use in aromatic medical therapy. Formerly used in clothes closets to repel moths; young buds reportedly used by Native Americans for dyeing porcupine quills.

Chromosome No. $2n = 16, 48, 96$; $n = 8, 24, 48$; $x = 8$

Similar Species. In Japan, it is known as *Myrica gale* L. subsp. *tomentosa* C. DC. The northern bayberry, *Myrica pensylvanica* Mirbel, occurs along the Atlantic coast from Newfoundland to North Carolina and locally inland to the Lake Erie region, very rarely (3 counties) in Michigan. The southern bayberry, *M. heterophylla* Raf., is similar, becoming evergreen southward, ranging from New Jersey to Florida, Arkansas, and Texas.

1. Leaf, × 1
2. Vegetative shoot, × 1
3. Twig with vegetative shoot and shoot with male catkins, × 1/2
4. Fruit, beaked nutlet, × 10
5. Winter lateral bud and leaf scar, enlarged
6. Winter male flower bud, enlarged
7. Winter twig with male flower buds, reduced

KEY CHARACTERS

- low shrub to 1.5 m with strongly ascending branches; all parts aromatic
- leaves with blade slightly toothed toward apex, both surfaces dotted with shiny yellow resin glands; aromatic when crushed
- fruit a nutlet borne in dry resinous catkins, enclosed by two persistent, winglike bracts
- habitat open, wet to wet-mesic sites, especially those with circumneutral to basic substrates

VITACEAE
Parthenocissus inserta
(A. Kerner) Fritsch

Woodbine, Thicket Creeper

Size and Form. Straggling, rambling, sprawling on the ground, or rarely high-climbing vine to 10 m, with a stem diameter of 20 cm or more; climbing by slender twining or coiling branches and few-branched tendrils; rooting at nodes on ground surface.

Bark. Gray, with older stems becoming thick and furrowed.

Leaves. Alternate, deciduous, palmately compound; leaflets usually 5, 5–12 cm long, elliptic to obovate; acuminate; cuneate; sharply toothed mainly beyond the middle; glossy, dark green, mostly glabrous above, paler and glabrous beneath; turning red in autumn; petiole glabrous, 5–15 cm long.

Tendrils. Opposite the leaves or leaf scars, absent from every third node, slender, long and sinuous, 4–15 cm long, with few branches, each ending in a blunt or pointed tip, lacking adhesive disks but may appear so by becoming club-shaped in bark crevices.

Stems-Twigs. Slender to moderate, glabrous, round, bark not shredding or flaking, brown, nodes swollen; climbing stems giving rise to few-branched tendrils; leaf scars elliptic, concave on top, bundle scars numerous, indistinct; pith large, continuous, whitish.

Winter Buds. Terminal bud absent; often collateral at a node, sessile, round-conical, blunt, brown, several scales exposed.

Flowers. June–July; bisexual; borne on umbellike clusters in widely branched terminal panicles that lack a main central axis; flowers small, pedicels glabrous and bright red at maturity, 2–5 mm long; greenish white; calyx minute, forming a shallow cup; petals 5, reflexed, about 2.5 mm long; stamens 5. Insect pollinated.

Fruit. Berry; September–October; pedicels in sunlight turning red; subglobose, dark blue to bluish-black with slight bloom, 5–8 mm across; formed low on plant to be plainly visible; seeds 1–3, 18,800/lb (41,360/kg).

Distribution. Relatively common throughout the state. County occurrence (%) by ecosystem region (Fig. 19): I, 63; II, 47, III, 43; IV, 75. Entire state 57%. QC, New England, Great Plains, MB s.w. to WY, UT, AZ, TX.

Site-Habitat. Occurs in a very wide range of landform, soil, and light conditions; wet to dry, neutral to usually basic soils or substrates; similar ecosystems and communities as *Parthenocissus quinquefolia*. Associates also similar.

Notes. Moderately shade-tolerant. As the name thicket creeper implies, it is not as effective a climber as *Parthenocissus quinquefolia*, but it has a greater ability to flower and fruit in moderately well-lighted forest understories and ground covers, as well as tree canopies. Similar in landscaping use to *P. quinquefolia* except not effective in holding to masonry due to lack of adhesive discs. The name *P. vitacea* (Knerr) Hitchc. was used for this plant in older literature.

Chromosome No. $2n = 40$, $n = 20$

Similar Species. The related Virginia creeper, *Parthenocissus quinquefolia*, is similar in morphology and ecology. See below for distinguishing characters. The similar East Asian Boston ivy, *P. tricuspidata* (Sieb. and Zucc.) Planchon, has simple, lustrous, 3-lobed leaves (rarely trifoliate), short tendrils (1.5–3 cm long), and glabrous stems that are channeled or grooved.

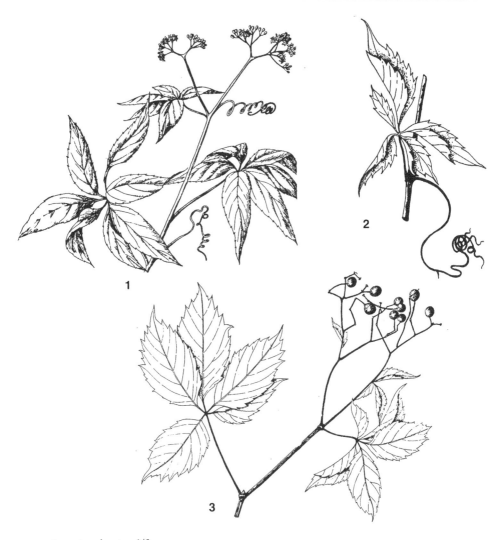

1. Flowering shoot, × 1/3
2. Vegetative shoot with tendril, × 1/3
3. Fruiting shoot, × 1/3

KEY CHARACTERS

- scrambling, sprawling, or rarely a high-climbing vine
- leaves palmately compound, leaflets usually 5, blades glossy green above, sharply toothed mainly beyond the middle, petioles long
- stems with few-branched tendrils that lack adhesive discs, lacking aerial roots
- flower inflorescence widely branching but lacking a distinct central axis
- fruit a dark blue to bluish-black berry

Distinguished from Virginia creeper, *Parthenocissus quinquefolia*, by leaves glossy green above and not glaucous beneath; climbing by twining stems and tendrils, lacking aerial roots; tendrils not developing adhesive disks; inflorescence wide branching but lacking a distinct central axis; often flowering in the forest understory and ground cover with fruit easily visible.

VITACEAE

Parthenocissus quinquefolia (L.) Planchon

Virginia Creeper

Size and Form. High-climbing or trailing vine; climbing >20 m by aerial roots and small adhesive disks attached to the ends of much-branched tendrils, the stem to 6 cm or more in diameter; rooting at nodes on ground surface; occasionally forming dense clonal patches.

Bark. Thin and tight, brown, not flaking or shredding; furrowed on very large individuals.

Leaves. Alternate, deciduous, palmately compound, leaflets mostly 5, but basal shoots and indeterminate shoots may be trifoliate; the lateral leaflets conspicuously inequilateral, the lower 2 much smaller; sessile or on petiolules to 15 mm, 4–15 cm long and nearly half as wide; elliptic to obovate or oblong, acuminate to acute; cuneate; coarsely toothed mainly beyond the middle; dull green above, paler and slightly glaucous beneath, glabrous to sparsely hairy; turning red to reddish-purple in late summer or early autumn; leaflets falling early in autumn; petioles hairy, 20 or more cm long.

Tendrils. Opposite the leaves or leaf scars, absent from every 3rd node, slender, 1–4 cm long, much branched with 3–12 branchlets, each ending in a small adhesive disk if it contacts a hard object.

Stems-Twigs. Slender to moderate, round, with bark not shredding or flaking, brown, terminal shoots usually pubescent, nodes swollen; climbing stems with coarse aerial roots, giving rise to branched tendrils; leaf scars in an ellipse, concave on top, bundle scars numerous, indistinct; stipule scars long and narrow; pith large, continuous, green to white.

Winter Buds. Terminal bud absent; often collateral at a node, sessile, round-conical, blunt, light brown, several scales exposed, outer scales often keeled.

Flowers. June–July; bisexual; borne in terminal, large compound cymes, longer than wide from the upper leaf axils, 6–12 cm long, opposite the leaves; the strong, zigzag, distinct central axis with divergent branches bears 50–200 flowers in terminal clusters; flowers small, about 6 mm across; greenish white; calyx minute, forming a shallow cup; petals 5, spreading or reflexed, about 2.5 mm long; stamens 5. Insect pollinated.

Fruit. Berry; September–October; pedicels in sunlight turning red; subglobose, dark blue to bluish-black, with slight bloom, 5–8 mm across; formed mostly high in forest canopy and not readily visible; seeds 1–3, 15,600/lb (34,320/kg).

Distribution. Common to abundant in southern lower Michigan and coastal counties in northern lower Michigan; rare in upper Michigan. County occurrence (%) by ecosystem region (Fig. 19): I, 66; II, 43; III, 43; IV, 25. Entire state 52%. Wide ranging in e. North America and the Great Plains, SK s. to TX.

Site-Habitat. Occurs in a very wide range of landform, soil, and light conditions; wet to dry, neutral to usually basic soils or substrates, open to moderately shaded sites; swamps, floodplains, dry to mesic forests, wooded dune and swales, near lakes Michigan and Huron, roadsides, fencerows. Associates include a great range of species, though not most characteristic acid-site ericaceous species.

Notes. Moderately shade-tolerant; thriving in full sunlight. Widely used in landscaping. The adhesive disks attach tightly to both smooth and rough surfaces. It is reported that a tendril with 5 disk-bearing branches can stand a force of 4.5 kg (10 lbs). Commonly associated with poison-ivy and differentiated from it by palmately compound leaves with 5 leaflets, shape of winter buds, and berries bluish-black instead of whitish. An excellent low-maintenance cover for walls and trellises. Grows in cultivation in about any kind of soil, full sun to full shade, exposed, windy conditions, and polluted situations; salt tolerant (Dirr 1990).

Chromosome No. $2n = 40$, $n = 20$

Similar Species. A related species is woodbine or thicket creeper, *Parthenocissus inserta* (A. Kerner) K. Fritsch (see comparison below). A native of East Asia, Boston ivy, *P.*

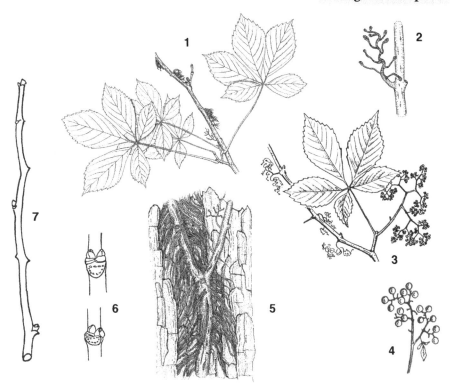

1. Climbing vine with aerial roots, determinate vegetative shoot, reduced
2. Shoot at node bearing tendrils with adhesive discs, × 1
3. Branch with tendrils and flowering shoot, reduced
4. Fruiting shoot, berry, reduced
5. Climbing vine secured to tree by aerial roots, × 1
6. Winter stem with leaf scar at 2 nodes with collateral vegetative (*above*) and flower buds (*below*), enlarged
7. Winter twig, × 1

KEY CHARACTERS

- high-climbing or trailing woody vine, climbing to >20 m by aerial roots and small adhesive disks attached to the ends of much-branched tendrils
- leaves palmately compound, leaflets usually 5, blades dull green above, coarsely toothed mainly beyond the middle, petioles long
- stems with bark not shredding or flaking, long, producing aerial roots and much-branched tendrils, each branch ending in a small adhesive disk if it contacts an object
- flower inflorescence with a distinct central axis with divergent branches
- fruit a dark blue to bluish-black berry formed high in the forest canopy

Distinguished from woodbine, *Parthenocissus inserta*, by leaves dull green above and glaucous beneath; climbing by coarse aerial roots; tendrils developing adhesive disks; inflorescence with a distinct central axis; flowering and fruiting in upper tree crowns.

tricuspidata (Sieb. and Zucc.) Planchon, is also widely used in landscape plantings. It has simple, lustrous, 3-lobed leaves (rarely trifoliate), short tendrils (1.5–3 cm long), and glabrous stems that are channeled or grooved.

VITACEAE
Parthenocissus tricuspidata
(Siebold & Zucc.) Planchon

Boston Ivy, Japanese Creeper

Size and Form. High-climbing or trailing woody vine to 18 m, the stem up to 8 cm or more in diameter, attaching to hard objects by adhesive disks; very closely appressed to the surfaces on which it grows.

Bark. At first thin, smooth, tan to reddish-brown, with conspicuous lenticels, becoming moderately thick, very rough, and grayish, with prominent grooves or channels.

Leaves. Alternate but appearing opposite on short-shoots; deciduous; adult shoots with simple leaves, leaf blades broadly ovate to heart-shaped, 5–30 cm long and nearly as wide; 3-lobed, with acute sinuses, apex acuminate; base cordate to deep and broadly U-shaped; unequally and very coarsely toothed, the teeth mucronate-tipped; glabrous, lustrous bright green above, paler beneath, main veins beneath with scattered, short, stiff hairs; palmately veined; petioles 5–30 cm long. Juvenile shoots with simple leaves, blades coarsely toothed but not 3-lobed, glabrous both sides. Juvenile basal shoots with simple, trifoliate leaves 10–28 cm long, leaflets short-stalked, the terminal obovate, the two lateral ones oblique at the base, all leaflets shallowly and coarsely dentate, the inner margin of the two lateral leaflets entire or nearly so, petioles 5–14 cm long.

Tendrils. Opposite the leaves or leaf scars, absent from every 3rd node, slender, short, 1.5–3 cm long, much branched, with each branch ending in a large, preformed adhesive disk.

Stems-Twigs. Slender to very stout, trailing stems slender, <0.05–1.5 cm at base of object supporting climbing stems; climbing stems to 8 cm in diameter; round to laterally compressed against walls adhering or separated from structure; slightly to deeply grooved or channeled; conspicuous vertically arranged lenticels on young stems; climbing stems >3 cm very rough, with many projections of broken tendrils; grayish, tan in the grooves; matrix of stems of all sizes crisscrossing walls; short-shoots (up to age 20 years and 20 cm long) arising from climbing stems >3 cm, very knobby, with conspicuous concave leaf scars, appearing opposite; flowering and young vegetative stems arising in axils of leaves of short-shoots; bundle scars in an ellipse, in-

conspicuous; pith continuous, greenish when young, becoming white.

Winter Buds. Terminal bud absent; small, sessile, round, conical, blunt, light brown, with 2–4 outer scales, often collateral at a node, outer scales often keeled.

Flowers. June–July; bisexual; borne in panicles on 2-leaved short-shoots, longer than wide, 6–15 cm long; inconspicuous, greenish-white, calyx minute; petals 5, reflexed; stamens 5. Insect pollinated.

Fruit. Berry; October–November, sub-globose, bluish-black with bloom, 5–8 mm across; seeds 1–4.

Distribution. A cultivated nonnative ornamental of urban areas and built environments. Not yet documented as an escape from cultivation. Cultivated widely in e. North America, New England, w. Great Lakes Region, mid-South, mid-Atlantic states.

Site-Habitat. Once planted it grows vigorously in open to lightly shaded, moist to dry-mesic and basic to slightly acidic soils.

Notes. Shade-intolerant, thriving in full sunlight. Used in landscaping as a tough, dense, low-maintenance, and attractive cover for walls and other objects. Tolerates smoke, dust, and ozone; attractive glossy green summer foliage, turning brilliant scarlet and orange in fall; especially vigorous on brick walls, wooden structures, and trees. A native of Japan and central China, introduced in 1862. Not as hardy as Virginia creeper but persisting in sheltered conditions. Cultivars selected for very small leaves, glossy green leaves, and leaf colors reddish-purple and brilliant shades of red.

Chromosome No. $2n = 40$, $n = 20$

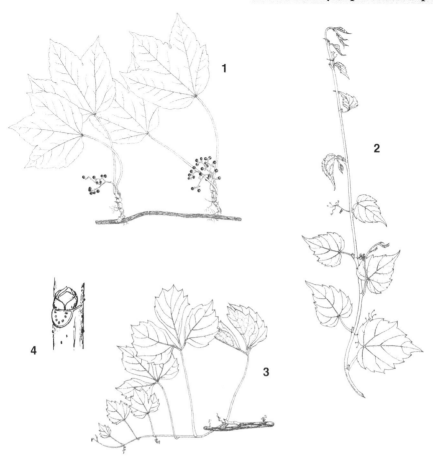

1. Climbing vine, fruiting shoots, simple 3-lobed leaves, berry, reduced
2. Climbing indeterminate shoot with simple, non-lobed leaves, tendrils, reduced
3. Basal shoot with simple trifoliate leaves, reduced
4. Winter lateral bud and leaf scar, × 1

KEY CHARACTERS

- high-climbing or trailing woody vine to 12 m, attaching to hard objects by adhesive disks
- leaves predominately simple, blades broadly 3-lobed with acute sinuses, base cordate to deep and broadly U-shaped, lustrous bright green above
- stems slender to very stout, those >3 cm very rough, grayish, grooved or channeled; tendrils short, much branched, each branch ending in a large adhesive disk
- fruit a bluish-black berry
- native of China and Japan, widely cultivated as a cover for walls and other structures, urban areas and built environments only

Similar Species. The native Virginia creeper, *Parthenocissus quinquefolia,* is also widely used in landscape plantings. It has mostly compound leaves with 5 leaflets and stems with aerial roots, as well as many-branched tendrils with adhesive discs. The native woodbine, *P. inserta,* also has 5 leaflets and spreads and climbs by twining stems and tendrils without adhesive discs or aerial roots.

ROSACEAE
Physocarpus opulifolius (L.) Maximowicz
Ninebark

Size and Form. Medium to tall, much-branched, erect, spreading shrub, 1–3.5 m; loosely branched, with stiff recurved branches forming a rounded, densely foliated crown; branches dense, tangled, giving a coarse, irregular appearance in winter; forming clonal clumps by basal sprouting.

Bark. Exfoliating conspicuously on larger branches in numerous, long, very thin strips or layers, exposing brown inner bark.

Leaves. Alternate, simple, deciduous, blades 3–8 cm long and about as wide; ovate-orbicular, usually 3-lobed, with smaller and unlobed blades on flowering branchlets; acute; truncate, subcordate, or cuneate; irregularly crenate-dentate, with 3 main veins from the base; glabrous, medium green above, more or less pubescent beneath in axils of main veins; yellowish to bronze in autumn; petioles glabrous or stellate-pubescent, slender, 1–2 cm long, with an elongate pair of stipules at base.

Stems-Twigs. Moderate to stout, zig-zag, young twigs yellowish-green, turning brownish, glabrous, round, strongly ridged downward from the nodes; twigs 2–3 mm in diameter starting to exfoliate; leaf scars raised, 3-sided or 3-lobed, bundle scars 3 (5), the lower one larger than the others; stipule scars present, pubescent, small, on an elevated ridge on each side at top edge of leaf scar; pith large, brown, round, continuous.

Winter Buds. Terminal bud present unless replaced by flower or fruit clusters; buds small, conical-oblong or ovoid, pointed, sometimes twisted, sessile, with about 5 scales visible; lateral buds appressed.

Flowers. May–July; bisexual; white to pinkish, showy, 0.5–1 cm across; borne in many-flowered (usually more than 25 on vigorous stems), terminal, umbellike corymbs 3–5 cm across, pedicels 1–2 cm; sepals 5, pubescent, with stellate hairs; petals 5, more or less pubescent; stamens 20–40. Insect pollinated.

Fruit. Follicle, borne as an aggregate, usually in 3s in terminal clusters of 10 to >25; August–September; a single follicle is inflated, resembling a bellows, glabrous or stellate-hairy, ovoid and longer than broad, united at the base, splitting along both margins, acute to very sharp-pointed, 0.7–1 cm long, red-tinted

in exposed situations; persistent in winter; seeds lustrous, light brown, 2–4 per follicle, about 2 mm long, 1,045,000/lb (2,299,000/kg).

Distribution. Frequent to common throughout the state, especially on river and stream floodplains; edges of wet meadows; interdunal wetlands; open, bedrock ecosystems, especially in upper Michigan along the shore-lines of lakes Huron and Michigan. County occurrence (%) by ecosystem region (Fig. 19): I, 66; II, 80; III, 100; IV, 100. Entire state 77%. Widely distributed in e. North America and the Great Plains, NS to MN, s. from New England to FL, AL, and from MN s. to OK, AR.

Site-Habitat. Predominantly open sites, soils circumneutral to basic; often with water table fluctuation from spring inundation to summer drought on bedrock sites; river flood-plains and stream banks, wet and wet-mesic prairies, prairie fens, ecosystems on bedrock with little or no soil development, including lakeshores and cliffs. Associates include primarily species of open, extreme, harsh sites with a preference for circumneutral to basic soils or substrates, willows, ground juniper, red-osier and silky dogwood,

Notes. Very shade-intolerant. Water plays a major role in dispersal. Tolerates extreme water-table changes from wet to dry on bedrock sites with little or no soil; commonly occurring in cracks in bedrock and near-vertical bedrock cliffs. In large, multistemmed clumps, a single, 1-year-old, unbranched basal sprout can exceed 2 m. High variation in degree of pubescence on many vegetative and flower parts. The genus was named from the Greek *physa,* "a pair of bellows," which is the arrangement of pubescent follicles in the mallow ninebark (see "Similar Species"). The

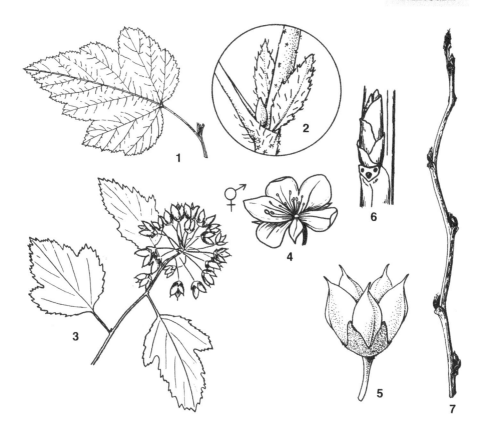

1. Leaf, × 1
2. Stipules, × 5
3. Flowering shoot, reduced
4. Flower, × 3
5. Fruit, follicle, × 2
6. Winter lateral bud and leaf scar, enlarged
7. Winter twig, × 1

KEY CHARACTERS

- medium to tall, much-branched, multistemmed spreading shrub to 3.5 m
- exfoliating on larger branches in numerous, long, very thin strips or layers
- leaf blades usually 3-lobed, ovate-orbicular
- fruit a follicle, borne as an aggregate of follicles in terminal clusters of many aggregates, brown, persistent in winter
- habitat open sites with circumneutral to basic soils; floodplains, stream banks, wet prairies, bedrock lakeshores and cliffs

common name derives from the belief that the bark has nine layers of exfoliating bark on the branches (Pojar and MacKinnon 1994).

Chromosome No. $2n = 18$, $n = 9$; $x = 9$

Similar Species. Several related species occur in the West, including mallow ninebark, *Physocarpus malvaceus* (Greene) Kuntze., of the northern and central Rocky Mountains; the Rocky Mountain ninebark, *P. monogynus* (Torr.) M. J. Coult., of the central and southern Rockies; and the Pacific ninebark, *P. capitatus* (Pursh) Kuntze., of coastal northern California, Oregon, Washington, and British Columbia. In northeastern China and Korea, the Amur ninebark, *P. amurensis* (Maxim.) Maxim., has larger leaves, flowers, and fruit.

ROSACEAE
Prunus americana Marshall

American Wild Plum

Size and Form. Tall shrub or small tree with crooked branches, 3–10 m, and thorn-tipped twigs; clone-forming by root sprouting, giving rise to dense thickets. Michigan Big Tree: girth 91.4 cm (36 in), diameter 29.1 cm (11.5 in), height 10.7 m (35 ft), Oakland Co.

Bark. Dark brown to black, tightly exfoliating when young and irregular plates with age; inner bark mildly bitter; lenticels horizontal, distinct, light brown.

Leaves. Alternate, simple, deciduous; blades thin, 4–12 cm long, 2–5 cm wide; obovate to oblong-ovate; abruptly acuminate to a prolonged tip; broadly cuneate to rounded; margin sharply serrate with forward-pointing bristle-tipped teeth; light green above, pubescent when young, becoming glabrous, glabrate or pubescent along veins beneath; veins prominent; petioles mostly glandless, occasionally with a few obscure glands near the blade base, 1 cm long.

Stems-Twigs. Moderately slender, reddish-brown or orange-brown, glabrous, partly or entirely covered with gray epidermis; armed with thorns at ends of twigs; lenticels distinct, light brown; pungent when bruised, bark mildly bitter; leaf scars small, broadly crescent-shaped or elliptical, with a fringe of hairs along the upper edge, bundle scars 3; pith round or angled, brown, continuous.

Winter Buds. Terminal bud absent; end and lateral buds sessile, ovoid, or broadly conical, with about 6 visible scales, 3–6 mm long, reddish-brown or chestnut brown; glabrous; lateral buds solitary or occasionally collateral (2–3) at nodes.

Flowers. April–May, before the leaves, bisexual, white; 2–4 in an umbel, 1.5–2.5 cm wide; very fragrant; calyx lobes glabrous or with some hairs near the base of the lobes, which are pubescent within and smooth or hairy without, often toothed; petals 5, narrowly obovate, 0.7–1 cm long. Insect pollinated.

Fruit. Drupe, August–September, usually subglobose, red to yellow, glaucous; 2–3 cm wide; skin tough, astringent, flesh juicy, pleasantly flavored; pit 1–1.5 cm wide, compressed; seeds 870/lb (1,914/kg).

Distribution. Occasional in the southern half of lower Michigan; rare in northern lower Michigan and upper Michigan. County occurrence (%) by ecosystem region (Fig. 19):

I, 63; II, 13; III, 43; IV, 25. Entire state, 40%. Widely ranging in North America, QC to SK, s. from Canada throughout e. North America, the Great Plains and Rocky Mountains, the Southwest, and Washington.

Site-Habitat. Open to lightly shaded, dry to dry mesic, fire-prone sites, slightly to moderately acidic upper horizons and typically with basic subsoil within the rooting zone; pine and oak barrens, gaps and edges of mixed oak forests and oak openings, as well as fence-rows and roadsides. Associates include up to 8 fire-dependent oak species, pignut hickories, sassafras, sumacs, sand cherry, prairie willow, gray dogwood, American and beaked hazels.

Notes. Very shade-intolerant, persistent in light shade. Seeds dispersed widely by birds and mammals. Germination is hypogeal. Ripe fruit is frequently eaten raw and was once widely used for making jams, jellies, and conserves/preserves. Ornamentally, it tends to thrive with neglect.

Chromosome No. $2n = 16$, $n = 8$; $x = 8$

Similar Species. Most similar in our region is *Prunus nigra* Aiton (Barnes and Wagner 2004, 174–75). The related *P. mexicana* S. Watson ranges northward from Mexico to the Great Plains, the Midwest, and east and south to the Gulf of Mexico and the Atlantic coastal plain. The blackthorn or sloe, *P. spinosa* L., is a wide-ranging counterpart in Europe, West Africa, and western Asia. In Michigan it is reported from Kalamazoo and Livingston Co.

Also similar is the rarely encountered Alleghany plum, *P. umbellata* Elliott. It is a small scraggly shrub to 2 m, relegated to dry, sandy, fire-prone sites and usually associated with jack pine–northern pin oak forests in District 8 (Fig. 19) of northern lower Michigan, where

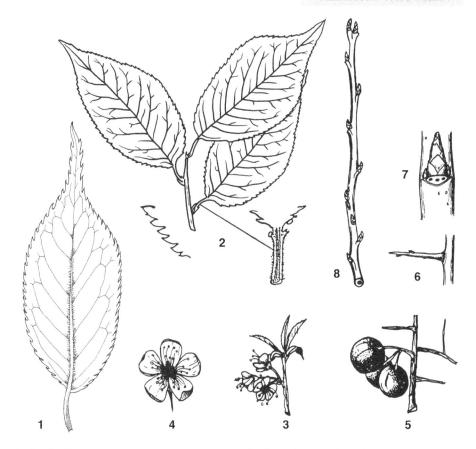

1. Leaf, × 1
2. Vegetative shoot, reduced; blade margin and petiole enlarged
3. Flowering shoot, reduced
4. Flower, enlarged
5. Fruit, drupe, enlarged
6. Winter stem with thorn, × 1
7. Winter lateral bud and leaf scar, enlarged
8. Winter twig, × 1

KEY CHARACTERS

- clone-forming shrub or small tree with crooked branches to 10 m
- stems reddish-brown or orange-brown, glabrous, more or less thorny, bark mildly bitter
- leaf blades obovate to oblong-ovate and abruptly acuminate to a prolonged tip, teeth sharply serrate or doubly serrate, pointing forward; petiole mostly glandless
- fruit a red to yellow drupe, juicy, edible when ripe
- habitat open, dry to dry-mesic fire-prone sites, surface horizons acidic, often with basic sub-soil within the rooting zone

Distinguished from the Canada plum, *Prunus nigra*, by the oblong-ovate leaf blade with abruptly acuminate and conspicuously elongated tip, sharply serrate or doubly serrate glandless teeth; mostly glandless petioles; flowers white; and petals longer.

it is reported from 10 counties. Besides its markedly different form, its leaf blade shape is broadly ovate, it lacks the abruptly acuminate tip of the American wild plum, margin with rounded serrations, and its fruits are markedly smaller and dark purple.

195

ROSACEAE
Prunus pumila Linnaeus
Sand Cherry

Size and Form. Low, erect or decumbent, loosely branched shrub, 0.5–1.5 m; forming clones by rhizomes and layering, rarely creeping except old stems.

Leaves. Alternate, simple, deciduous; blade relatively thick or leathery in full sun, 3–8 cm long, 0.6–2 cm wide; narrowly oblanceolate, oblong-elliptic to obovate; acute; cuneate; finely and remotely glandular-serrate above the middle; bright glossy green above, paler or gray-whitish beneath, with prominent veins, turning reddish in autumn, glabrous both sides; leaf glands often obscure, when present tending to occur near the blade base; petioles 6–12 mm long, glandless; stipules narrowly linear, 4–7 mm long, deciduous.

Stems-Twigs. Slender, angled, glabrous, reddish-brown when young, with small, pale lenticels, becoming gray and turning dark brown with age; pungent odor lacking; leaf scars small, half round, bundle scars 3; pith small, round, pale, continuous.

Winter Buds. Terminal bud absent; end and lateral buds sessile, small, subglobose or ovoid, with about 4 scales exposed.

Flowers. May–June, bisexual; white, solitary or in sessile umbels of 2–4 on pedicels 4–12 mm long, 1.2–1.6 wide; calyx lobes with irregular glandular-toothed margins; petals 5, 4–7.5 mm long. Insect pollinated.

Fruit. Drupe; July–September, subglobose, purplish to black, 1–1.5 cm wide, largest of Michigan's native cherries; edible, usually astringent; seeds 2,920/lb (6,424/kg).

Distribution. Common in northern lower Michigan and upper Michigan, frequent in southern lower Michigan. County occurrence (%) by ecosystem region (Fig. 19): I, 53; II, 80; III, 100; IV, 62. Entire state, 67%. Widely ranging from NS to SK s., including New England, w. Upper Great Lakes Region, Midwest, Great Plains, and central and northern Rocky Mountains.

Site-Habitat. Open, dry, fire-prone sites primarily of two kinds. *P. pumila* var. *susquehanae* occurs in very strongly to mildly acidic, excessively to well-drained inland ecosystems characterized by communities of pine, pine-oak, and oak barrens. Associates include jack and red pines; northern pin, black, and white oaks; prairie willow; low sweet blueberry; bearberry; sweetfern; and prairie redroot. In contrast, var. *pumila* occurs in coastal ecosystems with medium to mildly basic soils with well-drained to seasonally wet ecosystems such as open dunes, alvars, lakeshores, and beach ridges around the Great Lakes. Associates include ground and creeping junipers, willows, and red-osier.

Notes. Very shade-intolerant, persistent in light shade. Variable in form and morphology due to a combination of geography and physical conditions of ecosystems supporting it. The decumbent form of stems is caused by the weight of snow in the high plains of northern lower Michigan and blowing and shifting sand on dunes near the Great Lakes.

Chromosome No. $2n = 16$, $n = 8$; $x = 8$

Similar Species. Three varieties are recognized in Michigan (Rohrer 2000; Voss and Reznicek 2012). *Prunus pumila* var. *susquehanae* (Willd.) Jaeger is the most distinctive and widespread and has oblong-elliptic to obovate leaf blades (see "Site-Habitat"). It is also recognized as a species, *P. susquehanae* Willd, Appalachian dwarf cherry, and is geographically, in some works, considered wide ranging (Braun 1989; Catling, McKay-Kuja, and Mitros 1999; Smith 2008). *P. pumila* var. *pumila* is restricted to beaches and dunes of the Great Lakes. *P. pumila* var. *besseyi* (L. H. Bailey) Waugh is confined to the western Upper Peninsula and ranges as far west as North Dakota, Colorado, and WY. Natives of China *P. japonica* Thunb. and dwarf flowering almond, *P. glandulosa* Thumb.; the Japanese Fugi cherry, *P. incisa* Thumb.; and the central and eastern European ground cherry, *P. fruticosa* Pall., are prized for their horticultural value (Coates 1992; Dirr 1990).

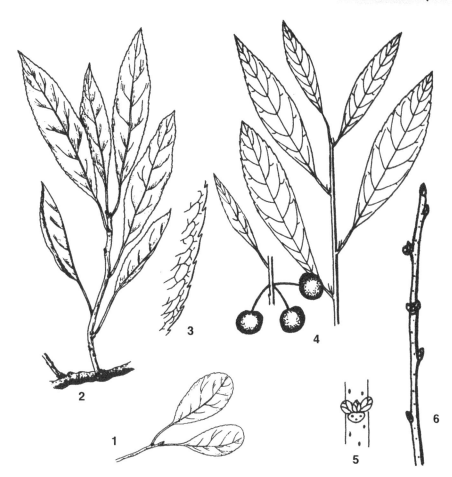

1. Vegetative shoot, var. *susquehanae*, × 1/2
2. Vegetative shoot, var. *pumila*, × 1/2
3. Leaf blade margin, × 2
4. Vegetative shoot, var. *pumila*, and fruiting shoot, drupe, reduced
5. Winter lateral bud and leaf scar, enlarged
6. Winter twig, reduced

KEY CHARACTERS

- low, erect or decumbent, loosely branched shrub
- leaf blade leathery in full sun, narrowly oblanceolate to obovate, bright glossy green above, margin finely and remotely glandular-serrate above the middle
- fruit a relatively large drupe, subglobose, purplish to black
- habitat open, dry, sandy, acidic, and basic sites depending on variety

ROSACEAE
Prunus virginiana Linnaeus

Choke Cherry

Size and Form. Tall shrub or small tree, 2–10 m and 10–15 cm in diameter. Trunk crooked, often leaning or twisted; forming a narrow, irregular, or somewhat rounded crown. Dense clonal colonies produced by sprouting from the root collar and root suckers. Michigan Big Tree: girth 170 cm (67 in), diameter 54.2 cm (21.3 in), height 23.4 m (77ft), Kent Co.

Bark. Thin, dark brown, slightly fissured or with fine scales; lenticels prominent but not horizontally extended.

Leaves. Alternate, simple, blades 5–10 cm long, half as wide; obovate to oblong-obovate or broadly oval; abruptly acuminate; rounded; margin finely and sharply serrate, teeth pointing away from blade or ascending, each tooth ending in a narrow point; very thin; dull dark green above, paler beneath; turning yellow in autumn; glabrous; petioles short, slender, glandular near the blade.

Stems-Twigs. Slender to moderately stout, at first light brown or greenish, becoming grayish-brown, finally dark brown; bitter almond odor and taste when crushed; leaf scars half round, small; bundle scars 3; pith white.

Winter Buds. Terminal bud 0.6–1.2 cm long, conical, sharp-pointed; scales rounded at the tip, dark brown with pale brown edges, giving a two-toned appearance (dark brownish-black at the tip and pale brown at the middle); lateral buds similar but smaller.

Flowers. May–June, when the leaves are nearly grown; bisexual; 0.8–1 cm across; borne on short, slender pedicels in many-flowered racemes, 7–15 cm long; calyx cup-shaped, 5-lobed; petals 5, white; stamens 15–20; stigma broad, on a short style. Insect pollinated.

Fruit. Drupe; July–August; globular, 7–9 mm in diameter, calyx deciduous except for minute remnant at base of each fruit, varying from deep red to dark purple, with dark red flesh; extremely astringent but edible; seeds 2,173/lb (4,780/kg).

Distribution. Abundant in the southern half of the Lower Peninsula; common throughout the rest of the state. County occurrence (%) by ecosystem region (Fig. 19): I, 97; II, 83; III, 100; IV, 100. Entire state 93%. Ranging throughout North America, including Mexico, with the exception of NU and YT in the far North and SC, FL, AL, MS, and LA in the deep South.

Site-Habitat. Open to moderately shaded, dry to wet-mesic, mostly circumneutral to basic but occasionally slightly to moderately acidic sites; common in 13 natural communities of Michigan (Kost et al. 2007), including oak-pine and oak barrens, dry and dry-mesic southern and northern forests (especially edges), and bedrock lakeshores and cliffs. Associates include oaks, maples, ashes, black cherry, sassafras, and a great many shrubs accompanying these trees in the communities noted above and many others.

Notes. Moderately shade-tolerant, slow-growing, individual stems short-lived; resprouting vigorously following fire. Unlike many *Prunus* species, germination is epigeal. An opportunistic species whose seeds are widely disseminated by birds and mammals very common in the understory of forests where light is sufficient for growth. One of the most widely distributed native shrubs and small trees of North America, extending from the Atlantic west to the Pacific Ocean and from central Canada (latitude 58° north) south into the Rocky Mountains and from California into Mexico.

Chromosome No. $2n = 20$, $n = 10$; $x = 10$

Similar Species. Two varieties are recognized: *Prunus. virginiana* var. *virginiana* L. of eastern North America and *P. virginiana* var. *demissa* (Nutt.) Torr. of western North America. Morphologically similar are *Prunus grayana* Maxim. in Japan and *Prunus padus* L. in central Europe, but both are trees to 15 m.

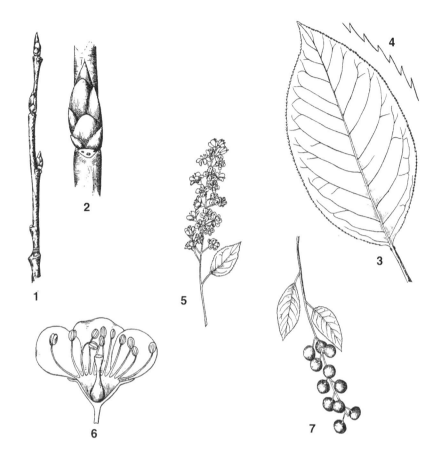

1. Winter twig, × 1
2. Portion of twig, enlarged
3. Leaf, × 1
4. Margin of blade, enlarged
5. Flowering shoot, × 1/2
6. Vertical section of flower, enlarged
7. Fruiting shoot with drupes, × 1/2

KEY CHARACTERS

- clone-forming shrub or small tree
- leaf blade broadly ovate, abruptly acuminate; fine narrow teeth point away from the blade; petioles with small glands near the blade base
- bark smooth, dark brown; twigs when crushed with aroma of bitter almond
- terminal buds conical, sharp-pointed, appearing two-toned dark and pale brown with a conspicuous dark brown tip
- fruit a reddish-purple drupe, borne on a raceme

Distinguished from pin cherry, *Prunus pensylvanica,* by its smaller and broader leaves; twigs grayish-brown; buds large, more sharp-pointed and conical, two-toned brown; fruit deep red to purple; and borne in racemes.

RUTACEAE
Ptelea trifoliata Linnaeus

Wafer-ash, Hop-tree

Size and Form. Tall shrub or small tree, 3–5 m; trunk short, crooked, often low-branched; crown irregular, rounded; root sprouting occasional. Michigan Big Tree; girth 83.8 cm (33 in), diameter 26.7 cm (10.5 in), height 10.7 m (35 ft), Kent Co.

Bark. Thin, smooth, reddish-brown, lenticels elongated, becoming roughened and warty; light to dark gray or brown, bitter to taste.

Leaves. Alternate, a few sometimes opposite or nearly so; deciduous; compound, trifoliate; leaflets sessile or nearly so, 5–17 cm long, the terminal larger than the laterals; ovate-oblong to obovate, acute to acuminate; laterals rounded at base, terminal tapers gradually to point of attachment; entire to crenulate; glabrous, dark green above, slightly pubescent and paler beneath, dotted with small translucent glands (visible when held up to sunlight); pungent to malodorous when crushed, bitter to taste; petioles stout, long, 6–10.

Stems-Twigs. Slender to moderate, unarmed, warty-dotted, glabrous, yellowish-brown to dark reddish-brown, aromatic when bruised (suggesting lemon peel or turpentine), bitter taste; leaf scars rather large, U-shaped when torn by the buds, bundle scars 3; pith large, white, continuous.

Winter Buds. Terminal bud absent; lateral buds small, flattened, low-conical, sessile, closely superposed in pairs, yellow, silvery silky-pubescent, without apparent scales, sunken or nearly surrounded by leaf scar and often covered by it.

Flowers. June; with the leaves; bisexual or unisexual (usually polygamous), borne in large, terminal compound cymes on slender pedicels; small, 0.8–1.3 cm, greenish-white; calyx lobes, petals, and stamens 4–5. Insect pollinated.

Fruit. Samara, borne in loose open cymes; flat and broadly winged, that is, a conspicuous circular, reticulate-veined, membranous wing surrounds the seed cavity (hence the name wafer-ash); 2-seeded; September–October; 1.8–2.5 cm across, with odor of hops; persistent; seeds 12,000/lb (26,400/kg).

Distribution. Occasional in the southern half of lower Michigan, north along Lake Michigan to Manistee Co.; Marquette Co. in upper Michigan. County occurrence (%) by ecosystem region (Fig. 19): I, 42; II, 10; III, 0; IV, 12. Entire state 27%. Ranging widely throughout e. North America, the central Great Plains, and the Southwest.

Site-Habitat. Open to moderately shaded, circumneutral to basic, seasonally wet, sandy to alluvial sites; tolerant of calcareous substrate; sandy beaches, dunes, river floodplains, stream banks; particularly common near the Lake Michigan shore and on crests of dunes; may be perceived as tolerating dry sites, but the water table is within reach of roots. Associates include balsam poplar, willows, ground juniper, choke cherry, fragrant sumac, American bittersweet, sand cherry, pasture rose, riverbank grape.

Notes. Moderately shade-tolerant. Leaves similar to those of poison-ivy. All parts when bruised produce an odor usually regarded as disagreeable, rank, akin to that of a polecat, citruslike, or rarely pleasant. The fruit was once used as a substitute for hops in beer brewing. The hop-tree is adaptable for planting in full sun or shade. The wood is hard, close-grained, and bright yellow. *Ptelea* is the Greek name for an elm and was given to this species by Linnaeus because its fruit resembles that of an elm.

Chromosome No. $2n = 42$, $n = 21$; $x = 21$

Similar Species. In eastern North America, the hop-tree belongs to the variety *trifoliata*. A few collections from the Lake Michigan area have been referred to as the southern var. *mollis* Torr. & A. Gray, woolly common hop-tree, which has densely pubescent twigs, leaves, and inflorescences (Voss and Reznicek 2012). The counterpart in California is *Ptelea crenulate* Greene, California hop tree, and in Baja California it is *P. aptera* Parry (Bailey

1. Flowering shoot, × 1/2
2. Flower, × 2
3. Fruiting shoot, × 1/2
4. Fruit, samara, × 1
5. Winter lateral bud surrounded by leaf scar, enlarged
6. Winter twig, reduced

KEY CHARACTERS

- tall shrub or small tree, trunk short, often low-branched, crown irregular
- leaves alternate, compound, trifoliate, both sides with translucent dots, pungent or rank odor when crushed
- fruit a samara with a conspicuous circular, membranous wing surrounding the seed cavity; borne in persistent drooping clusters
- habitat river floodplains, stream banks, common on sandy, nutrient-rich sites

1962). Several other taxa are described from the Southwest (see Vines 1960; Bailey 1962 and USDA Plants, http://plants.usda.gov).

FAGACEAE
Quercus prinoides Willdenow

Dwarf Chinkapin Oak

Size and Form. Low, spreading clonal shrub, 2–4 m (6–12 ft) high; occasionally a small tree to 6 m (20 ft) high and 8–12 cm (3–5 in) in diameter. Trunk slender, usually crooked and leaning, supporting a thin, rounded, straggly crown. Typically occurring in small clonal groups. Deep-rooting. Michigan Big Tree: girth 58 cm (23 in), diameter 19 cm (7.3 in), height 14.0 m (46 ft), Berrien Co.

Bark. Smooth and thin, bitter, light brown to ashy gray; becoming roughened and flaky with age.

Leaves. Alternate, simple, blades 6–12 cm long and 4–7 cm wide; obovate to oblong lanceolate, often somewhat lance-shaped; acute; cuneate; margin wavy and coarsely toothed with 4–8 rounded teeth; 5–8 veins per side; thick and firm; somewhat lustrous green above, pale-pubescent beneath, turning reddish-brown in autumn; petioles short, 1–2 cm long.

Stems-Twigs. Slender, rusty, fine-pubescent on new growth, becoming glabrous and reddish-brown, lenticels pale, inconspicuous; leaf scars small, concave above; bundle scars numerous; pith continuous, white.

Winter Buds. Terminal bud small, 1–3 mm long, ovate, rounded at apex, light brown, glabrous; clustered at ends of twigs.

Flowers. Late May, with the leaves; bisexual; male catkins hairy, 2.5–7 cm long; female flowers solitary or in pairs, sessile, stigmas yellowish-red. Wind pollinated.

Fruit. Acorn, ripening in autumn of first season; sessile or short-stalked; cup 9–12 mm deep and 13–17 mm wide, with small, appressed, somewhat swollen scales, becoming knobby toward the base, pale woolly outside, downy inside, thin, deep, enclosing 1/3 to 1/2 of the nut; nut ovoid to ellipsoid, 1.2–2 cm long, chestnut brown, pale-downy at apex, blunt-pointed; kernel relatively sweet.

Distribution. Uncommon in the southern portion of lower Michigan. County occurrence (%) by ecosystem region (Fig. 19): I, 29; II, 0; III, 0; IV, 0. Entire state 13%. Widely distributed in the e. USA: New England w. to s.e. ON, s. to AL, MS, LA, n. to OK, e. MO, s.e. NE.

Habitat. Characteristic of open or thinly forested dry sites with calcareous soils, including sandy-gravelly hillsides, knobs, kames, and eskers. Associates include black and white oaks; pignut hickory; ground juniper; smooth, staghorn, and fragrant sumacs; gray dogwood; northern dewberry.

Notes. Shade-intolerant; slow-growing. Germination hypogeal in autumn. Develops small clones by vigorous sprouting at the root collar of living and dead stems whose roots remain alive. In addition, small suckers arise from rhizomes. Sometimes, what appears to be a single colony is clusters of root-collar sprouts, each cluster circling a core of a former tree. The leaves are notably variable in size, shape, and margins. The teeth vary from long and pointed to extremely shallow and rounded.

Chromosome No. $2n = 24$, $n = 12$; $x = 12$

Similar Species. Closely related to and often not easily distinguished from chinkapin oak, *Quercus muehlenbergii*.

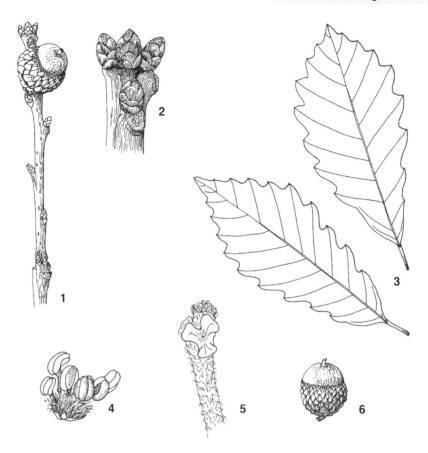

1. Winter twig, × 1
2. Terminal portion of twig, × 3 1/2
3. Leaves, × 1/2
4. Male flower, enlarged
5. Female flower on stalk, enlarged
6. Fruit, acorn, × 1

KEY CHARACTERS

- low shrub or scraggly small tree, usually forming a small clone, trunk crooked, leaning
- leaves with coarse, rounded teeth
- acorn cup thin, deep, with knobby scales toward the base
- occurs only in southern lower Michigan (Region I)

Distinguished from chinkapin oak, *Quercus muehlenbergii*, by shrub or small tree form, crooked trunk, and scraggly crown; clonal habit; smaller leaves with fewer, more rounded teeth and fewer veins; acorn cup with knobby scales toward the base.

RHAMNACEAE
Rhamnus alnifolia L'Héritier
Alder-leaved Buckthorn

Size and Form. Low, few-branched shrub 0.3–0.6 m; unarmed, clone-forming by layering and often spreading, with decumbent branches to 2 m, shallow-rooted.

Bark. Thin, gray to dark gray, smooth.

Leaves. Alternate, simple, deciduous; blades 4–10 cm long, 2–5 cm wide, ovate to elliptic; acute; cuneate; margin unevenly crenate-serrate; dark green, glabrous above, slightly pubescent on veins beneath; veins pinnate and slightly arcuate, 6–8 per side; stipules conspicuous in spring, 6–9 mm long, deciduous; petioles grooved, 0.5–1.2 cm long.

Stems-Twigs. Slender, unarmed, ridged below the nodes, red or brown, minutely downy, becoming glabrous and gray, leaf scars small, crescent-shaped; bundle scars 3; stipule scars small; pith white, continuous.

Winter Buds. Terminal bud present; buds small, not beaked, solitary, sessile, dark red-brown, 4–5 mm long; 2–4 imbricate scales exposed.

Flowers. May–June, appearing with the leaves; functionally unisexual, plants dioecious; greenish-yellow, borne in axils of lower leaves; solitary or 2–3 together; small, 2–3 mm across, pedicels slender, 4–5 mm long; 5-parted; hypanthium saucer-shaped, sepals 5, about 1 mm long; petals lacking; stamens very short in female flowers. Insect pollinated.

Fruit. Drupe; August–September, subglobose, black, 6–8 mm in diameter; pit surface roughened with 2 grooves on back; usually 3 seeds.

Distribution. Locally common throughout the state. County occurrence (%) by ecosystem region (Fig. 19): I, 68; II, 77; III, 100; IV, 88. Entire state 76%. Transcontinental range northern and boreal, n. Great Plains to BC, s. to UT and CA; in the East, s. to NC and TN.

Site-Habitat. Open to lightly shaded, wet to wet-mesic sites with slightly acidic to basic surface soil and calcareous substrates within the rooting zone; wet lake plain prairies, prairie fens, hardwood-conifer swamps, limestone bedrock glades, edges of lakes and streams. Associates include a wide array of species characteristic of ecosystems with calcareous, nutrient-rich substrates: northern white-cedar, tamarack, black ash, willows, blue-beech, shrubby cinquefoil, bog birch, poison-sumac, ninebark, meadowsweet, spicebush.

Notes. Shade-intolerant, persisting in light to moderate shade. Usually in the more open microsites of swamps caused by periodic windthrow and uprooting of trees. In Minnesota loose but continuous clones up to 15 m across have been reported (Smith 2008).

Chromosome No. $2n = 24$, $n = 12$; $x = 12$

Similar Species. A similar but tall shrub of southern and Great Plains range with definite overlapping bud scales is the lance-leaf buckthorn, *Rhamnus lanceolata* Pursh. In Japan a similar shrub is *Rhamnus ishidae* Miyabe & Kudo, a rare species of Hokkaido Island (Satake et al. 1993).

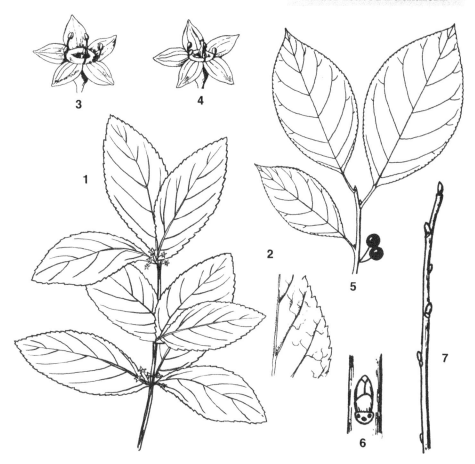

1. Flowering shoot, × 1/2
2. Leaf blade margin, × 2
3. Male flower, × 5
4. Female flower, × 5
5. Fruiting shoot, drupe, reduced
6. Winter lateral bud and leaf scar, enlarged
7. Winter twig, reduced

KEY CHARACTERS

- low, erect, few-branched shrub <1 m, often spreading clonally by layering
- stems slender, ridged below the nodes, unarmed
- leaves with blade margin crenate-serrate, veins pinnate
- fruit a drupe, subglobose, black
- habitat wet-mesic to wet sites with calcareous substrates

Distinguished from glossy buckthorn, *Frangula alnus,* by low, spreading form, green upper leaf surface, terminal bud not naked or hairy; a native species occurring in wetlands throughout the state.

RHAMNACEAE
Rhamnus cathartica Linnaeus

Common Buckthorn, European Buckthorn

Size and Form. Tall shrub or usually small tree with a single stem to 7 m. Trunk crooked, low-branching, supporting a rounded, bushy crown of crooked branches and thorn-tipped twigs. Michigan Big Tree: girth 114.3 cm (45 in), diameter 36.4 cm (14.3 in), height 26.3 m (61 ft), Washtenaw Co.

Bark. Thin, dark brown to blackish, shiny, smooth, becoming rough and scaly, bitter.

Leaves. Mostly opposite or subopposite, occasionally alternate; simple, deciduous, blades 4–9 cm long and about half as wide; elliptic to ovate, acute; rounded to subcordate; crenate-serrate, occasionally appearing entire; thin and firm; dull green above, paler and yellowish-green beneath, remaining green until shed in early winter; glabrous; usually 3 veins per side, parallel-arcuate; petioles hairy, 3–4 cm long.

Stems-Twigs. Slender, gray, glabrous, lenticels prominent, often tipped with a short, needlelike thorn as long as the end lateral buds; leaf scars crescent-shaped, bundle scars 3; pith tan in small twigs <1 cm to dark brown and spongy in twigs >1.5 cm.

Winter Buds. Terminal bud absent, replaced with a thorn; lateral buds brownish-black, appressed, with several glabrous bud scales exposed.

Flowers. May, with the leaves; functionally unisexual, plants dioecious; borne in 2- to 5-flowered sessile umbels in the axils of lower leaves of the current season's shoot; small, inconspicuous; sepals 4, 2–3 mm long, petals 4, erect, lanceolate, 0.6 mm in female flowers and 1–1.3 mm in male flowers; yellowish-green; fragrant. Insect pollinated.

Fruit. Drupe; August–September, black, round, about 6 mm across; nauseating and reported to be poisonous to some people; contains usually 4 nutlike pits, each enclosing a seed; enormously prolific in seed production annually in open or lightly shaded sites.

Distribution. Native in Europe and western and northern Asia; naturalized in eastern North American from initial cultivation as an ornamental plant. Common to locally abundant in southern lower Michigan, rare in northern lower and eastern upper Michigan, infrequent in western upper Michigan. County occurrence (%) by ecosystem region (Fig. 19):

I, 45; II, 7; III, 29; IV, 50. Entire state 30%. Widely distributed in e. North America except the deep South; also occurs in Great Plains, Rocky Mountains, CA.

Site-Habitat. Establishing and persisting on any disturbed and open to moderately shaded upland site and many wetlands, including periodically flooded bottomlands and swamps. Characteristic of urban and suburban environments, sometimes cultivated but also vigorously colonizing homesites, adjacent lots, old fields, roadsides, fencerows and cutover forest understories.

Notes. Shade-tolerant as a juvenile, becoming moderately shade-tolerant; moderately fast-growing; short-lived. Unusual for its variable leaf arrangement. A tough, persistent shrub or small tree adapted to urban, suburban, and nearby disturbed lands. Establishing nearly everywhere that birds disperse seeds. Occurrence and abundance diminish with distance from its seed source. Widely regarded as a major invasive species whose shade prevents the establishment of native trees and shrubs. Difficult to eradicate by any means because it sprouts vigorously from the root collar following cutting, girdling, burning, or poisoning. Only repeated treatment reduces its occurrence. The genera *Frangula* and *Rhamnus* of the Rhamnaceae are closely related; see Voss and Reznicek 2012 for distinguishing features.

Chromosome No. $2n = 24$, $n = 12$; $x = 12$

Similar Species. Similar native species of primarily tree habit include Carolina buckthorn, *Frangula caroliniana* Walter, of the South and the cascara buckthorn, *F. purshiana* DC., of western North America. Known as "the all-American laxative," the bark of cascara (containing hydroxymethyl-anthraquinones)

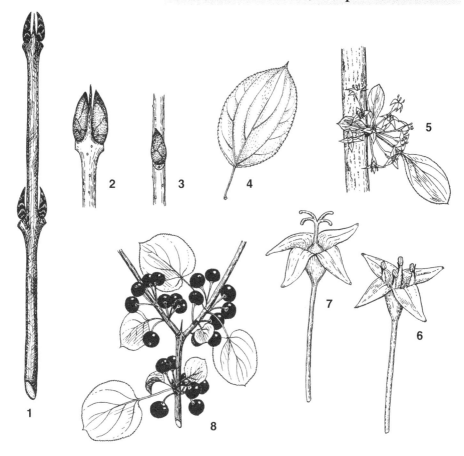

1. Winter twig, × 1 1/2
2. Terminal portion of twig, enlarged
3. Axillary bud, × 1 1/2
4. Leaf, × 1/2
5. Male flowering shoot, × 1 1/2

6. Male flower, enlarged
7. Female flower, enlarged
8. Fruit (drupe) in umbels in axils of leaves, × 1/2

KEY CHARACTERS

- tall shrub or usually small tree to 7 m
- leaves opposite or subopposite, dark green, with arcuate venation, remaining green into winter
- twigs with stout, needlelike thorns instead of terminal buds
- fruit a black, round drupe, borne in umbels; nauseating
- occurrence primarily in urban and suburban environments and adjacent disturbed areas of southern lower Michigan

was boiled and used as a strong laxative by native peoples of the Pacific Northwest. The similar East Asian species *Rhamnus utilis* Decne. is known from 3 counties in southern Michigan (Voss and Reznicek 2012). Other similar species in China and Japan include *R. davurica* Pall.; in Japan *R. japonica* Maxim. Common buckthorn is also related to the Eurasian glossy buckthorn, *Frangula alnus* Miller, another major invasive species in southern Michigan.

ERICACEAE
Rhododendron groenlandicum (Oeder) Kron & Judd
Labrador-Tea

Size and Form. Upright, straggly, evergreen shrub to 1 m, often forming dense clonal patches by layering and rhizomes.

Leaves. Alternate, simple, evergreen, sessile or subsessile, crowded toward the ends of the shoots, pointing downward; blades 2–5 cm long and about 1/4 as wide, thick and leathery; oblong to linear-oblong; obtuse; cuneate; entire, strongly revolute; bright green and finely rugose above, midrib depressed and finely hairy, white the 1st year, later densely rusty-brown, woolly beneath; fragrant when crushed; petioles short, 2 mm long.

Stems-Twigs. Moderately slender, brittle, rounded, brown-woolly; leaf scars half elliptic or cordate; bundle scar 1; pith small, somewhat 3-sided, brownish, spongy.

Winter Buds. Terminal flower buds large, round or ovoid, with about 10 broad, glandular-dotted outer scales; lateral vegetative buds small, solitary, sessile, ovoid, somewhat compressed, with about 3 outer scales.

Flowers. May–June; bisexual; borne on slender, puberulent, threadlike pedicels 1.5–2 cm long in umbellike terminal clusters; small, 6–7 mm across; white; calyx very small, 5-toothed; petals 5, distinct, narrowly obovate, about 5 mm long, spreading; stamens 5–7; style persistent. Insect pollinated.

Fruit. Capsule; August–September; narrowly oblong, 5–6 mm long, hairlike style persistent, 5-parted, splitting from the base; many-seeded, seeds threadlike; persisting for several years, the pedicels gradually curving downward.

Distribution. Common to locally abundant in northern lower and upper Michigan, rare or absent in southern lower Michigan. County occurrence (%) by ecosystem region (Fig. 19): I, 11; II, 77; III, 100; IV, 100. Entire state 51%. A circumboreal species; it occurs in Greenland and thus the name *groenlandicum*.

Site-Habitat. Wet, open, acidic, nutrient-poor sites; bogs, boggy lake and swamp margins, peat mounds in fens, acidic conifer swamps and interdunal swales; "on shaded sandy bluffs along Lake Superior, in rock crevices on Isle Royale" (Voss and Reznicek 2012). Associates include tamarack, white and black spruces, bog-laurel, leatherleaf, blueberries, mountain holly, bog-rosemary, and many other species of acidic sites.

Notes. Shade-intolerant, persisting in light shade. Ideally suited to its northern and boreal occurrence; a thick woolly cover protects its stems and the lower surface of the leaves, which helps prevent loss of water via evapo-transpiration in cold and dry weather and oxygen in wet soils. The common name tea is derived from the use of the leaves. Fresh or dried leaves can be boiled to make an aromatic tea that was a favorite among native peoples and early settlers of eastern and western North America. Tea should be boiled well before drinking and consumed in moderation to avoid drowsiness. Excessive doses are reported to act as a strong diuretic and cathartic and to cause intestinal disturbances (Pojar and MacKinnon 1994). The medicinal uses and properties are described by Marles et al. (2000). In older literature its name was *Ledum groenlandicum* Oeder.

Chromosome No. $2n = 26$, $n = 13$; $x = 13$

Similar Species. Trapper's tea, *Rhododendron neoglandulosum* Harmaja (formerly *Ledum groenlandicum* Nutt.), occurring in the Cascade Range of Washington, Oregon, and British Columbia and the mountains of AK to the southern Rocky Mountains, is distinguished primarily by its whitish-hairy leaf undersides. In northern Europe and Asia, wild rosemary, *R. tomentosum* Harmaja (formerly *L. palustre* L.), is a small upright shrub to 1 m. The marsh or northern Labrador-tea, *R. tomentosum* ssp. *decumbens* Elven & D. F. Murray (formerly *L. palustre* L. ssp. *decumbens* [Ait.] Hultén), is of smaller stature and occurs in the northern parts of the circumpolar boreal forest, subarctic, and arctic tundra (Merles et al. 2000).

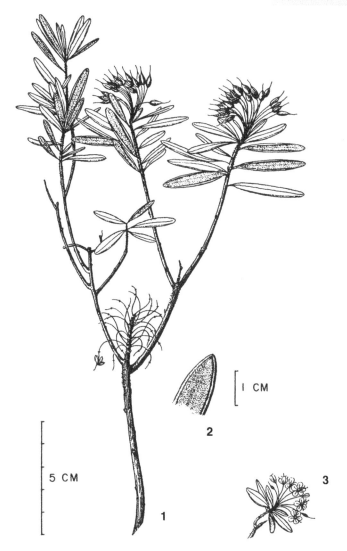

1. Individual with vegetative and fruiting shoots, capsule
2. Lower leaf blade surface
3. Flowering shoot, reduced

KEY CHARACTERS

- upright, straggly, evergreen shrub to about 1 m
- leaves evergreen, crowded toward the ends of shoots, thick and leathery, oblong to linear-oblong, entire, strongly revolute, bright green above, at first white beneath, becoming densely rusty-brown and woolly; fragrant when crushed
- stems brown and woolly, pith brownish
- flowers in terminal clusters, small, white
- fruit a capsule, persisting for several years, pedicels curving downward
- habitat open, wet, acidic bogs, swamps, and swales of northern lower and upper Michigan

THE SUMACS—*RHUS*

Worldwide, there are approximately 100 species of *Rhus*, mostly in Asia and southern Africa. Four shrub species of *Rhus* are native in Michigan, and the 2 most common reach small tree size, up to 10 m: staghorn and smooth sumac (*R. typhina* and *R. glabra*, respectively). Shining sumac, *R. copallina*, is typically a medium to tall shrub with a characteristic "winged" leaf rachis. It has a related counterpart in East Asia, *R. chinensis*. Fragrant sumac, *R. aromatica*, is a shrub found in dry, open, nutrient-rich sites of lower Michigan; *R. trilobata* is its counterpart species in western North America.

The genus *Rhus* is a member of the sumac (sumach) and poison-ivy family: Anacardiaceae. It encompasses some 70 genera and 600 species, mostly confined to the warmer parts of the Earth, especially mediterranean and tropical regions. This important family includes trees that produce the mango fruit and the cashew and pistachio nuts. Also notable in the family is the varnish-tree of China, Japan, and the Himalayas, *Toxicodendron vernicifluum*, whose sap makes a black lacquer used to coat fine furniture. This brilliant paintlike coating of oriental furniture was most fashionable in the late seventeenth and early eighteenth centuries.

Species of the Anacardiaceae have leaves that are usually alternate and pinnately compound but sometimes trifoliate or unifoliate. Vertical resin canals are well developed in the inner bark and associated with the larger leaf veins and often also leaf tissues. Flowers are almost always unisexual (plants are usually dioecious); the fruit is typically a flattened asymmetrical drupe. Note that we have included the genus *Toxicodendron* as part of this introduction to the sumac group. *Toxicodendron* was once included as a subgenus of *Rhus*, but today it is maintained as a separate genus with 30 species worldwide. *Toxicodendron* distinguishes the group of species in North America and Asia that have axillary inflorescences and glabrous, greenish-whitish fruit compared to species of *Rhus* with terminal inflorescences and reddish, glandular-pubescent fruit.

The resins of *Rhus* are not poisonous, whereas those of *Toxicodendron* cause mild to severe dermatitis. In Michigan, this includes poison-sumac (*Toxicodendron vernix*) and poison-ivy (*Toxicodendron radicans* and *T. rydbergii*), whereas none of the shrub sumacs of North America—cause dermatitis. Occurring in 99% of Michigan counties (lacking in Sanilac Co.), poison-ivy, in its vine (*T. radicans*) and shrub (*T. rydbergii*) habits, is the most widely reported woody plant organism in Michigan.

The rash from both poison-sumac and poison-ivy is easily avoided with an ounce of prevention. Be sure to wear long pants and a long-sleeved shirt— no sandals! Bring along a bottle of rubbing alcohol (isopropyl alcohol) on field trips where they are likely to occur. During or after the trip, generously wash with your hands and other possibly affected areas with rubbing alcohol.

It effectively removes the phenolic resin, 3-n-pentadecycatechol, that causes the rash.

Medium-height shrubs, aromatic and winged sumac, are best distinguished during the growing season by their clonal form and distinctive compound leaves. The taller staghorn and smooth sumacs are easily identified by their typical growth in compact clones of multiple stems, large, compound leaves, thick twigs with large brown pith, and antlerlike fruit clusters. The individual stems are repeatedly forked, giving a much-branched or "twiggy" appearance. This distinctive form is caused by the terminal position of the inflorescence and therefore lack of a terminal bud (Gilbert 1966). New extension growth the following year is therefore from one of the lateral buds below the inflorescence. At almost all times of the year, the upright, dense flowering or fruiting panicles are easily spotted. The antlerlike form of both species is a distinguishing feature, although that of staghorn sumac is distinctly more erect.

Another characteristic of many genera of the Anacardiaceae is the brilliant coloration of leaves or fruits in autumn. This trait is exemplified in all Michigan shrub or small-tree species of *Rhus* and *Toxicodendron*. Their leaves turn brilliant red and orange hues. Because all leaves of all stems in a clone turn the same color, the clones are easily identified by a distinctive color. The sumacs spread clonally by sprouts that arise from shallow roots, and a compact patch or thicket is typically formed.

Sumac species are very shade-intolerant and require full sunlight for their development and clonal spread. Therefore, they are characteristic of open, disturbed, well-drained sites that afford little competition, especially roadsides, old fields, open hillsides, and forest edges.

KEY TO SPECIES OF *RHUS*

1. Low to medium, erect or spreading shrub to 2 m; stems, buds, and leaves aromatic when crushed; leaflets usually 3, crenate-serrate . . . *R. aromatica,* p. 212
1. Medium erect shrub to 2 m or small tree >5 m; stems, buds, and leaves not aromatic when crushed; leaflets many (7–31), serrate or entire
    2. Medium shrub or small tree; leaf rachis winged between leaflets; leaflets entire or few-toothed . . . *R. copallina,* p. 214
    2. Tall shrub or small tree; leaf rachis not winged; leaflets serrate
        3. Young stems and current shoots glabrous; leaflets glabrous beneath (or with a few hairs); petioles and winter twigs glabrous or nearly so; hairs on fruit less than 0.5 mm long; panicle of mature fruit ± drooping . . . *R. glabra,* p. 216
        3. Young stems and current shoots densely pubescent; leaflets pubescent beneath, at least on veins; petioles and winter twigs densely soft-pubescent, with long hairs; hairs on fruit mostly 1–2 mm long; panicle of mature fruit ± erect . . . *R. typhina,* p. 218

ANACARDIACEAE
Rhus aromatica Aiton
Fragrant Sumac

Size and Form. Medium, erect, spreading clonal shrub, to 2 m; many branches spontaneously sprout from root collar to form a shrub of irregular habit.

Bark. Thin, smooth, brownish to gray.

Leaves. Alternate, deciduous, compound, 5–12 cm long; leaflets usually 3, sessile, 2.5–7 cm long; ovate, terminal sometimes 3-lobed or with 3 large teeth; coarsely and irregularly toothed above the middle, with rounded or abruptly pointed teeth; thin and soft; green, becoming orange to red in autumn; pubescent both sides, becoming glabrate; aromatic when crushed; petioles slender, 1–3 cm long.

Stems-Twigs. Slender, glabrate, unarmed, brown to reddish-brown, dotted with small, brown, slightly elongate lenticels; aromatic when bruised; in autumn and winter clusters of 2–3, dormant male flowers occur in conspicuous spikes, 0.6–1 cm long, at the end of lateral and terminal twigs; leaf scars raised, circular; bundle scars 5, 7, or 9; pith light brown to orange-brown, continuous.

Winter Buds. Terminal bud absent, lateral buds small, about 1 mm long, yellowish-hairy, ciliate, embedded in the persistent base of the petiole, which is joined to the upper edge of the leaf scar and appears as the upper ridge or "flap" of leaf-scar tissue.

Flowers. April–May, before the leaves; usually unisexual, plants dioecious; yellow; male flowers borne in scaly spikes, 0.6–1 cm long, that are formed in summer of previous year and conspicuous at ends of twigs in autumn and winter; female flowers borne in dense, spikelike clusters from axillary buds in late autumn. Insect pollinated.

Fruit. Drupe; July–August; persistent; produced sparingly in dense, compact clusters, subglobose, reddish, hairy, small, 6–8 mm across; 1 bony pit.

Distribution. Frequent in the southern part of the Lower Peninsula; infrequent northward; rare to absent in the Upper Peninsula. County occurrence (%) by ecosystem region (Fig. 19): I, 37; II, 13; III, 14; IV, 0. Entire state 23%. Widely distributed in e. North America, QC and ON to FL, w. to TX, n. to SD and MN.

Site-Habitat. Open sites in a variety of habitats; most vigorous on dry to dry-mesic sites and basic conditions; sand dunes and limestone flats along lakeshores. Due to widespread bird dispersal, it is found on diverse sites where lack of competition allows establishment. Associates include black oak, white oak, dwarf chestnut oak, pignut hickory, shagbark hickory, white ash, black cherry, sassafras, choke cherry, New Jersey tea, downy arrow-wood.

Notes. Shade-intolerant; relatively fast-growing; forming dense clones on open-disturbed sites. Sprouting vigorously following fire. Tolerant of drought and calcareous soil. A good indicator of dry-mesic to dry sites. Leaves similar to those of poison-ivy, *Toxicodendron radicans*, but distinguished by smaller, prominently crenate-serrate leaflets; terminal leaflet not stalked; fruits pubescent and reddish, not white and glabrous.

Chromosome No. $2n = 30$, $n = 15$; $x = 15$

Similar Species. Squawbush, *Rhus trilobata* Nutt., is similar in habit, leaves, fruit, and all parts being aromatic or ill-scented when crushed. It grows in open sites in foothills and along streams in the Rocky Mountains from Alberta to New Mexico; known as skunkbush in California; habit more upright and flowers less conspicuous than those of *R. aromatica*.

1. Leaves, × 1/2
2. Male flowering shoot, reduced
3. Male flower, × 5
4. Female flowering shoot, × 1/2
5. Female flower, × 5
6. Fruiting shoot, drupe, × 1/2
7. Male flowering shoot in winter, × 1
8. Winter side view, persistent base of petiole and lateral bud, enlarged
9. Winter front view, persistent base of petiole with leaf scar, enlarged
10. Winter twig, reduced

KEY CHARACTERS

- medium, erect, and spreading clonal shrub to 2 m
- all parts aromatic
- leaves compound, trifoliate, turning orange or red in autumn
- twigs of male clones with conspicuous terminal clusters of dormant male catkins in autumn and winter, lateral buds embedded in leaf scar, appearing absent
- fruit red, hairy drupes in compact clusters

ANACARDIACEAE
Rhus copallina Linnaeus
Shining Sumac, Winged Sumac

Size and Form. Low to medium shrub to 2 m, rarely a small tree to 10 m; similar to smooth and staghorn sumacs in form and clonal habit but usually smaller. Michigan Big Tree: girth 50.8 cm (20 in), diameter 16.2 cm, (6.4 in), height 10.1 m (33 ft), Kalamazoo Co.

Bark. Thin, smooth, greenish-brown to grayish-brown.

Leaves. Alternate, deciduous, pinnately compound, 15–40 cm long; rachis pubescent, more or less winged between each pair of leaflets, wings 1–5 mm wide; leaflets 7–21, laterals sessile, 4–8 cm long; 1–3 cm wide; ovate-lanceolate; acute or acuminate; terminal cuneate, laterals more rounded; inequilateral; entire or few-toothed near the apex; very lustrous, dark green, glabrous above, usually pubescent and paler beneath, turning red, crimson, scarlet, or reddish-purple in autumn; petioles 3–8 cm long, more or less pubescent, wingless.

Stems-Twigs. Slender to moderately stout, reddish-brown, downy-pubescent, older, larger twigs becoming glabrous; lenticels prominent, raised, small and vertical on current shoots, narrow and horizontal on older-larger twigs, orange-red; sap watery and resinous with odor resembling that of turpentine; leaf scars U-shaped, concave; bundle scars 7; pith brown, continuous.

Winter Buds. Terminal bud absent; lateral buds small, globose, 1–2 mm long and equally wide, light brown to reddish-brown, pubescent.

Flowers. June–July; usually unisexual, plants dioecious; tiny, about 4 mm across, in elongate terminal panicles, greenish-yellow. Insect pollinated.

Fruit. Drupe; September; borne in dense erect or drooping panicles, subglobose, dark red, glandular-hairy, 3–4 mm across, persistent in winter; 1 bony pit; seeds 57,000/lb (125,400/kg).

Distribution. Frequent in the southern half of the Lower Peninsula north to Benzie and Grand Traverse Co.; absent in the Upper Peninsula. County occurrence (%) by ecosystem region (Fig. 19): I, 55; II, 27; III, 0; IV, 0. Entire state 35%. Widely distributed in e. North America, ON to FL, w. to TX, n. to NE.

Site-Habitat. Open, disturbed, dry to dry-mesic, slightly to moderately acid, and seasonally wet sites; common on beach ridges of lake plains in southeastern lower Michigan. Associated with scarlet, swamp white, white, and pin oaks; sassafras; black cherry; witch-hazel; low sweet blueberry; black huckleberry; gray dogwood; meadowsweet; northern dewberry; pasture rose.

Notes. Shade-intolerant, persisting in light shade. Sprouting from a wide-spreading root system following fire. Desirable ornamentally because of its very lustrous foliage and autumn colors but rarely cultivated. Species name *copallina* means "lustrous upper-blade surface" as in copal gum or resin.

Similar Species. Little-leaf sumac, *Rhus microphylla* Engelm., is a clonal shrub of dry rocky hillsides and gravelly mesas from 610 to 1,830 m in Texas, New Mexico, and Arizona and widespread in Mexico. Closely related are *Rhus javanica* L. in Japan and *R. chinensis* Mill., *R. wilsonii* Hemsl., and *R. teniana* Hand.-Mazz. in China.

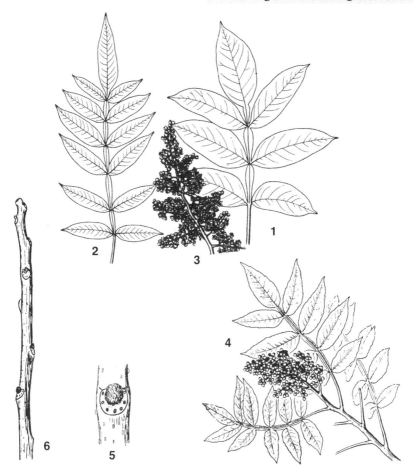

1. Leaf with 7 leaflets, × 1/4
2. Leaf with 13 leaflets, × 1/4
3. Flowering shoot, × 1/2
4. Fruiting shoot, drupe, reduced
5. Winter lateral bud and leaf scar, enlarged
6. Winter twig, × 1/2

KEY CHARACTERS

- low to medium clonal shrub or small tree to 2 m
- leaves compound with 7–12 lustrous leaflets, leaf rachis winged between leaflets
- twigs stout, pubescent-glabrous, with U-shaped leaf scars
- fruit a dark red, glandular hairy drupe, borne in dense terminal panicles

ANACARDIACEAE
Rhus glabra Linnaeus
Smooth Sumac

Size and Form. Tall clonal shrub or small tree to 7 m; branches few, crooked; spouting from a shallow, wide-spreading root system to form an open or dense clone to >4 m across; crowns of many stems often merge, giving a rounded or flat-topped appearance. Clonal habit described by Gilbert (1966). Michigan Big Tree: girth 33.0 cm (13 in), diameter 10.5 cm (4.1 in), height 5.5 m (18 ft), Hillsdale Co.

Bark. Thin, smooth, brownish to gray.

Leaves. Alternate, pinnately compound, deciduous, 10–50 cm long; leaflets 11–31 laterals subsessile, middle leaflets larger than those at both ends, 5–12 cm long, 1.5–3 cm wide; lanceolate to oblong, acuminate, rounded; serrate to rather remotely serrate; bright green above, paler to conspicuously white beneath, turning a striking orange, scarlet, crimson, or purple in autumn; glabrous; petioles glabrous or nearly so, enlarged at the base and surrounding and enclosing the bud, 3–10 cm long.

Stems-Twigs. Very stout, forking repeatedly, somewhat 3-sided below the nodes, glabrous, light brown to reddish-brown, glaucous, often with waxy bloom; exuding a milky yellowish sap when cut; leaf scars horseshoe-shaped; bundle scars many; pith thick, yellowish-brown, continuous.

Winter Buds. Terminal bud absent; lateral buds ovoid, 5–7 mm long, light brown; leaf scars horseshoe-shaped, almost completely encircling the bud, 3 groups of bundle scars.

Flowers. June, after the leaves; usually unisexual, plants dioecious; borne in large, erect terminal panicles, 10–25 cm long; similar to those of a staghorn sumac except for glabrous stem and branchlets of panicles. Insect pollinated.

Fruit. Drupe; September–October, 3–4 mm across, covered with dense, sticky, short hairs 0.2 mm long, borne densely clustered in large terminal or axillary panicles, which become loose and drooping by spring; juicy, sour-lemony to taste, single-seeded; seeds 49,000/lb (107,800/kg).

Distribution. Abundant in the southern half of southern lower Michigan; common in lake-moderated counties in northern lower Michigan; absent in eastern upper Michigan; occasional in western upper Michigan. County occurrence (%) by ecosystem region (Fig.

19): I, 50; II, 27; III, 0; IV, 62. Entire state 39%. Distributed throughout the conterminous USA and adjacent Canada, QC to BC.

Site-Habitat. Similar to that of *Rhus typhina*. For pre-European settlement communities and associates, see Kost et al. 2007.

Notes. Very shade-intolerant; early successional; fast-growing; ramets of clones are short-lived (<25 years), but clones may be long-lived if regenerated by disturbance. Gilbert (1966) found clones in southeastern Michigan ranging in size up to 22 ×40 m (880 m2). Tolerant of droughty soils and those with calcareous substrates. Sprouting from roots following fire. Good for mass plantings in dry, infertile soils and roadsides. Rich in tannic acid and formerly used for tanning leather. Leaves and leaflets variable among clones; generally similar to *Rhus typhina* but overall with leaflets less sharply serrate. Hybridizes readily with *R. typhina* to form *R.* × *pulvinata* Greene. They are commonly found wherever the plants occur together. Hybrids may be recognized by their finely hairy twigs and the intermediate nature of the hairs on the fruit. Four morphological variants are described by Voss and Reznicek (2012).

Chromosome No. $2n = 30$, $n = 15$; $x = 15$

Similar Species. Closely related to staghorn sumac, *Rhus typhina* L. A related pubescent dwarf clonal shrub is Michaux's sumac or false poison sumac, *Rhus michauxii* Sargent, an endangered species of sandy or rocky woods of the piedmont and coastal plain of Virginia, North Carolina, South Carolina, Georgia, and northern Florida.

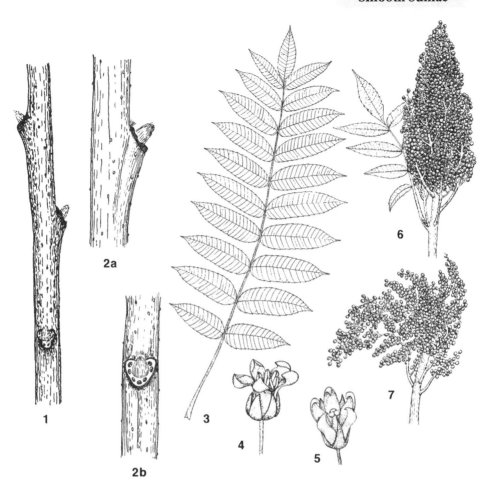

1. Winter twig, × 3/4
2a, 2b. Twig, enlarged, × 1 1/2
3. Leaf, × 1/4
4. Male flower, × 4
5. Female flower, × 4
6. Fruiting body (infructescence) in fall, × 1/4
7. Fruiting body (infructescence) in early spring, × 1/4
8. Fruit, drupe, × 2 1/2

KEY CHARACTERS
- tall shrub or small tree to 7 m
- leaves pinnately compound, 11–31 leaflets, glabrous, turning a striking orange, scarlet, crimson, or purple in autumn
- twigs stout, glabrous, light brown to reddish-brown, glaucous, often with waxy bloom
- fruit a small hairy drupe, borne in large, dense terminal panicles, erect in autumn, becoming loose and drooping in spring

ANACARDIACEAE
Rhus typhina Linnaeus
Staghorn Sumac

Size and Form. Tall clonal shrub or small tree to 9 m; branches few, crooked; sprouting from a shallow, wide-spreading root system from an open or dense clone; often crowns of many stems merge, giving a dense, rounded, or flat-topped appearance. Clonal habit described by Gilbert (1966). Michigan Big Tree: girth 101.6 cm (40 in), diameter 32.3 cm (12.7 in), height 7.9 m (26 ft), Oakland Co.

Bark. Thin, gray to dark brown, smooth and occasionally separating into small, square scales; lenticels horizontal, cherrylike, showing a rich reddish-brown color.

Leaves. Alternate, pinnately compound, deciduous, 20–60 cm long; leaflets 11–31, laterals nearly sessile or short-stalked, opposite or lower leaflets slightly alternate, middle 3–4 pairs considerably longer than those at both ends, 5–12 cm long, 1.5–3.5 cm wide; lanceolate to oblong or rarely lanceolate long-pointed; rounded or subcordate; sharply serrate to rather remotely serrate; covered above at first with red caudate hairs, bright yellowish-green until half grown, dark green above at maturity, pale or often nearly white below, glabrous with the exception of short, fine hairs on the underside of the stout midrib, primary veins forked near the margins; turning bright scarlet with shades of crimson, purple, and orange in autumn; petioles stout, covered with soft, dense pubescence, including many long (mostly straight 1–2 mm) hairs, enlarged at base and surrounding and enclosing the bud, 3–10 long.

Stems-Twigs. Very stout, forking repeatedly, brittle, rounded; brown to reddish, densely pubescent, with long hairs suggesting a young stag's velvety antlers, exuding a milky yellowish sap when cut; leaf scars large, horseshoe- or C-shaped, nearly enclosing the bud; bundle scars many; pith large, yellowish-brown, aromatic, continuous.

Winter Buds. Terminal bud absent; lateral buds small, conic, about 6 mm long, brown; velvety-pubescent, with long, silky, pale brown hairs, appearing naked; leaf scars not elevated, somewhat C-shaped and almost encircling the bud, 3 groups of bundle scars.

Flowers. June, after the leaves, opening gradually and in succession, the female a week or 10 days later than the male; usually unisexual, plants dioecious; borne on slender pedicels from the axils of small, acute pubescent bracts, in large terminal panicles, with pubescent stem and branchlets, 20–30 cm long and 12–15 cm across, with wide-spreading branches, male panicle nearly 1/3 larger than the more compact female; calyx lobes acute, covered on the outer surface with long, slender hairs, much shorter than the petals in the male flower and nearly as long in the female flower; petals of male flower yellowish-green, strap-shaped, apex rounded; petals of female flower green, narrow, and acuminate; disk bright red and conspicuous; stamens slightly exserted, with slender filaments and large, bright orange anthers; ovary ovoid, pubescent, the 3 short styles slightly united at base, with large capitate stigmas, in the male flower glabrous, much smaller, usually rudimentary. Insect pollinated.

Fruit. Drupe; September-October; 3–4 mm across, covered with long crimson hairs; juicy, sour-lemony to taste; densely clustered in large, terminal, cone-shaped panicles, persistent most of the winter and more erect than in smooth sumac; single-seeded; seeds 53,300/ lb (117,260/kg).

Distribution. Common to abundant in lower Michigan; locally common in upper Michigan. County occurrence (%) by ecosystem region (Fig. 19): I, 89; II, 57; III, 71; IV, 88. Entire state 76%. Widely distributed in e. North America, NS to ON, s. to Gulf states; KS, SD.

Site-Habitat. Open, upland sites of diverse landforms and soils; roadsides, old fields, fencerows, swamp margins, large forest openings. For occurrence in near-natural communities and associates, see Kost et al. 2007.

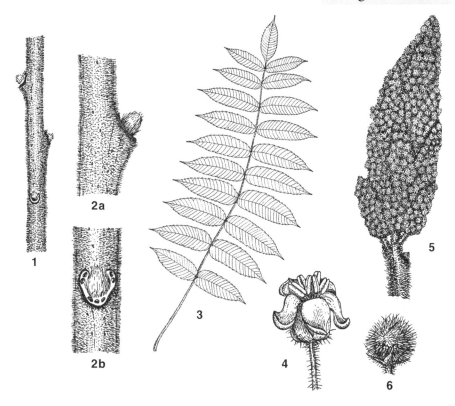

1. Winter twig, × 1/2
2a, 2b. Twig enlarged, × 1 1/2
3. Leaf, × 1/4
4. Male flower, × 8

5. Fruiting body (infructescence),
   fall, × 1/3
6. Fruit, drupe, × 5

KEY CHARACTERS

- tall, many-stemmed clonal shrub or small tree to 9 m
- leaves pinnately compound, 1–31 leaflets, leaves pubescent on leaflet midrib and main veins beneath; petioles and rachis conspicuously pubescent, turning bright scarlet with shades of crimson, purple, and orange in autumn
- twigs stout, densely velvety-pubescent, with long hairs
- the dense, erect cluster of drupes suggests a young stag's velvety antlers; persistent in winter

Notes. Very shade-intolerant; early successional; fast-growing; ramets of clones are short-lived, but clones may be long-lived if regenerated by disturbance. Leaves flush late in spring and are shed in early fall. The habit of many-branched stem covered with velvety pubescence often appearing as a stag's antlers in velvet. Tolerant of droughty soils and those with calcareous substrates. Sprouting from roots following fire. Good for mass plantings in dry, infertile soils, roadsides. Hybridizes readily with *R. glabra* to form *R.* × *pulvinata* Greene. See "Notes" under *R. glabra* for further information.

Chromosome No. $2n = 30$, $n = 15$; $x = 15$

Similar Species. Similar to *Rhus glabra*. A related pubescent dwarf clonal shrub is Michaux's sumac or false poison sumac, *Rhus michauxii* Sargent, an endangered species of sandy or rocky woods of the piedmont and coastal plain of Virginia, North Carolina, South Carolina, Georgia, and northern Florida.

THE CURRANTS AND GOOSEBERRIES—*RIBES*

Species of *Ribes*, about 160, are widely distributed in temperate and boreal landscapes of the Northern Hemisphere and in the Andes south to Patagonia. Phylogenetically based biogeographic studies are available in Schultheis and Donoghue 2004. Ten species are native in Michigan. As erect or low creeping-sprawling, mainly deciduous shrubs, they occur from low elevations to alpine sites and from inundated floodplains and swamps to dry uplands. Two groups comprise *Ribes*, the currants (subg. *Ribes*) and gooseberries (subg. *Grossularia*). Five gooseberry and 3 currant species are native in Michigan; all have leaves that are alternate, simple, and deciduous. Year round the 2 groups are probably most easily distinguished by the presence (gooseberries) or absence (currants) of armament prickles and spines. A notable exception in the upper Great Lakes Region and adjacent Canada is the bristly black currant, *Ribes lacustre*. Of adaptive significance, young stems, in the juvenile phase of physiological development, are markedly more prickly/spiny than adult-phase stems, which are typically out of the reach of small mammals. Furthermore, the native landscape home of currants is usually floodplains, swamps, and other wetlands where historically animal concentrations may have been lower than in uplands. The contrast between the very prickly, prickly gooseberry, *Ribes cynosbati*, of upland sites and the unarmed wild black currant, *R. americanum*, native in wetlands, is particularly instructive. However, the spread of *Ribes* species to adjacent open or cutover forests due to massive disturbances in modern landscapes and the wide dissemination of germinable seeds by birds have blurred these distinctive site differences.

The flowers have a conspicuous floral tube extending well above the inferior ovary. Major floral distinctions between the two groups are whether flowers are solitary or corymblike clusters of 2–3 as in gooseberries or borne in many-flowered clusters as in currants and whether the flower/fruit pedicel is not jointed at the base (gooseberries) or jointed (currants). European species (*Ribes nigrum, R. rubrum*) and varieties were once widely planted for their attractive flowers and edible fruit.

Seedlings of white pine infected with the white pine blister rust fungus, *Cronartium ribicola*, were introduced from European tree nurseries to American tree nurseries on both the East and West Coasts in the first decade of the twentieth century. Because *Ribes* species are the alternate host for the fungus, the disease spread in epidemic proportions in populations of eastern and western white pines (*Pinus strobus, P. monticola*) and other five-needled white pines. Because the fungus requires *Ribes*, as well as white pines, to persist, the planting of any *Ribes* species in the range of these pines is strongly discouraged. Enormous amounts of time and money were devoted to eradicating *Ribes* bushes in the East and West by pulling them out of the ground or chemical spraying—both of which were ineffective because of the sprouting ability of the species and the fact that the fungal spores that infect pines had traveled much farther than once thought. In the West, a blister-rust-resistant strain of western white pine has been successfully developed (Mahalovich 2010)

KEY TO SPECIES OF *RIBES*

1. Stems unarmed.
    2. Stems erect to 1.5 m; leaves resin-dotted above and beneath; fruit a black berry . . . *R. americanum*, p. 222
    2. Stems low, creeping, straggling, or ascending to <1 m; leaves not resin-dotted; fruit a red berry . . . *R. triste*, p. 228
1. Stems armed with persistent prickles and spines.
    3. Stems densely prickly, the nodal spines about the same length as internodal prickles; stems and leaves with unpleasant odor when bruised or crushed; flowers in racemes of usually 5 or more; fruit purplish-black at maturity; wet to wet-mesic habitats . . . *R. lacustre*, p. 226
    3. Middle and upper stems usually without prickles between nodes; stems and leaves lacking unpleasant odor when bruised or crushed; flowers solitary or in corymblike clusters of usually 2–3; fruit reddish-purple at maturity; usually upland habitats . . . *R. cynosbati*, p. 224

GROSSULARIACEAE
Ribes americanum Miller

Wild Black Currant

Size and Form. Low, erect, unarmed shrub, 1–1.5 m; branches few, slender and spreading; roots shallow, forming small clones by tip rooting and rhizomes.

Bark. Thin, smooth, dark brown.

Leaves. Alternate, simple, deciduous, borne several in a cluster on short shoots or separately with distinct internode elongation (0.5–3 cm) on long shoots; blades thin, 3–8 cm long and equally wide, blades of indeterminate long shoots 2 to 3 times larger than leaves of short-shoots; broadly ovate, palmately 3-lobed, sometimes 5-lobed when the lower lobes are deeply cleft, lobes triangular to ovate; acute; subcordate to nearly truncate; margin dentate to doubly serrate; glabrous, yellowish-green above, pubescent beneath, yellow resin-dotted above and beneath; petioles pubescent, often glandular-dotted, 1–3 cm long on short-shoot leaves and 4–8 cm long on vigorous indeterminate long-shoots.

Stems-Twigs. Both long- and short-shoots present. Long-shoots slim, rounded, with 4-sided ridges extending downward from the nodes, pale, yellowish-green when young, becoming dark brown; glabrate, with large, conspicuous resin glands; short current shoots, with typically 2–3 small leaves, give rise to virtually all flower-bearing peduncles; short-shoots rarely older than 5 years; leaf scars broad, triangular, bundle scars 3; pith relatively large, pale, round, continuous.

Winter Buds. Terminal bud present, 2–4 mm long, conical, pubescent; lateral buds 1–3 mm long, ovoid to conical; exposed bud scales keeled with occasional hairs and conspicuous sessile resin glands.

Flowers. April–May; bisexual; many flowers borne in axillary or terminal racemes on leafy short-shoots, drooping, 4–8 cm long, peduncles pubescent; flowers greenish to yellowish-white; 8–10 mm in diameter; about 1 cm long; pedicels downy, bracts longer then pedicels, persistent; floral tube campanulate, 3–5 mm, calyx tubular, sepals obovate, spreading, 5-lobed; petals 5, oblong; stamens 5; ovary glabrous, only half inferior. Insect pollinated.

Fruit. Berry; July–August; globose, smooth, black, 0.6–1 cm across, many seeded; palatable when cooked; seeds 313,000/lb (688,600/kg).

Distribution. Common to locally abundant in lower and upper Michigan. County occurrence (%) by ecosystem region (Fig. x19): I, 97; II, 87; III, 100; IV, 75. Entire state 92%. Distribution ranges from NS to AB, s. to VA, MO, NM.

Site-Habitat. Wet to wet-mesic to seasonally wet, open to lightly shaded, slightly acidic to basic sites; deciduous swamps, nutrient-rich conifer swamps, swampy borders of lakes and streams; shrub thickets, seasonally wet river bottomlands, rocky sites, meadows; rare or lacking in upland forests. Associates include northern white-cedar, spruces, white pine, black ash, American elm, red maple, yellow birch, willows, poison-sumac, common and red elders, sweet gale, wild red raspberry, and meadowsweet, among others.

Notes. Shade-intolerant, persisting in light shade; tolerant of oxygen deficiency in ecosystems of floodplains and swamps. Relatively tolerant of inundation and sedimentation. Short-shoots are rarely older than 4–5 years. It is an alternate host for the blister rust disease organism (*Cronartium ribicola*) of eastern white pine and other 5-needle white pines. Fruits are a choice food of many birds; 31 species are listed by DeGraaf and Whitman (1979). Fruit (100 g fresh weight) contains 89 mg of vitamin C.

Chromosome No. $2n = 16$, $n = 8$; $x = 8$

Similar Species. Closely related is the northern black currant, *Ribes hudsonianum* Richardson, distinguished by leaves glandular-dotted on the lower surface only and several other minor morphological characteristics. It is common in moist woods and rocky slopes from Alaska and British Columbia to western Quebec and in the western, midwestern, and upper Great Lakes states, including Michigan.

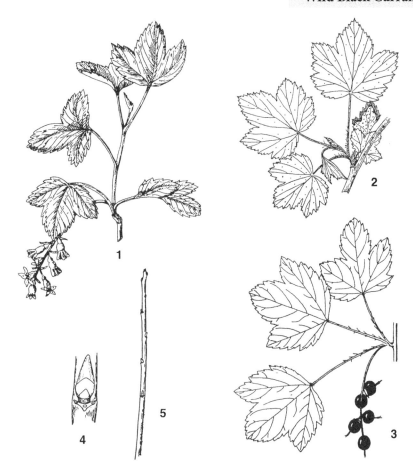

1. Vegetative long-shoot with flowering shoot at base, × 1/2
2. Vegetative short shoot, × 1/2
3. Fruiting shoot, berry, reduced
4. Winter lateral bud and leaf scar, enlarged
5. Winter twig, reduced

KEY CHARACTERS

- low, erect, unarmed shrub to 1.5 m with slender, spreading branches
- leaves glabrous, yellowish-green above, pubescent beneath, yellow resin-dotted above and beneath
- stems with many flower-bearing short-shoots, long-shoots slim, with conspicuous resin glands, unarmed
- fruit a globose berry, black, unarmed
- habitat wet to wet-mesic sites, swamps, shrub thickets, floodplains

Also closely related is the European black currant, *R. nigrum* L., which is widely distributed in Europe and Asia and was once widely planted in American gardens. Fresh foliage of *R. americanum* lacks the strong, sweet "tomcat" odor of *R. hudsonianum* and *R. nigrum*, which is unpleasant to some people.

GROSSULARIACEAE
Ribes cynosbati Linnaeus

Wild Gooseberry,
Prickly Gooseberry

Size and Form. Low, erect, or spreading shrub, 0.5–1.2 m; solitary or forming small clones by layering, tip rooting, and sprouting from the root collar.

Bark. Larger branches with gray shredding outer bark, inner bark brownish-purple.

Leaves. Alternate, simple, deciduous, borne several in a cluster on short-shoots or separately, with distinct internode elongation on long-shoots; blades thin, 2–6 cm long and equally wide; broadly ovate, palmately 3- to 5-lobed, lobes oblong to rhombic, often sharply cleft, acute or obtuse; cordate to truncate; margin irregular crenate-dentate; green, finely pubescent above, velvety pubescent and paler beneath, with gland-tipped hairs along the veins; petioles pubescent, shorter than the blade, 1–4 cm long.

Stems-Twigs. Both long- and short-shoots present. Long-shoots slender; yellowish to medium brown with 1–3 slender spines 0.5–1 cm long at each node; juvenile indeterminate long-shoots arising from the root collar and the lower part of older stems range from extremely prickly along internodes to sparsely prickly to absent near the apex (Fig. 11); prickles 5–12 mm long; current short-shoots, with typically 2–3 small leaves, give rise to virtually all flower-bearing peduncles; short-shoots may be >10 years old and > 1 cm long; leaf scars partly raised, very narrow, U-shaped to crescent-shaped; bundle scars 3; pith relatively large, pale, round, continuous, becoming porous with age.

Winter Buds. Terminal bud present, conical, 2–8 mm long, 6–8 scales exposed; lateral buds very slender, narrowly ovoid, elongate, pointed, 3–7 mm long; golden brown to reddish-brown, somewhat hairy, silky, or entirely glabrous, 4–6 scales exposed, the lower ones keeled.

Flowers. April–May; bisexual; borne in small clusters on current short-shoots, with1–3 flowers on pubescent peduncles 7–18 mm long; flowers greenish-yellow or greenish-white; floral tube campanulate, 5-lobed, 3–6 mm long and 5–7 mm wide, pedicels 5–12 mm long; sepals broadly oblong, shorter than the tube, soon reflexed; petals 5, obovate, shorter than the reflexed calyx lobes; stamens 5, not exserted; ovary with stalked glands that become stiff prickles on the fruit; inferior. Insect pollinated.

Fruit. Berry; borne on long pedicels; July–September; globose, 0.8–1.7 cm across, not including prickles; light green, turning dark reddish-purple when ripe, usually with dense, still, long prickles; many-seeded; edible; not persistent in winter; seeds 205,000/lb (451,000/kg).

Distribution. Common to abundant in lower and upper Michigan. County occurrence (%) by ecosystem region (Fig. 19): I, 95; II, 90; III, 86; IV, 88. Entire state 92%. Wide ranging in e. North America from QC, ON, s. to GA, AL, and from MN and ND to OK, AR.

Site-Habitat. Occurs in a wide range of open to moderately shaded upland sites, common in dry-mesic oak-hickory forests; mesic beech–sugar maple forests; hemlock–northern hardwoods forests; boreal forests, floodplains; rare in wetlands except on drier microsites. Associates include a great many trees and shrubs of the diverse range of forests mentioned above.

Notes. Moderately shade-tolerant. It is the most common *Ribes* species in Michigan. Unlike other *Ribes* species, it is characteristic of mesic beech–sugar maple and hemlock–northern hardwoods forests where it leafs out and flowers before overstory and understory deciduous trees leaf out. Seeds are disseminated widely by birds and mammals. *R. cynosbati* is an alternate host for the blister rust disease organism (*Cronartium ribicola*) of eastern white pine and other 5-needle white pines. Fruit is sometimes used for jellies and pies.

Chromosome No. $2n = 16$, $n = 8$; $x = 8$

Similar Species. Closely related to the Missouri gooseberry, *Ribes missouriense* Nutt.,

1. Flowering short shoot, × 1 1/2
2. Fruiting shoot, berry, × 3/4
3. Winter lateral bud, spine, and leaf scar, enlarged
4. Winter twig, juvenile shoot, × 1
5. Winter twig, adult shoot, × 1

KEY CHARACTERS

- low shrub, erect or spreading
- leaves with 3–5 lobes, sharply cleft, pubescent above and especially below, lacking resinous glands
- stems with many flower-bearing short-shoots, long-shoots with spines at nodes, and prickles along internodes
- fruit a globose berry, prickly, dark reddish-purple at maturity, borne on long pedicels

which is rare in Michigan but common south and west of it. Also similar is the swamp gooseberry, *R. hirtellum* Michx., which is primarily a swamp species and infrequent to common throughout Michigan. The berries lack prickles, nodal spines are absent or poorly developed, and flowering peduncles are short.

GROSSULARIACEAE
Ribes lacustre (Pers.) Poiret

Swamp Black Currant

Size and Form. Low, bristly-spiny shrub, 0.4–1 m; erect or arching slender stems.

Bark. Loose, shredding outer bark, inner bark blackish.

Leaves. Alternate, simple, deciduous, unpleasant odor when bruised or crushed; borne several in a cluster on short-shoots or separately with distinct internode elongation on long-shoots; blades thin, 3–7 cm long and equally wide; nearly orbicular, palmately and mostly 5-lobed, deeply incised lobes further divided and coarsely toothed, tip acute, base cordate; dark green, nearly glabrous above, paler beneath, with scattered gland-tipped hairs; petioles with spreading gland-tipped hairs, 3–5 cm long.

Stems-Twigs. Both long- and short-shoots present. Long-shoots slender, weak, rounded but decurrently ridged below the nodes; yellowish-brown, glossy, covered with dense to scattered bristles and clusters of longer, slender spines at the nodes; leafy flower-bearing short-shoots may be at least 15 years old and >1 cm long; unpleasant odor when bruised or crushed; leaf scars somewhat raised, very narrow, bundle scars 3; pith relatively large, pale, round, continuous, becoming porous with age.

Winter Buds. Terminal bud present, conical, sharp-pointed, 3–8 mm long, light brown, 4–6 scales exposed; lateral buds small, solitary, sessile, conical-elongated, 2–6 mm long.

Flowers. May–June; bisexual; several flowers borne on drooping racemes, 2–5 cm long on current short-shoots, peduncle puberulent and glandular-bristly; flowers yellowish-green to purplish, about 7 mm across, pedicels puberulent and glandular, floral tube flat to saucer-shaped, not a campanulate tube, sepals 5, short and broad, spreading, 2 mm; petals 5, pinkish, fan-shaped, 1.5 mm; stamens 5; ovary densely covered with gland-tipped bristles, inferior. Insect pollinated.

Fruit. Berry; July–August; July–September; globose, 6–12 mm across, reddish or purplish-black, glandular-bristly; disagreeable taste; many-seeded.

Distribution. Absent in southern lower Michigan, infrequent to common in northern lower Michigan, common and locally abundant in upper Michigan. County occurrence (%) by ecosystem region (Fig. 19): I, 0; II, 33; III, 100; IV, 88. Entire state 29%. Wide ranging

from NL to AK, s. to CA and throughout the n. and middle Rocky Mountains; New England s. to VA and western Great Lakes Region.

Site-Habitat. Wet to wet-mesic, primarily circumneutral to mildly basic sites; relatively nutrient-rich conifer and conifer-hardwood swamps and exposed rocky ledges, cliffs and talus. Associates include northern white-cedar, spruces, tamarack, red maple, black ash, American elm, trembling aspen, balsam poplar, mountain maple, speckled alder, winterberry, swamp red currant, Canadian fly honeysuckle, velvetleaf blueberry.

Notes. Moderately shade-tolerant. Stems with dense bristles and nodal spines make it one of the most heavily armed of *Ribes* species. More of a boreal than northern species, it is characteristic of the Canadian boreal forest region to ca. latitude 56° north. *R. lacustre* is an alternate host for the blister rust disease organism (*Cronartium ribicola*) of eastern white pine and other 5-needle white pines. Fruit (100 g fresh weight) contains 58.2 mg of vitamin C.

Chromosome No. $2n = 16$, $n = 8$; $x = 8$

1. Vegetative shoot, × 1/2
2. Flower, × 3
3. Fruiting shoot, × 1/2
4. Fruit, berry, × 2
5. Winter lateral bud and leaf scar, enlarged
6. Winter twig, reduced

KEY CHARACTERS

- low, bristly-spiny shrub to 1 m
- leaves deeply 5-lobed, nearly glabrous above, paler beneath, lacking resinous glands, unpleasant odor when bruised or crushed
- stems with many flower-bearing short-shoots, long-shoots yellowish-brown, glossy, covered with dense bristles with longer spines at nodes, unpleasant odor when bruised or crushed
- fruit a globose berry, reddish or purplish-black, glandular-bristly
- habitat relatively nutrient-rich swamps and rocky sites of northern lower and upper Michigan

GROSSULARIACEAE
Ribes triste Pallas

Swamp Red Currant

Size and Form. Low, creeping, straggling shrub, 0.5–1 m; unarmed; initially with short ascending stems, becoming prostrate on the forest floor with age and often rooting at the nodes (i.e., layering) in moist soil or when covered with moss or leaf litter; shallow-rooted.

Bark. Loose, shreddy, or peeling outer bark, inner bark reddish-purple to blackish.

Leaves. Alternate, simple, deciduous; borne several in a cluster on short-shoots or separately with distinct internode elongation (0.5–3 cm) on long-shoots; blades thin, 5–10 cm wide, 4–8 cm long; suborbicular to subcordate with wide sinuses; palmately 3–5 lobed, the lateral lobes acute, pointing forward; base subcordate to broadly truncate; margin coarsely serrate from sinus to tip; green and glabrous above, nearly glabrous to pubescent to tomentose beneath; petioles sparsely hairy, 2.5–6 cm long.

Stems-Twigs. Both long- and short-shoots present. Long-shoots rounded but decurrently ridged below the nodes, unarmed; young shoots slightly pubescent, slightly glandular, brownish to dark gray; flowering stems ascending from reclining main stems; short current shoots, with typically 2–3 small leaves, give rise to flower-bearing peduncles; short-shoots rarely older than 3 years; leaf scars somewhat raised, broadly crescent-shaped; pith relatively large, pale, round, continuous but becoming porous with age.

Winter Buds. Terminal bud present, oblong, 2–6 mm long, 4 pubescent scales exposed; lateral buds solitary, sessile ovoid, 1–3 mm long, gray-puberulent, without glands.

Flowers. May, with the leaves; bisexual; several flowers borne on drooping racemes, 3–6 cm long, peduncles and pedicels usually puberulent and glandular, bracts very small; flowers ranging from greenish-purple to pinkish-red, 4–5 mm across, floral tube saucer-shaped, <1 mm, sepals 5, spreading, obtuse; petals 5, truncate or notched, 1 mm; stamens 5; ovary glabrous, inferior. Insect pollinated.

Fruit. Berry; July-August; bright red, translucent, globose, smooth, 6–8 mm across; many-seeded; tart.

Distribution. Relatively common throughout the state except for the southernmost three tiers of counties. County occurrence (%) by ecosystem region (Fig. 19): I, 47; II, 70; III, 100; IV, 100. Entire state 65%. A circumboreal and northern species found across the USA and Canada, including nonboreal ecosystems in s. interior BC.

Site-Habitat. Moderately shade-tolerant, persisting in heavy shade; moist to wet, moderately acidic to mildly basic wetlands; deciduous and conifer swamps; wet depressions, seepages, stream banks and terraces, clearings, and rocky slopes. Associates include northern white-cedar, tamarack, white and black spruces, red maple, black ash, American elm, trembling aspen, balsam poplar, mountain maple, speckled alder, autumn willow, winterberry, Canadian fly honeysuckle, blueberries, elderberries, leatherleaf, and mountain holly, among many others.

Notes. Moderately shade-tolerant, occasionally persisting in deep shade. Large, shade-grown leaves resemble those of mountain maple, *Acer spicatum*, or thimbleberry, *Rubus parviflorus*. *R. triste* is an alternate host for the blister rust disease organism (*Cronartium ribicola*) of eastern white pine and other 5-needle white pines. The fruit (100 g fresh weight) contains 51.5 mg of vitamin C. Fruits are eaten fresh or cooked to make jelly.

Chromosome No. $2n = 16$, $n = 8$; $x = 8$

Similar Species. Closely related to the skunk currant, *Ribes glandulosum* Grauer, and red currant, *R. rubrum* L., of central Europe. Other similar species occurring in swamps include wild black currant, *Ribes americanum*, swamp black currant, *R. lacustre*, and northern black currant, *R. hudsonianum* Richardson.

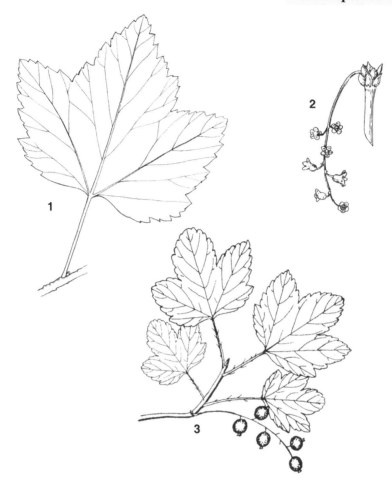

1. Leaf, × 1
2. Flowering shoot, × 1/2
3. Fruiting shoot, × 1/2

KEY CHARACTERS

- low, straggling, unarmed shrub to 1 m
- leaves simple, blade suborbicular, 3–5 lobed, the lateral lobes acute, pointing forward, margin coarsely serrate from sinus to tip; lacking resin glands
- stems with many flower-bearing short-shoots, long-shoot stems lacking resin glands, un-armed
- fruit a berry, bright red, smooth
- habitat wet to wet-mesic, relatively nutrient-rich sites, especially swamps

THE ROSES—*ROSA*

There are about 120 (100–150) native roses in the Northern Hemisphere, of which 7 are native in Michigan (Voss and Reznicek 2012); worldwide there are 2,050 species (Judd et al. 2008). *Rosa* is a member of the subfamily Rosoideae of the Rosaceae. Although roses of the garden are familiar to everyone, less well known are the native wild roses of Michigan and the western Great Lakes Region. They are all "single-flowered," with 5 petals, whereas most cultivated garden roses are "double," that is, some or all of the stamens form extra petals. The universal name *rose* is derived from the ancient Greek word for the rose flower; later, in England, the word *rose* was derived from a Celtic word meaning "red" (Coates 1992). However, rose breeding throughout the world, especially in England, has produced an enormous variety of colors as well as fragrances. Thus, not only the rosarian of today has cause to ponder the statement "A rose is neither red nor sweet, though we may think it so."

The first record of a cultivated English rose dates from the reign of William Rufus (1087–1100), but the roses Sir William sought in the nunnery at Romsey most certainly were not derived from wild rose stock. According to English garden historian Alice Coates (1992), there were only 5 English wild briars. These native roses played only a minor role in the development of English garden forms. The spectacular advance in English rose growing in the early nineteenth century was due to the arrival of various Chinese garden roses, which brought with them the property of perpetual blooming. By 1826 an English catalog listed 1,393 species and varieties, and they continued to increase greatly.

Seven wild roses are native in Michigan, and we describe 4 of the most widely distributed together with the multiflora rose, *Rosa multiflora*, a native of Japan and Korea. It was introduced before 1868 (Rehder 1940) for wildlife habitat and erosion control and has spread out of control. Prior to European settlement, native rose species likely had more typical or distinctive ecological homes (e.g., *R. palustris* in wetland ecosystems or *R. carolina* in primarily dry to dry-mesic upland ecosystems). However, great disturbances to natural landscapes by humans, together with the ease of dispersal of seeds by birds and mammals and hybridization among native species, led to their wide occurrence today in open and disturbed sites of diverse kinds (see "Site-Habitat" sections under individual species).

Although rose species grow today in a range of soil-site conditions from wet to dry, from basic to acid, and in open to moderately shaded situations, they show a preference for rooting and persistence in circumneutral to basic soils and substrates. Such situations may be microsites where favorable seasonal water table fluctuations occur or where roots reach capillary fringes supplying circumneutral water in sites with acidic upper horizons. Rose species, like those of numerous other genera, occur together with species tolerating or requiring acidic soils together with those that typically grow in circumneutral or basic soils.

Native rose species are characterized by upright, arching, or trailing stems, usually armed with prickles. The alternate deciduous leaves are pinnately compound with serrate leaflets. The native species have narrow but prominent stipules; stipules of all species are fused to the petiole for most of their length. Stems are circular in cross section, with prickles commonly in pairs below the nodes; bristles often occur on flowering and nonflowering shoots. The pith is relatively large and brown. Winter buds are small and solitary, with 3 or 4 scales exposed; the terminal bud is present. Leaf scars are very narrow, extending about halfway around the stem; there are 3 bundle scars. Flowers are "single" (not "double," as in most garden roses), with 5 petals of pale pink to deep rose color. A reddish, fleshy receptacle or hypanthium, the hip, encloses several to many achenes and is usually persistent long into winter. Sepals often crown the apex of the hip and are quickly deciduous or persistent into winter (*R. blanda*). Asexual reproduction and clone-forming is by rhizomes, roots, and layering. The reader is directed to excellent treatments of native and introduced rose species and hybrids in Michigan (Voss and Reznicek 2012), Minnesota (Rosendahl 1965), Ontario (Soper and Heimburger 1982), Ohio (Braun 1989), and the northeastern United States and adjacent Canada (Gleason and Cronquist 1991).

KEY TO SPECIES OF *ROSA*

1. Wetland habitat, wet to wet-mesic, poorly to somewhat poorly drained sites; flowers very fragrant . . . *Rosa palustris*, p. 240
1. Upland habitat, dry to dry-mesic, well-drained sites; flowers not or markedly less fragrant.
    2. Medium shrub, 0.7–3 m; an introduced species occurring primarily in southern lower Michigan; stout recurved prickles at and between the nodes; stipules conspicuously fringed or comblike-serrate (i.e., deeply pinnatifid); stems spreading to form impenetrable thickets . . . *Rosa multiflora*, p. 238
    2. Low shrub, 0.2–1.5 m; native species; prickles not stout, straight; stipules entire, not conspicuously fringed or comblike-serrate; erect shrub not forming impenetrable thickets.
        3. Stems with few prickles, only on nodes of lower branches; frequent to locally common throughout the state . . . *Rosa blanda*, p. 234
        3. Stems with numerous prickles; not frequent throughout the state.
            4. Stems densely covered with prickles; leaflets often double serrate; rare in southern lower Michigan . . . *Rosa acicularis*, p. 232
            4. Stems with fewer prickles; leaflets coarsely serrate; common to abundant in southern lower Michigan, rare in northern lower Michigan, absent in upper Michigan . . . *Rosa carolina*, p. 236

ROSACEAE
Rosa acicularis Lindley
Prickly Wild Rose

Size and Form. Low, erect, bushy shrub, 0.3–1 m, taller in shade; plant very prickly, clones of few stems formed by rhizomes.

Leaves. Alternate, deciduous; pinnately compound, leaflets 5–7, short-stalked; elliptic to ovate or obovate, oval, leaflets 1.5–5 cm long, 1–2.5 cm wide; usually acute; cuneate or rounded to subcordate; margin coarsely serrate, usually doubly serrate toward apex, frequently glandular-serrulate, teeth pointed outward; dull green and glabrous above, paler and somewhat pubescent and usually more or less resinous-glandular beneath; petiole and leaf rachis usually minutely pubescent and glandular; stipules simple, narrow at base, adnate and broadened at the free end, stipule pair to 1.5 cm wide, pubescent, the margins glandular.

Stems-Twigs. Slender, flexible, round, greenish, becoming somewhat reddish, densely covered with slender, straight, unequal-sized prickles, 3–6 mm long, those near the nodes often longer and stouter than others; flower-bearing shoots prickly; leaf scars long, narrow, and linelike, extending about halfway around the stem; bundle scars 3; pith relatively large, round, brown, continuous.

Winter Buds. Terminal bud present; buds small, 2–3 mm long, solitary, sessile, ovoid, with several scales exposed.

Flowers. Late May–July; bisexual, pink to deep rose, 5–7 cm wide, fragrant, usually solitary on short, prickly lateral shoots from stems of the previous year, occasionally in clusters of 2 or 3, pedicels glabrous, rarely glandular-hispid, 1–2 cm long; sepals usually simple, somewhat broadened at the tip, 2–3 cm long, glandular outside, often finely pubescent within, erect and persistent; petals 2–3 cm long. Insect pollinated.

Fruit. A fleshy hip enclosing an aggregate of achenes; the achenes are bony, thick, very light-colored; August–September; hip fleshy, ellipsoid, pear-shaped or nearly globular, calyx persistent, red and translucent, glabrous, about 2 cm long, 1.5 cm wide; seeds ca. 17,250/lb (37,950/kg).

Distribution. Common in northern lower Michigan and upper Michigan, rare in southern lower Michigan. County occurrence (%) by ecosystem region (Fig. 19): I, 8; II, 43; III, 86; IV, 75. Entire state 34%. It has the widest range of North American rose species, from NS and QC to AK, s. throughout the Rocky Mountains to NM, w. Great Lakes Region to New England. In the Southwest, it ranges between altitudes of 1,370 to 3,050 m.

Site-Habitat. Almost any upland site with high light irradiance and circumneutral to mildly basic soils or substrates; rocky lakeshores and cliffs, talus accumulations, forest and field edges, cutover forests, roadsides, fencerows, meadows, prairies, dunes. Of particular prominence in cobble, sandy, and bedrock sites and lakeshores of northern lower and upper Michigan, especially the harsh shorelines and cliffs of Lake Superior. It occurs in sites with little to no soil development in cracks, joints, depressions of bedrock, and talus. In the latter sites, it occurs with stunted tree conifers and hardwoods, willows, ninebark, mountain alder, wild red raspberry, thimbleberry, and blueberries.

Notes. Shade-intolerant, persisting in light shade. Flowers earlier than *R. blanda*. *R. acicularis* is circumboreal in distribution. As a native of Eurasia, it is an octoploid ($2n = 56$), whereas our native species, a hexaploid ($2n = 42$), is considered a subspecies and differs from the Eurasian taxon in various morphological characteristics (Voss and Reznicek 2012). Many varieties and forms have been described. It readily forms hybrids with *Rosa blanda*. *Acicularis* means "needlelike," that is, with many needlelike prickles. It is Alberta's provincial floral emblem.

Chromosome No. $2n = 42$, $n = 21$; $x = 7$

Similar Species. In Michigan and the upper Great Lakes Region, the smooth wild rose, *Rosa blanda* Ait., is the most similar. The sim-

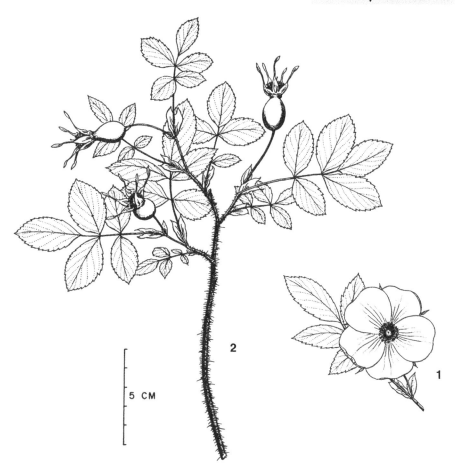

1. Flowering shoot, × 1/2
2. Fruit, the hip encloses the true fruit, an achene, × 1/2

KEY CHARACTERS

- low, erect, bushy shrub, usually <1 m
- leaves pinnately compound; leaflets usually 5; margin rather coarsely serrate, blades glabrous above, minutely pubescent beneath; petiole and rachis usually minutely pubescent and glandular; stipules with glandular margins
- stems densely covered with slender, straight prickles; flowering shoots prickly
- flowers pink to deep rose with glabrous pedicels
- fruit an achene enclosed in a red hip, ellipsoid, elongated at base

Distinguished from the smooth rose, *Rosa blanda,* by its low form, usually <1 m; leaflets usually 5; margin rather coarsely serrate, whole plant very prickly; twigs with dense, straight prickles, lateral floral shoots prickly, hip ellipsoid; of rare occurrence in southern lower Michigan.

ilar prairie rose, *R. arkansana* Porter, reported from 15 Michigan counties, is dominant in the Great Plains south to Texas and is widely distributed by varieties and hybrids from New York across the northern United States and adjacent Canada to British Columbia.

233

ROSACEAE
Rosa blanda Aiton

Smooth Wild Rose, Meadow Rose

Size and Form. Low, erect, bushy shrub 0.2–1.5 m; spreading widely by long-lived rhizomes and forming clonal clumps by root-collar sprouting.

Leaves. Alternate, deciduous; pinnately compound, leaflets 5–7; short-stalked; elliptic-oblong to oval or obovate, 1–4.5 cm long, about half as wide; acute to rounded; cuneate to rounded; margins coarsely serrate to just below the middle; dull green and nearly glabrous above, pale and finely pubescent beneath; petiole and rachis finely pubescent; stipules generally pubescent above and beneath, margins entire, the adnate portion gradually broadened toward the free end.

Stems-Twigs. Slender, flexible, round, reddish-purple; larger stems at the base armed with a few slender, straight prickles; vigorous indeterminate current shoots often densely prickly on the lower part, becoming unarmed or with few prickles on the upper vegetative and flowering shoots; leaf scars long and narrow, linelike, extending about halfway around the stem; bundle scars 3; pith relatively large, round, brown, continuous.

Winter Buds. Terminal bud present; buds small, 2–3 mm long, solitary, sessile, ovoid, red, with 3–4 scales exposed.

Flowers. May, peaking in June; bisexual, pale or bright pink, solitary or in pairs or clusters of 3–5 at the ends of current shoots; pedicels glabrous, generally 1–2 cm long; calyx tube glabrous, calyx lobes lanceolate, erect and persistent on hip; petals, 2–3 cm long. Insect pollinated.

Fruit. A fleshy hip enclosing an aggregate of achenes; the many achenes slender, light brown, about 4 mm long; August–October; hip fleshy, subglobose, more or less erect, round or somewhat elongated, sepals persistent; scarlet to red, glabrous, 0.8–1.5 cm wide; seeds 45,000/lb (99,000/kg).

Distribution. Frequent to locally common throughout the state. County occurrence (%) by ecosystem region (Fig. 19): I, 61; II, 77; III, 100; IV, 100. Entire state 73%. Ranging in North America from NS to AK and NT, s. to MT, the Great Plains, w. Great Lakes Region and New England, s. to VA.

Site-Habitat. Almost any upland site with high light irradiance, mostly circumneutral to basic sites; open forests, meadows, rock outcrops, gravelly shores of lakes, dunes, roadsides, fencerows. Associates include a wide variety of pioneer species in communities of these sites.

Notes. Highly shade-intolerant, persisting in light shade. Highly variable morphologically and readily hybridizing with *Rosa acicularis* and *R. palustris* (Voss and Reznicek 2012). Only the basal part of the stems is prickly because it remains in the juvenile phase of physiological development. The name *blanda* denotes the scarcity of prickles on the leafy flower-bearing shoots.

Chromosome No. $2n = 14$, $n = 7$; $x = 7$

Similar Species. In Michigan and the upper Great Lakes Region, the prickly wild rose, *Rosa acicularis,* is the most similar.

1. Flowering shoot, × 1/2
2. Fruiting shoot, hip × 1/2

KEY CHARACTERS

- low, erect, bushy shrub 0.2–1.5 m
- leaves pinnately compound, leaflets 5–7, margin coarsely serrate, rachis finely pubescent; stipules generally pubescent, margins entire
- stems prickly at base whereas upper part of stems and leafy flowering branches unarmed or nearly so
- hip scarlet to red, subglobose, sepals persistent on the hip into winter

Distinguished from the prickly wild rose, *Rosa acicularis*, by its larger habit to 1.5 m; leaflets 5–7, margin coarsely serrate, stems generally smooth except prickly on lower part of stems, lateral flower-bearing shoots unarmed or nearly so; hip subglobose; frequent in southern lower Michigan.

ROSACEAE
Rosa carolina Linnaeus
Pasture Rose

Size and Form. Low, slender, little-branched shrub, usually 0.3–0.6 m (1); forming clones by basal sprouting and rhizomes.

Leaves. Alternate, deciduous, pinnately compound; leaflets 5–7, sessile or short-stalked; elliptic-lanceolate to nearly orbicular, 1.5–5 cm long, 0.8–2 cm wide; acute or rounded; rounded or cuneate; margin coarsely and sharply serrate with ascending teeth; dull green and glabrous above, paler and glabrous or pubescent on main veins beneath; leaf rachis more or less pubescent, glandular, and prickly; stipules narrow, flat, not fringed, 3 mm wide, 12–10 mm long, nearly glabrous, entire or with a few glandular teeth, attached to the petiole for nearly their entire length.

Stems-Twigs. Slender, green to reddish-brown, glabrous; prickles arranged singly around the stem at nearly right angles, at nodes often paired, slender, straight, needle-like, round to the base, to 10 mm long, prickles numerous between the nodes, more or less deciduous; leaf scars long and narrow, linelike, extending about halfway around the stem; bundle scars 3; pith relatively large, brown, round.

Winter Buds. Terminal bud present; buds small, solitary, sessile, red, generally glabrous, with 3–4 scales exposed.

Flowers. Mid-June to July; bisexual, pink, mostly solitary, up to 7 cm wide; pedicels generally 1–2 cm long, glandular-hispid; calyx tube glandular-hispid, calyx lobes lanceolate, often expanded at the tip, either entire or more or less lobed, tardily deciduous from mature fruit; petals 2–3 cm long. Insect pollinated.

Fruit. A fleshy hip enclosing an aggregate of achenes; the many achenes are bony, light-colored; September, persisting into winter; hip subglobose; bright red, glandular-hispid, 1–1.5 cm wide.

Distribution. Common in southern lower Michigan, rare in northern lower Michigan; absent in upper Michigan. County occurrence (%) by ecosystem region (Fig. 19): I, 79; II, 20; III, 0; IV, 0. Entire state 43%. Widely distributed in e. North America from NS to ON, s. from New England and the w. Great Lakes Region s. to FL , w. to TX.

Site-Habitat. Mainly upland sites. Open to lightly shaded, fire-prone, dry and dry-mesic sites with mostly circumneutral to basic substrates but also occurring in natural communities with seasonally fluctuating water tables and wet-mesic soils, prairies, oak and pine barrens, dunes. In addition, spreading to other open and disturbed sites: forest and field edges, meadows, roadsides, railroad rights-of-way, fencerows. Many associated tree and shrub species of seasonally wet to dry ecosystems; the oak barrens alone support 21 shrub associates rooting in either acidic or circumneutral soils or substrates.

Notes. Shade-intolerant, persisting in light shade; tolerating fluctuating hydrology from seasonally wet to summer drought. A variable species with two subspecies reported.

Chromosome No. $2n = 28$, $n = 14$; $x = 7$

Similar Species. In eastern North American, *Rosa palustris* is similar in some morphological characteristics, though not in habitat.

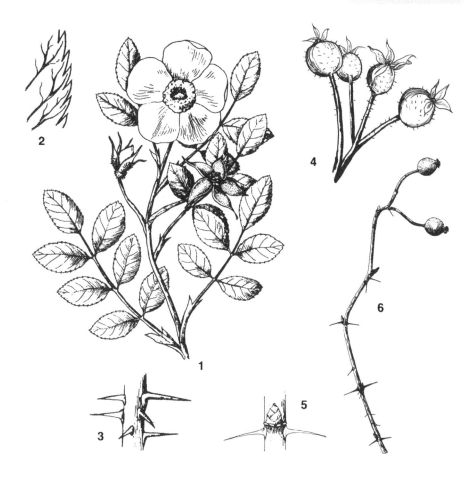

1. Vegetative and flowering shoots, × 1/2
2. Leaf blade margin, × 2
3. Stem with prickles, × 2
4. Fruiting shoot, hip, reduced
5. Winter lateral bud, leaf scar, prickles, enlarged
6. Winter twig with fruit, hip, × 1/2

KEY CHARACTERS

- low, slender, little-branched shrub, 0.3–0.6 m
- leaves pinnately compound, leaflets 5–7, elliptic-lanceolate to nearly orbicular, margins coarsely serrate; stipules narrow, about 1 cm long, flat, not fringed
- stems with many prickles arranged singly around the stem or paired at the nodes; slender, straight, needlelike, round
- flowers pink, hips red, persisting into winter

237

Rosa multiflora Murray

Multiflora Rose, Japanese Rose

Size and Form. Vigorous, erect, shrub with long reclining, climbing, arching, or spreading stems to 3 m; forming dense, nearly impenetrable clones by rhizomes, tip rooting, and layering.

Leaves. Alternate, late deciduous, pinnately compound, leaflets usually 7 (5–9), sessile or short-stalked, elliptic to obovate 1.5–4 cm long and about half as wide; acute; cuneate to rounded; margin sharply serrate, teeth with red tips; dull green or slightly lustrous and glabrous above, paler and soft pubescent beneath; stipules long, slender, conspicuously fringed or comblike-serrate and glandular-ciliate; petiole and rachis often prickly and softly pubescent.

Stems-Twigs. Stout, vigorous, long, flexible, round, reddish to green, glabrous, prickles on older main stems stout, recurved, those at nodes 5 mm long; prickles on indeterminate shoots solitary, slender; leaf scars long and narrow, linelike, extending about halfway around the stem; bundle scars 3; pith relatively large, brown, round, continuous.

Winter Buds. Terminal bud present; buds small, 2–3 mm long, solitary, sessile, blunt pointed, glabrous, red; located 2–4 mm distal to the paired prickles at a node, occasionally at right angles to the stem; 4–5 scales exposed.

Flowers. May–June; bisexual, usually white, fragrant, several to many in large flat-topped clusters with usually many flowers; pedicels glandular, sepals 5–7 mm long, glandular, the outer with toothlike lobes, eventually deciduous; petals 1–2 cm long. Insect pollinated.

Fruit. A fleshy hip enclosing an aggregate of achenes; the achenes are bony, light-colored; August through winter; hip small, about 3.5 mm long and 5 mm wide, ellipsoid to obovoid, red, glossy, smooth, in drooping or erect clusters; seeds ca. 66,000/lb (145,200/kg).

Distribution. Increasingly common and locally abundant in southern lower Michigan, rare elsewhere. County occurrence (%) by ecosystem region (Fig. 19): I, 58; II, 23; III, 14; IV, 12. Entire state 34%.

Site-Habitat. Widely planted for wildlife habitat (see "Notes") and spreading to open, dry-mesic upland sites with circumneutral to moderately basic soils, forest and field edges, naturally open or cutover oak or oak-pine forests, roadsides, fencerows, meadows, pastures, prairies, dunes. Associates include a great many shade-intolerant and midtolerant species over its wide planted and naturalized range.

Notes. Shade-intolerant, persisting in light shade; requiring higher soil fertility than any native rose species (Gill and Healy 1974). A native in Japan and Korea, it was introduced before 1868 for bank stabilization, as "living fences," and for wildlife food and cover. In a study of food-producing plants in Michigan, it received more use by wildlife than any other plant (Gysel and Lemmien 1964). It escaped cultivation and spread widely in the Midwest and eastern United States and adjacent Canada. Closed forests and shaded understories restrict its invasion except along edges. At one time distributed by state nurseries, it has become a serious invasive pest in some areas. The combination of dense, clonal thickets and stout prickles prevents penetration beyond its edges (Fralish and Franklin 2002).

Chromosome No. $2n = 14$, $n = 7$; $x = 7$

Similar Species. In Japan a variety from China with pink to deep rose flowers is cultivated, *Rosa multiflora* var. *adenochaeta* (Satake et al. 1993). In China, among varieties reported, *R. multiflora* var. *cathayensis* Rehder & E. H. Wilson has pink flowers and *R. multiflora* var. *multiflora* has white petals (Flora of China 2014).

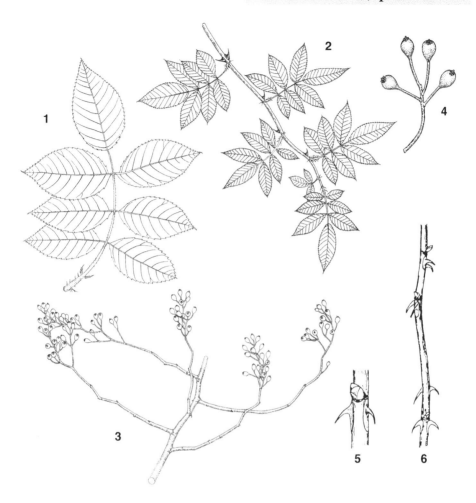

1. Leaf, ×1
2. Shoot with leaves, reduced
3. Winter branch with fruiting shoots, hip, reduced
4. Hips, × 1
5. Winter lateral bud and leaf scar, enlarged
6. Winter twig, reduced

KEY CHARACTERS

- vigorous shrub with long, climbing, arching, or spreading branches to 3 m; forming dense, nearly impenetrable thickets
- leaves late deciduous, pinnately compound, leaflets elliptic to obovate rachis often prickly; stipules narrow, lobes conspicuously fringed or comblike and glandular-ciliate
- twigs glabrous with stout, recurved prickles
- flowers white, fragrant, in flat-topped clusters with few to many flowers

ROSACEAE
Rosa palustris Marshall
Swamp Rose

Size and Form. Erect, much-branched, medium shrub, 0.7–2 m; clone-forming by rhizomes and layering.

Leaves. Alternate, deciduous, pinnately compound, leaflets usually 7 (5–9), narrow elliptic to oblanceolate, 2–7 cm long and <3 cm wide; usually acute, sometimes rounded; long-cuneate; finely and evenly serrate almost to the base, 20 or more teeth per side; dull green or slightly lustrous and glabrous above, pale and minutely pubescent on main veins beneath; petiole and leaf rachis softly pubescent, often with small prickles; stipules long, narrow, 1–3 cm long, with toothlike lobes only diverging 4–6 mm from the petiole.

Stems-Twigs. Slender, round, green to deep red, glabrous; internodal prickles none, usually a pair of recurved, broad-based, stout prickles at each node, 3–6 mm long; lacking internodal bristles; flower-bearing shoots often unarmed; leaf scars long and narrow, linelike, extending about halfway around the stem; bundle scars 3; pith relatively large, brown, round, continuous.

Winter Buds. Terminal bud present; buds small, 2–3 mm long, solitary, sessile, red, smooth or sparingly glandular-bristly, 3–4 scales exposed.

Flowers. June, peaking late July; bisexual, pink, very fragrant, in clusters of 2–5, sometimes solitary; pedicels 1–2 cm long; calyx tube glandular-hispid, calyx lobes attenuate, 1.5–2.5 cm long, often pinnately lobed, tomentose within and on outer margins, reflexed and wide-spreading after flowering, late deciduous; petals 2–3 cm long. Insect pollinated.

Fruit. A fleshy hip enclosing an aggregate of achenes; achenes bony, thick, light-colored; August–September; hip depressed-globose to oblong, red, glandular-hispid, 7–12 mm wide.

Distribution. Common throughout the state. County occurrence (%) by ecosystem region (Fig. 19): I, 95; II, 67; III, 100; IV, 62. Entire state 82%. Widely distributed in e. North America, NS to ON, s. from New England to FL and from w. Great Lakes Region to LA, MS, AL.

Site-Habitat. A wetland species. Open to moderately shaded, wet to wet-mesic sites with circumneutral to moderately basic soils or substrates; wetlands; fens, conifer and conifer-hardwood swamps, wet-mesic lake and stream margins; southern shrub carr, inundated shrub swamp, roadside ditches. Associates include a wide variety of pioneer species in open and lightly shaded wet and wet-mesic sites. For example, there are 20 likely associates in the southern shrub carr and 30 trees and shrubs in the rich tamarack swamp.

Notes. Moderately shade-tolerant. Shallow-rooted and tolerant of low oxygen availability in wetlands. Hybrids with *Rosa blanda* may be called *R. × palustriformis* Rydb (Voss and Reznicek 2012).

Chromosome No. $2n = 14$, $n = 7$; $x = 7$

1. Flowering and fruiting shoots, × 1/2
2. Leaf blade margin, × 2
3. Stem with prickles, × 1
4. Fruiting shoots, hip, × 1/2
5. Winter lateral bud, leaf scar, prickles, enlarged
6. Winter twig, reduced

KEY CHARACTERS

- erect, much-branched, medium shrub to 2 m
- leaves pinnately compound, leaflets narrow elliptic to oblanceolate, very finely and evenly serrate almost to the base
- stems lacking internodal prickles, usually a pair of recurved, broad-based prickles at each node, lacking internodal bristles
- flowers pink, calyx lobes wide-spreading after flowering, deciduous
- fruit an achene, many enclosed in a red hip
- habitat open to lightly shaded, circumneutral to medium basic wetlands

THE BRAMBLES: RASPBERRIES, DEWBERRIES, THIMBLEBERRIES, AND BLACKBERRIES—*RUBUS*

Rubus is one of the most interesting and important shrub genera in Michigan and the upper Great Lakes Region. Easily recognized, individuals currently occupy or have the potential to occupy almost every kind of terrestrial landform, landscape ecosystem, forest community, and prairie of the region. *Rubus* is a large genus, 400–740 species (Judd et al. 2008), with more than 400 reported from eastern North America, as well as many hybrids, both diploids ($x = 7$) and polyploids. Species of *Rubus* are common to locally abundant throughout the state—there are more than you can shake a stick at. With high sexual and asexual reproductive capacity and attractive and edible fruit, they are very widely dispersed by birds and animals. Being shade-intolerant to midtolerant, they have taken advantage of enormous human disturbances in our region to occupy a continuum of sites from dry to wet. Having the ability to remain dormant on the forest floor, even in beech–sugar maple forests, they are able to take immediate advantage of partial- or clear-cutting to form nearly impenetrable stands of raspberry-blackberry brambles. In doing so, they recouple the nutrient cycling system, prevent erosion, and provide wildlife habitat and light cover for regeneration of some tree species. Furthermore, species of *Rubus* regenerate vegetatively by sprouting from rhizomes, rooting at stem nodes, and rooting at the tips of arching stems that reach the ground. Therefore, once established and provided with abundant light they may persist in place for years or decades. Ecologically, *Rubus* species provide a range of site indicator values from useless (i.e., ubiquitous in open, disturbed places) to specific conditions of soil-water (dry to wet) and nutrient availability.

The six species we describe or contrast are relatively easy to identify in the field and place in one of their major groups using first the soil-site conditions where they grow and then characteristics of the whole plant: habit, flowers, stems, leaves, and fruits. These six species occur in three subgenera of the genus. The largest subgenus is *Rubus*, the blackberries and dewberries; we describe three species. The subgenus *Idaeoblatus* includes the raspberries; we describe two species. In the subgenus *Anaplobatus* we describe the thimbleberry. The subgenera may be distinguished in fall as follows.

Idaeoblatus: Plants armed, ripe fruit separating easily from the receptacle on the plant: raspberries
Rubus: Plants armed, ripe fruit not separating easily from the receptacle on the plant: dewberries and blackberries
  Old, 2nd-year stems (primocanes) prostrate, trailing: dewberries
  Old, 2nd-year stems (primocanes) erect or arching: blackberries

Anaplobatus: Plants unarmed, ripe fruit separating easily from the receptacle on the plant: thimbleberries

Using the field key and descriptions that follow, the species may be distinguished by using multiple characteristics, but due to natural variation within the genus, hybridization, and apomixis patience is required. The uppercase letters following the scientific name indicate: subgenus *Anaplobatus*, thimbleberry, T; subgenus *Idaeoblatus,* raspberry, R; subgenus *Rubus*, Blackberry, B, and Dewberry, D. Although it is not described, the herbaceous dwarf raspberry, *Rubus pubescens*, is included in the key because of its low habit and similarity to the swamp dewberry, *Rubus hispidus*. More detailed keys and descriptions for Michigan species by Voss and Reznicek (2012, 15 species or species complexes) and others (Braun 1989; Soper and Heimburger 1982; Gleason and Cronquist 1991; Smith 2008) provide the *Rubus* enthusiast with ample resources.

Rubus is most abundant in the temperate to boreal and alpine areas of North America and Eurasia; few species occur in the tropics and Southern Hemisphere. Primarily deciduous or evergreen shrubs, they exhibit a prostrate, trailing, erect, or arching habit, and many have stems with spines, prickles, or bristles. Most species send up biennial stems (canes) from perennial roots. Stems arising from seed, rhizome, or root collar the first year are typically unbranched, do not flower, and are termed *primocanes.* From axillary buds on the primocanes arise short lateral stems, *floricanes*, early in the second growing season, which bear flowers and fruit (Fig. 18). Old canes are short-lived, 2–4 years, and those that remain erect add to the woody network of a prickly bramble thicket along with the new primocanes that arise each year. Leaves are alternate, simple or 3–5-foliate, with stipules often persistent on the petiole remnant. Leaves are typically deciduous above the base of the petiole, leaving a shriveled stub covering the leaf scar. Leaves of primocanes are usually compound, whereas some or all of those of the floricanes are simple. Roselike flowers are usually white with a 5-parted calyx, 5 deciduous petals, numerous stamens, and many carpels, which ripen into drupelets attached to a fleshy receptacle. The fruit is an aggregate of drupelets.

Occasionally used in landscaping for borders or ground cover, *Rubus* species are very well known for their edible fruits, including the red raspberry (from the European *R. idaeus*, red raspberry) and the boysenberry and loganberry (from *R. ursinus*, California dewberry). The fruit is eaten raw and has long been used in jellies, jams, and preserves. Native peoples, especially in the Pacific Northwest, used stems, flowers, fruits, leaves, and bark medicinally for many complaints and ailments. Young shoots were peeled and eaten raw or steamed. Fruits were eaten raw or cooked, dried and stored for winter, and used to flavor liquor and make red dye (Moore 1979; Pojar and MacKinnon 1994; Marles et al. 2000).

1. Upright shrub; stems upright, not prostrate or decumbent.
    2. Stems unarmed; plants not biennial (canes able to produce flowering shoots the 1st year); leaves simple, broad, palmately lobed, and deeply cleft . . . *R. parviflorus* (T), p. 254
    2. Stems armed; plants biennial (vegetative primocanes produce floricanes in the 2nd year); primary leaves compound; leaflets 3–7.
        3. Stems bristly or prickly, not strongly armed; leaves densely white-tomentose beneath.
            4. Stems prickly, covered with whitish bloom; pedicels and peduncles with glandless prickles; mature fruit purplish-black . . . *R. occidentalis* (R), p. 252
            4. Stems densely bristly, bloom not conspicuous; pedicels and peduncles with gland-tipped bristles; mature fruit red . . . *R. strigosus* (R), p. 256
        3. Stems strongly armed, prickles stout, abundant, straight or recurved, broad based; leaves not densely white-tomentose beneath . . . *R. alleghaniensis* (B), p. 246
1. Low, decumbent shrub; stems trailing on the ground with ascending floricanes (flowering   shoots that arise the 2nd growing season from the vegetative primocanes).
    5. Stems herbaceous or nearly so; stems unarmed or occasionally with a very few weak bristles . . . *R. pubescens* (R).
    5. Stems slender but woody, hispid, bristly, or prickly.
        6. Stems with slender, straight, or barely recurved bristlelike prickles, scarcely or not enlarged at the base; leaves essentially evergreen; flowers small, petals 4–9 mm long; usually in a wet to wet-mesic site . . . *R. hispidus* (D), p. 250
        6. Stems with stiff, commonly recurved or hooked, broad-based prickles; bristles none; leaves clearly deciduous; flowers larger, the petals 1–1.5 cm long; usually in a dry to dry-mesic site . . . *R. flagellaris* (D), p. 248

Fig. 18. Early in the second growing season primocanes of black raspberry, *Rubus occidentalis*, give rise to floricanes. A, a primocane is shown with a single floricane that developed from a bud in the axil of a primocane leaf of the first growing season. At the same node a vegetative bud gave rise to the single, compound primocane leaf. B, a vigorous and arching primocane gives rise to multiple, upward-extending floricanes. Flowers have now formed on axillary and terminal stalks of the floricane. Due to this biennial shoot structure the flowers are well positioned for pollination and fruit development.

ROSACEAE
Rubus alleghaniensis Porter

Common Blackberry,
Highbush Blackberry

Size and Form. Erect, high-arching shrub, 1–2 m, canes long, to 3.5 m, rarely reaching the ground; clone-forming by root-collar sprouting and root suckering.

Leaves. Alternate, deciduous; palmately compound or simple; primocane leaves, 15–30 cm long; usually 5 leaflets, terminal leaflet 8–15 cm long, 4–10 cm wide, broadly ovate to broadly lanceolate, widest near or below the middle; long acuminate; rounded or subcordate; long-stalked, finely and sharply serrate to doubly serrate; lateral leaflets smaller, shorter-stalked or sessile (basal pair); floricane leaves with 3 leaflets or a simple blade in the upper part of the inflorescence, elliptic to oblong-ovate; all leaflets pubescent when young, becoming glabrous or sparsely hairy above to velvety beneath, with prickles along the midvein, green both sides; stipules linear, to 14 mm long, pubescent and glandular, persistent; petioles and petiolules densely pubescent, with stalked glands and scattered, broad-based, hooked prickles; petioles 5–16 cm long, base persistent.

Stems-Twigs. Biennial; primocanes stout, sylleptic branching common especially on vigorous stems; strongly angled or ridged (3- to 6-sided), often sparsely glandular when young; green, becoming reddish-brown to purplish with exposure and age; covered with gland-tipped hairs at least near the tip, prickles stout, much flattened at base, straight or recurved; floricanes brownish to purplish-red, somewhat ridged, with scattered broad-based prickles and small glandular hairs; leaf scar covered by persistent petiole base, bundle scars 3; pith large, creamy-white, continuous.

Winter Buds. Terminal bud absent; lateral buds pointing outward from stem, often superposed, 3–7 mm long, the lower and smaller one covered by petiole base; 4–6 scales exposed, scales ciliate, pubescent, especially at the tip, green to reddish-brown depending on exposure and position on stem.

Flowers. May–June; bisexual; white, 2–3 cm wide, borne on few- to many-flowered glandular-pubescent racemes, elongate, 10–20 cm long, the pedicels copiously tomentose and stipitate-glandular, and with scattered prickles, the lower 1–3 flowers subtended by leaves, the others by stipules only; sepals ovate, abruptly cuspidate-acuminate, 6–8 mm long; petals cuneate and separate at base 1–2 cm long; stamens many. Insect pollinated.

Fruit. Drupe, borne as an aggregate of drupes on a receptacle; July–September; the aggregate not separating from the receptacle when ripe, breaking from the stalk together with the receptacle, black, hemispherical to elongate, 1–1.5 cm long, edible; single-seeded; seeds 262,000/lb (576,400/kg).

Distribution. Locally common throughout the state. County occurrence (%) by ecosystem region (Fig. 19): I, 82; II, 80; III, 71; IV, 75. Entire state 80%. Widely distributed in e. North America except FL, LA, TX.

Site-Habitat. Open to moderately shaded, dry to wet-mesic sites, slightly acid to circumneutral soils; disturbed sites, especially forest edges, roadsides, clearings, fencerows; all but the wettest ecosystems. Associates of many pioneer trees and shrubs of open to moderately shaded, disturbed areas.

Notes. Shade-intolerant, persistent in moderate shade. Sylleptic branches borne in axils of leaves on the ascending and upper curved part of vigorous stems; 6 such branches observed on a stem 3.5 m long and still extending in mid-November (Washtenaw Co.; see Fig. 9*B*). Stems highly prickly, dense thickets painful to walk through. This is the most common of the complex tall blackberry species (Gleason and Cronquist 1991; Smith 2008). Cultivars derived from this and related species provide the blackberries of commerce. Thornless and more or less thornless plants are observed in the field.

Chromosome No. $2n = 14, 28, n = 7, 14;$ $x = 7$

1. Primocane leaf, × 1/2
2. Primocane stem with floricane flowering shoots, × 1/2
3. Fruiting shoot, aggregate of drupes, reduced
4. Winter lateral bud and leaf scar, enlarged
5. Winter twig, reduced
6. Primocane stem and broad-based prickles, × 3
7. Floricane stem with small glandular hairs, × 3

KEY CHARACTERS

- erect, high-arching shrub with long primocanes; biennial
- stems strongly angled or ridged (3- to 6-sided), green, becoming purplish-red
- leaves palmately compound, usually 5 leaflets on primocane, base of petiole persistent
- fruit a drupe, borne as an aggregate of black drupelets on a receptacle, aggregate not separating from the receptacle

ROSACEAE
Rubus flagellaris Willdenow

Northern Dewberry

Size and Form. Decumbent, creeping, low-arching shrub; woody primocanes to 4 m long, ascending floricanes to 30 cm; clone-forming by rooting at nodes and shoot tips; shallow-rooting.

Bark. Shreddy or peeling in long strips.

Leaves. Alternate, mostly deciduous, trifoliate or simple; primocane leaflets 3–5, terminal 3–7 cm long, 1–5 cm wide, ovate to lanceolate-elliptical, widest distinctly below the middle, sharply acute to acuminate and often with small lobes above the middle; rounded to subcordate; prominently doubly and sharply serrate, petiolule 1–2 cm; lateral leaflets smaller, ovate, sessile or nearly so, often somewhat asymmetrical; floricane leaflets 3 or with a simple blade in the upper part of the inflorescence; short-stalked; elliptic to rhombic or obovate, usually broadest above the middle; all leaflets thin, dull green both sides, glabrous or finely hairy above, pubescent on veins beneath; midrib prickly beneath; stipules linear, persistent; petioles 3–5 cm long, finely pubescent, with scattered hooked prickles beneath.

Stems-Twigs. Biennial; slender, wirelike, strongly woody, mostly creeping and rooting at tips; primocanes at first short-ascending, soon becoming prostrate; young stems circular or only slightly angled, greenish when young, becoming brownish to purplish-red; glabrous or nearly so with scattered stiff, recurved or hooked, broad-based prickles; floricanes erect, 10–30 cm high, often reddish.

Winter Buds. Terminal bud absent; lateral buds usually superposed, the lower, smaller one covered by petiole base; upper bud oblong, 1–5 mm long, several overlapping scales exposed, scales pubescent, margins ciliate, especially near apex, lower scales keeled.

Flowers. May–June; bisexual; solitary or borne on 2–4-flowered cymes; white; 2.5 cm across; calyx 5-parted; petals 5, 1–1.5 cm long, 0.4–0.8 cm across; stamens and carpels numerous. Insect pollinated.

Fruit. Drupe, borne as an aggregate of drupes on a receptacle; July–August; the aggregate not separating from the receptacle when ripe, short-cylindric to thimble-shaped, 1–1.5 cm long, black, juicy, edible; single-seeded; seeds 131,000/lb (288,200/kg).

Distribution. Common throughout the state except in eastern upper Michigan. County occurrence (%) by ecosystem region (Fig. 19): I, 74; II, 60; III, 43; IV, 75. Entire state 66%. Widely distributed throughout e. North America, w. of the Mississippi River from NE s. to TX.

Site-Habitat. Open, predominantly dry and drought- and fire-prone, sandy ecosystems with basic to acidic surface soil; prairies, oak and pine savannas, southern to northern dry to dry-mesic oak savannas and forests, old fields, meadows, swamp edges, shores, roadsides. Associates include a very diverse range of tree and shrub species of open, fire-prone, and disturbed prairies, savannas, and forests.

Notes. Shade-intolerant, persisting in light shade. Leaflets thinner than those of *Rubus hispidus*. *Flagellaris*, "whiplike," refers to the long, slender floricanes. Shallow-rooted and drought-tolerant; sometimes forming clonal mats.

Chromosome No. $2n = 28$, $n = 14$; $x = 7$.

Similar Species. Closely related to *Rubus multifer* L. H. Bailey, a plant of sandy acidic soils of oak savannas and woodlands of western Wisconsin and eastern Minnesota.

1. Primocane leaf, × 1/2
2. Floricane flowering shoot, × 1/2
3. Primocane stem with leaf, × 1/2
4. Primocane stem with floricane fruiting shoots, aggregate of drupes, × 1/2
5. Primocane stem with prickles, × 1

KEY CHARACTERS

- decumbent, creeping, low-arching shrub; biennial
- leaves compound with 3–5 leaflets, thin, dull green both sides
- stems with stiff, recurved or hooked, broad-based prickles
- fruit a drupe, borne as an aggregate of black drupelets on a receptacle, the aggregate not separating from the receptacle when ripe
- habitat open, predominantly dry to dry-mesic, fire-prone, disturbed areas

ROSACEAE
Rubus hispidus Linnaeus
Swamp Dewberry

Size and Form. Low, creeping shrub with prostrate primocanes and ascending, low-arching floricanes to 1.5 m; clone-forming by rooting at stem tips and nodes; shallow-rooting.

Bark. Shreddy or peeling in long strips.

Leaves. Alternate; often evergreen, trifoliate or simple; primocane leaflets 3, (5), 2–7 cm long; 1.5–4 cm wide, coriaceous, somewhat leathery, terminal leaflet short-stalked, obovate to oblong-ovate, obtuse or acute; cuneate; laterals sessile or nearly so; rhombic-ovate to nearly orbicular, obtuse; rounded to cuneate; floricane leaflets 3 or with a simple blade in the upper part of the inflorescence, elliptic obovate, usually broadest above the middle, smaller than primocane leaves; all leaflets coarsely singly or doubly serrate above the middle, teeth blunt, lower margin often entire; glabrous and glossy dark green above, green and glabrous or with pubescence on veins beneath; stipules linear, persistent; petioles with bristles and prickles similar to those on stems,1.5–3.5 cm long.

Stems-Twigs. Biennial; primocanes creeping or low-arched, very slender, wirelike, woody, round or angular; prickles almost absent to many, slender, weak, bristlelike, straight or barely recurved, scarcely enlarged at the base, 2–4 mm long; glandular hairs usually present, 2–4 mm long; floricanes numerous, herbaceous, ascending, 6–20 cm long.

Winter Buds. Terminal bud absent; lateral buds usually superposed, the lower one covered by petiole base; the upper bud 1–3 mm long, 4 pubescent overlapping scales exposed.

Flowers. June–July; bisexual; borne on 2- to 6-flowered corymbs, peduncles finely hairy, with or without stiff bristles; white; 1.5–2 cm across; calyx 5-parted; petals 5, about 4–6 (9) mm long; stamens and carpels numerous. Insect pollinated.

Fruit. Drupe, borne in an aggregate of drupes; August; the aggregate not separating from the receptacle when ripe, reddish-purple, slowly turning black at maturity, sour, edible; single-seeded; seeds 185,500/lb (408100/kg).

Distribution. Common throughout the state; occasional in upper Michigan. County occurrence (%) by ecosystem region (Fig. 19): I, 63; II, 70; III, 57; IV, 38. Entire state 63%. Ranging in e. North America from NL to ON, s. through the mid-South to TN and SC, also LA and KS.

Site-Habitat. Open to moderately shaded, wet to wet-mesic, seasonally wet, acidic surface soil and circumneutral to mildly basic sites; wet-mesic sand prairie, shrub thickets, fens, swamps. Associates include speckled alder, willows, red maple, American elm, black ash, swamp white oak, red-osier and silky dogwood, winterberry, swamp rose.

Notes. Shade-intolerant, persisting in moderate shade. Typical of ecosystems with fluctuating water table that brings basic water to the surface (i.e., seasonally wet sites). Leaves of primocanes commonly persisting through the winter.

Chromosome No. $2n = 28$, $n = 14$; $x = 7$

Similar Species. In habit and leaf morphology, dwarf raspberry, *Rubus pubescens* Rafinesque-Schmaltz, is similar, but stems are herbaceous or nearly so and unarmed or occasionally with a very few weak bristles. The following key characters distinguish *R. hispidus*. Primocanes woody not herbaceous; prickles many rather than prickles lacking; leaves firm and leathery, often evergreen rather than deciduous; fruit not separating from the receptacle rather than easily separating; fruit color reddish-purple, turning black at maturity, rather than bright red. Both occur in open to moderately shaded wet to wet-mesic habitats. *R. pubescens* is frequent to locally common throughout the state. County occurrence (%) by ecosystem region (Fig. 19): I, 89; II, 90; III, 100; IV, 88. Entire state 92%.

250

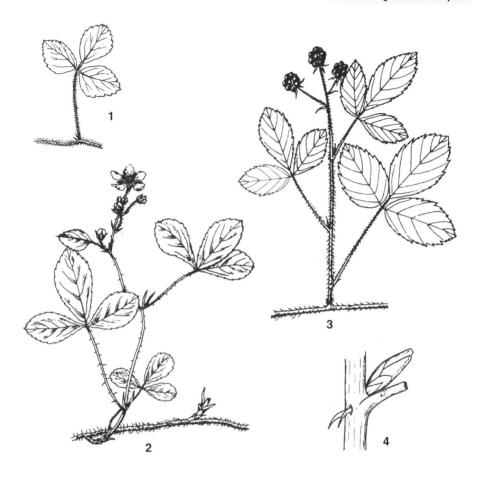

1. Primocane vegetative shoot, × 1/2
2. Floricane flowering shoots, × 1/2
3. Floricane fruiting shoot, aggregate of drupes, reduced
4. Winter lateral bud, enlarged

KEY CHARACTERS

- decumbent, trailing shrub with low-arching stems; biennial
- leaves compound with 3 leaflets, lustrous dark green above, mostly evergreen
- stems with slender, weak, bristlelike prickles, scarcely enlarged at the base, glandular hairs usually present
- fruit a drupe borne as an aggregate of reddish-purple drupelets on a receptacle, turning black when mature, the aggregate not separating from the receptacle when ripe
- habitat relatively open, moist to wet-mesic sites

ROSACEAE
Rubus occidentalis Linnaeus
Black Raspberry

Size and Form. Erect shrub to 2 m; strongly arching primocanes, 1–4 m long; clone-forming, rooting at tips of primocanes.

Bark. Brownish, shredding or exfoliating in patches

Leaves. Alternate, deciduous, trifoliate; primocane leaflets 3–5, floricane leaflets 3; all leaflets 4–10 cm long, 2–7 cm wide, primocane terminal to 10 cm long, petiolule 1–2.5 cm long; lower 2 leaflets sessile or nearly so, often lobed; broadly ovate to ovate-lanceolate (the lower 2 narrowly ovate), acuminate; rounded or subcordate; doubly serrate; dull green and more or less pubescent above, densely white-tomentose beneath; stipules linear, awl-shaped, deciduous; petioles glabrous, slightly pubescent or sparsely prickly, 3–8 cm long; base persistent.

Stems-Twigs. Biennial, slender, becoming stout with age and rooting at the tips; young primocane stems round, light purple to reddish, becoming purple the 2nd year; glabrous, covered with whitish bloom that is easily rubbed off; armed with short, broad-based prickles; floricanes very glaucous, appearing whitish; prickly; petiole base persistent; leaf scars crescent-shaped; bundle scars 3; pith large, tan to light orange-brown, continuous.

Winter Buds. Terminal bud absent; lateral buds oblong-ovoid, 3–8 mm long, pointing outward from stem, reddish-brown, usually superposed, the lower and smaller one often flattened, partially covered by petiole base; 4–6 overlapping pubescent scales exposed, tips of bud scales ciliate, reflexed on large buds.

Flowers. May–June; bisexual; white, borne on 3- to 7-flowered, tomentose, prickly corymbs, terminal or axillary, 1–1.5 cm across, pedicels 1–2.5 cm long, pedicels and peduncles with glandless, broad-based, slightly recurved prickles; calyx tomentose, sepals 5, triangular, 4–9 mm long, with ovate-lanceolate reflexed lobes; petals 5, elliptical, shorter than sepals, soon deciduous; stamens and carpels numerous. Insect pollinated.

Fruit. Drupe, borne in an aggregate of drupes; July–August; the aggregate separating easily when ripe from the white receptacle, which remains on the stalk; bright red at first, becoming purplish-black with a bloom, with belts of white tomentum between the drupelets, hemispherical, 0.8–1.5 cm wide; edible, sweet, juicy; single-seeded; seeds 334,000/lb (734,000/kg).

Distribution. Locally frequent to common in southern lower Michigan; infrequent to rare in northern lower Michigan; rare to absent in upper Michigan. County occurrence (%) by ecosystem region (Fig. 19): I, 74; II, 17; III, 29; IV, 0. Entire state 42%. Widely distributed in e. North America and Great Plains; lacking in FL, LA, TX.

Site-Habitat. Open to partially shaded areas, dry to mesic, basic to slightly acidic sites; almost any open, natural or human disturbed site, especially edges of forests, fields, clearings, roadsides. Associates very diverse due to widespread ecological occurrence.

Notes. Shade-intolerant, persisting in light shade. In open sites, the combination of vigorous tip-rooting primocanes, floricanes, and their intermingling may combine to create impenetrable thickets. According to Smith (2008), black raspberry was probably a savanna species, occurring between the mosaic of prairies and woodlands. Source of the many cultivars of black raspberry. Wild or cultivated, the fruits are prized for jams and jellies. Putative hybrids with red raspberry, *Rubus strigosis,* sometimes occur where the species grow together (*R.* × *neglectus* Peck; Voss and Reznicek 2012).

Chromosome No. $2n = 14$, $n = 7$; $x = 7$

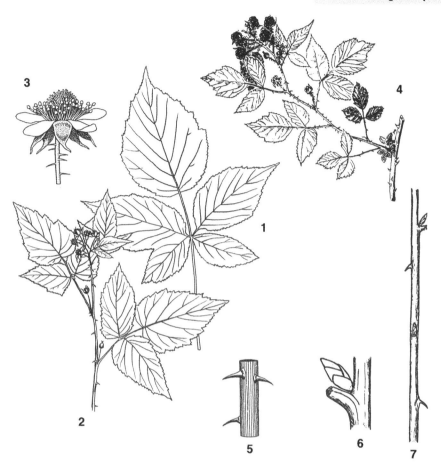

1. Primocane leaf, × 1/2
2. Floricane flowering shoot, × 1/2
3. Flower, × 1 1/2
4. Floricane fruiting shoot, aggregate of drupes, reduced
5. Primocane stem, prickles, × 2
6. Winter lateral bud and persistent petiole base, enlarged
7. Winter twig, reduced

KEY CHARACTERS

- erect, strongly arching shrub to 2 m; biennial
- leaves pinnately compound, leaflets 3–5, leaflets doubly serrate, somewhat pubescent above, densely white-tomentose beneath
- stems purple to reddish, covered with whitish bloom, especially on floricanes, armed with short broad-based prickles
- fruit a drupe, borne as an aggregate of drupelets on a receptacle, red, becoming purplish-black, the aggregate separating from the receptacle when ripe

ROSACEAE
Rubus parviflorus Nuttall
Thimbleberry

Size and Form. Upright, medium, un-armed shrub to 2.5 m, often much-branched; forming clones in clumps and thickets by sprouting from the root collar and a network of rhizomes.

Bark. Grayish-brown flaky, papery, peeling in long strips.

Leaves. Alternate, deciduous, simple, large, maplelike, blades rotund in outline, mostly 9–20 cm long and equally as wide and deeply cleft; base cordate or deeply cleft; usually palmately 5-lobed, lobes broadly triangular, apex acute, the upper 3 about the same length, small 3-lobed leaves occur on flowering shoots; margin conspicuously and irregularly coarsely serrate or toothed, teeth mucronate; bright green above, paler beneath; upper surface sparsely pubescent to glabrate, lower surface pubescent, at least along the veins; palmately veined; stipules lanceolate, glandular; petioles glandular-hispid, 5–15 cm long.

Stems-Twigs. Not biennial; slender, at first green, becoming light brown, young shoots lacking prickles but densely covered with brownish hairs; older branches stout with grayish, exfoliating bark. Stems live 2–4 years, increasing in branching complexity and thus in flower and fruit production.

Winter Buds. Terminal bud usually lacking; lateral buds ovoid to elongate, 2–6 mm long, light brown, densely hairy.

Flowers. June–July; bisexual; few, white, 3–5 cm wide, borne in long-peduncled, terminal, 2- to 9-flowered clusters; 3–5 cm across; flowering shoots borne on stems of the current year, not on floricanes of the following year; sepals ovate, long-tipped, with orange-yellow, very short, gland-tipped hairs; petals elliptic-ovate, 1.5–3 cm long; stamens and carpels numerous. Insect pollinated.

Fruit. Drupe, borne as an aggregate of drupes on a receptacle; August–September; the aggregate separating easily when ripe from the white receptacle, which remains on the stalk, bright to dull red; broad, low-hemispherical or thimble-shaped, hairy, 1.5–2 cm across, edible and opinions vary: sweet, sour, tart, tasteless, delicious, rather dry and insipid (likely dependent on season, habitat, and individual taster).

Distribution. Locally common in the northern tip of lower Michigan; common to locally abundant in upper Michigan, especially the Porcupine Mountains, Keweenaw Peninsula, and Isle Royale. County occurrence (%) by ecosystem region (Fig. 19): I, 0; II, 30; III, 86; IV, 100. Entire state 28%. Distributed throughout the upper Great Lakes Region; in the West occurring in all mtn. ranges from AK s. to BC and AB, s. to CA, NV, AZ, NM.

Site-Habitat. Open, rocky or sandy, exposed harsh to sheltered, northern to boreal sites following windstorm or fire, usually near the northern Great Lakes, especially Lake Superior; acidic and basic microsites; forest openings; bedrock shores and cliffs, dunes. Associates include a great variety of species that inhabit both basic and acid soils, red and white pines, hemlock, northern white-cedar, dwarf and creeping junipers, trembling aspen, speckled alder, blueberries, roses, wintergreen, mountain-ashes, ninebark.

Notes. Shade-intolerant, persisting in light shade. Reported in Minnesota to form dense clones by rhizomes, 30 m or more across (Smith 2008). Wide ranging from foothill to montane habitats from Alaska to New Mexico. Native peoples in the West used the large leaves as plates, containers, basket liners, and toilet paper (Kershaw, MacKinnon, and Pojar 1998). Indians ate berries with salmon eggs, accounting for a western common name, salmonberry. The name *parviflorus*, "small-flowered," is inappropriate because the flower is relatively large, largest among all species of *Rubus*. More often used in landscaping, especially in European gardens, than other armed *Rubus* species.

Chromosome No. $2n = 14$, $n = 7$; $x = 7$

Similar Species. Another thimbleberry, the purple-flowering raspberry, *Rubus odoratus* L.,

254

1. Twig with leaves and flowering shoot
2. Fruiting shoot, aggregate of drupes

KEY CHARACTERS

- upright branching shrub to 2.5 m
- leaves simple, large, maplelike, palmately 5-lobed and veined, base cordate or deeply cleft, petioles long
- stems lacking prickles, densely covered with brownish hairs
- fruit a drupe borne as an aggregate of drupelets on a receptacle, the aggregate when ripe separating easily from the receptacle, bright to dull red, thimble-shaped, edible
- habitat open, diverse, harsh, rocky, sandy sites of upper Michigan and the tip of northern lower Michigan

is similar in its tall thicket-forming habit, to 1.5 m; simple leaves; and the absence of spines and prickles. It is easily distinguished by its rose-purple to deep purple flowers and acidic and unpalatable fruit. It overlaps the range of thimbleberry in northeastern lower Michigan and hybridizes with it.

ROSACEAE
Rubus strigosus Michaux
Wild Red Raspberry

Size and Form. Upright shrub, ascending, rarely arching, 0.5–2 m high; clone-forming by sprouting from roots, not tip rooting.

Bark. Brownish, shreddy or peeling in long papery strips.

Leaves. Alternate, deciduous, trifoliate or pinnately compound; primocane leaflets 3–5 (7), 3–8 cm long, sessile except the terminal; terminal broadly ovate, sometimes 3-lobed, 5–10 cm long, 3–7 cm wide; acuminate; rounded to cordate; petiolule bristly, 0.5–3 cm long; lateral leaflets smaller, sessile, ovate-oblong to lanceolate; acute to acuminate; rounded to cordate; irregularly or coarsely doubly serrate; floricane leaflets mostly 3, short-stalked, sometimes simple; all leaflets green, glabrous to sparsely pubescent above, white-tomentose beneath; margins irregularly or doubly serrate; stipules slender, soon withering; petioles 2–7 cm long, with gland-tipped hairs and small, stiff bristles, leaf base persistent.

Stems-Twigs. Biennial; primocane upright, young stems round, brown to reddish, densely glandular-bristly, 1–2 mm long, ashy-woolly and appearing pink beneath the stiff, bristlelike prickles, older twigs with a few small, slender-based, hooked prickles or lacking prickles.

Winter Buds. Terminal bud absent; lateral buds oblong-ovoid, sessile, usually superposed, the lower and smaller one covered by petiole base; upper bud elongate, 1–3 mm long, several pubescent overlapping scales exposed.

Flowers. May–June; bisexual; white, about 1 cm wide, borne on 2- to 5-flowered, terminal clusters, solitary in upper leaf axils, pedicels and peduncles glandular-bristly; sepals lanceolate, long-acuminate, glandular-hispid, the grayish lobes with dark tips; petals spatulate to obovate, erect, shorter than the sepals, 4–6 mm long; stamens numerous. Insect pollinated.

Fruit. Drupe, borne as an aggregate of drupes on a receptacle; July–August; the aggregate separating easily when ripe from the white receptacle which remains on the stalk; red, elongate-hemispherical, about 1 cm across, edible; single-seeded; seeds 328,000/lb (721,600/kg).

Distribution. Locally frequent to abundant throughout the state. County occurrence (%) by ecosystem region (Fig. 19): I, 79; II, 77; III, 100; IV, 75. Entire state 80%. Widely ranging throughout North America except for coastal plain states GA to TX.

Site-Habitat. Open, mesic to wet sites and basic to acidic soils; best development on circumneutral to basic, nutrient-rich soils; virtually any substrate or physical process that provides wet to moist conditions; wet shrub thickets; swamps; floodplains; lakeshores; bedrock shores, glades, and cliffs. Associates highly diverse, ranging from northern white-cedar, speckled alder, bog birch, and poison-sumac to blueberries and bearberry.

Notes. Shade-intolerant, persisting in light shade. The name *strigosus* refers to its stiff bristles. *Rubus strigosus* and its Eurasian counterpart are major sources for jellies and jams, as well as raw fruit.

Chromosome No. $2n = 14$, $n = 7$; $x = 7$

Similar Species. The closely related Eurasian species is *R. idaeus* L., red raspberry or Himbeere. The North American *R. strigosis* is often treated as *R. idaeus* var. *strigosus* (Michx.) Maxim. Many cultivars of *R. idaeus* L. and *R. strigosus* supply the red raspberries of commerce.

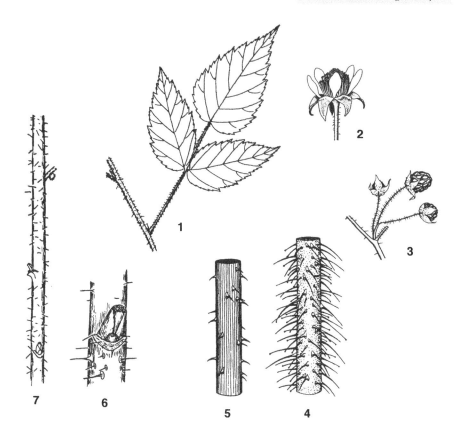

1. Primocane leaf, reduced
2. Flower, × 2
3. Floricane fruiting shoot, aggregate of drupes, × 1/2
4. Primocane stem, sprout or juvenile phase, × 2
5. Primocane stem, adult phase, × 2
6. Winter lateral bud and leaf scar, enlarged
7. Winter twig, reduced

KEY CHARACTERS

- upright shrub with arching branches; biennial
- leaves compound, 3–5 leaflets, white-tomentose beneath
- stems brown to reddish or pink, densely glandular-prickly, ashy-woolly beneath the bristly hairs
- fruit a drupe, red, borne as an aggregate of drupelets on a receptacle and separating from the receptacle when ripe
- habitat open, mesic to wet, often nutrient-rich sites

THE WILLOWS—*SALIX*

Salix and *Populus* are the main temperate zone genera of the large family Salicaceae. At one time, trees intermediate in characteristics between these genera were considered a separate genus, *Chosenia*, native of East Asia, but now they are included in *Salix*. The name *Salix* is derived from the Greek *helix*, "coil," referring to the pliant branches; similarly *willow* is from the Anglo-Saxon *welig* or *wican*, to "give way" or "bend" (Macleod 1952). However, according to Newsholme (2002) *willow* derives its scientific name from the Celtic word *salis, sal,* meaning "near," and *lis,* meaning "water"—an appropriate name for most of the 450 species and many hybrids that comprise the genus. Nearly worldwide in natural distribution, *Salix* species predominate in the Northern Hemisphere from temperate to arctic latitudes.

Willows occur naturally in open, moist to wet, disturbance-prone habitats. Trees, to 30 m tall, predominate along rivers. Shrubs, in addition to floodplains, stream sides, and lakeshores, predominate in peatlands; moist, rocky, mountain sites; and the Arctic tundra. Shrub willows are highly successful ecologically in occupying inhospitable sites for trees and herbs. They rapidly recouple nutrient cycles and develop extensive root systems following disturbances of fire, flooding, and landslides, which is significant in preventing erosion. Not surprisingly, they are widely used in watershed and soil stabilization and ecosystem restoration besides their well-known uses in habitat and food provision for wildlife, landscaping, basketry, furniture, cricket bats, and short-rotation forestry. The osiers, subg. *Vetrix*, have characteristic rod-like or flexible twigs, which are ideally suited for basketwork.

The 12 shrub willow species we describe, like most willows, have slender leaf- and catkin-bearing shoots and twigs. Crowns are typically bushy because current shoots forage extensively for light, virtually it seems without tidy hormonal control. Branches developing from current shoots are common (sylleptic growth), and crowns in open sun are marvelously intricate and totally unlike those of the stately red pine or tulip tree. They are more like American elm or white oak crowns compressed and displayed 2–4 m above ground. Thus indeterminate shoots are the rule, bearing leaves and flower buds in profusion. These shoots, 1–5 mm in diameter, grow vigorously until they are killed by freezing conditions of late fall and therefore lack terminal buds. The uppermost living lateral bud develops within it the leader shoot for next year. The preformed "early" leaves are often morphologically unlike in size and shape compared to the neoformed "late" leaves and the small leaves of flowering shoots. Thus for identification purposes, avoid the leaves of flowering shoots and the latest-formed leaves of indeterminate shoots. Winter buds have a single hood-shaped outer scale and one or more inner bud scales.

The pine cone willow gall, caused by the gall midge, *Rhobdophaga stro-*

biloides, is commonly found on the ends of shoots of willow species (e.g., *Salix eriocephala*). The midge lays an egg in the end bud of the willow shoot in early spring. The larva releases a chemical that causes the immature leaves to broaden and harden, forming a conelike gall rather than the typical shoot development with willow leaves.

Shrub willows are successful ecologically in open and disturbed sites because of their shade-intolerance, fast growth, ability to colonize and adapt to diverse and extreme habitats, and sexual and vegetative reproduction. The flowers are unisexual, with separate plants bearing male and female flowers (dioecious), which are densely arranged in catkins (aments). Catkins are usually preformed in winter buds on long indeterminate shoots in portions of the upper crown that are exposed to high light irradiance. Catkins may be borne densely, to a minimum of 1 cm apart, on slender shoots, 1–5 mm in diameter, indicating that huge amounts of pollen and seeds may be produced annually by a well-lighted plant. Catkins flush in the early spring before the leaves, with the leaves, or after young leaves are formed. In the upper Great Lakes Region, *Salix serissima* is the representative species, with ripe fruits and dispersal in late fall. The striking spring displays, especially of male catkins, are a major reason for using shrub willows in horticulture (e.g., pussy willow, *Salix discolor*).

Catkins terminate short, often leafy shoots, the leaves on female branchlets becoming larger and persisting longer than their male counterparts. Along the densely hairy rachis of the catkin are borne tiny individual flowers, each subtended by a flower bract with fine silky hairs, which extend well beyond the bract. These function in part to facilitate pollination by holding and bringing the often sticky pollen grains in contact with the stigma lobes as wind moves and slaps the long hairs against them. The flowers are highly reduced, lacking a perianth; they have nectar glands and an odor, which attract various insects, although wind pollination is important as well. Tiny seeds are produced 3–4 weeks after pollination and dispersed in spring or fall depending on the time of flowering. The seed coat of the tiny seeds, up to 2 million/lb), is covered with hairs (i.e., "cotton"), which facilitate wind dispersal. Seeds dispersed in spring and early summer germinate in 12–24 hours under moist conditions, whereas seeds dispersed in fall overwinter and germinate quickly following snowmelt early in the following spring.

Vegetative reproduction by basal sprouting and clump formation, layering, root sprouting, and fragmentation is extremely important in the maintenance and persistence of shrub willow clones. Unlike their root-sprouting and spatially spreading relatives, the aspens and balsam poplars of *Populus*, asexual reproduction is mainly by basal sprouting and layering. Clonal clumps spread locally by layering of branches and given open conditions form bushy clones, for example, prairie willow (*Salix humilis*) and bog willow (*S. pedicellaris*) (Smith 2008). Root sprouting is characteristic of *S. interior,* and dense clones form along sandbars and sandy-gravelly river margins.

Shrub willows are successful in Michigan and the upper Great Lakes Region not only because of their enormous reproductive ability but because they are adapted to colonize and persist in diverse kinds of wetland habitats: marshes, fens, bogs, river and stream edges and floodplains, swales, and lakeshores. Given their abundance and wide seed dispersal, they opportunistically colonize open and disturbed dryland sites wherever human disturbances provide favorable conditions. Of the 74 natural communities of Michigan (Kost et al. 2007, excluding caves and sinkholes), the 12 species we describe are included in 17—primarily wet sites of shrub thickets, fens, floodplains, and swamps where open conditions prevail.

Because willow species are generally shade-intolerant, they establish best in open or very lightly shaded conditions. Once established, they can withstand light to moderate shade of encroaching vegetation. However, their ability to reproduce declines markedly as the amount of overhead cover increases. Because they are fast growing and potentially enormous seed producers, they thrive in and are most characteristic of ecosystems well provided with water, nutrients, and oxygen. Such sites are those of moderate to high nutrient availability such as circumneutral or basic substrates where their roots penetrate or a seasonal rise in the water table provides the nutrients required for extension growth into the open. Moving water with high or adequate oxygen is the most favorable condition. However, virtually all willows, unlike most trees and shrubs of our region, have adaptations to tolerate low oxygen availability in (1) floodplains of rivers and streams when roots and lower stems are inundated by flooding *during the growing season* or covered by fine particles of silt and clay (i.e., silting, sedimentation), and (2) in wetlands with saturated soils such as swamps, bogs, fens, and muskegs (Barnes 1991, 319–20).

Willows tolerate surface acidic conditions of Michigan ecosystems but are not characteristic of ecosystems or microsites that are highly acidic, which are dominated by ericaceous species. In the boreal forest and tundra, they predominate along the warmer, open, and more nutrient-favorable riverine ecosystems. The dwarf and creeper forms are adapted to the climatically and edaphically harsher conditions but nevertheless full-sun environments of tundra.

Hybridization is reported to be more prevalent in willows than in any other woody plant group, although relatively little is known about its extent (150–200 species). Willows were once used widely for medicinal purposes because of the presence in the bark of 12 different salicylates, especially salicin and its esters, the starting point for the synthesis of acetylsalicylic acid (ASA) of aspirin. When willow bark is consumed, the salicylates are transformed in a series of processes into salicylic acid, which has the pain-relieving, fever-reducing, and anti-inflammatory effects that we associate with ASA pills (Marles et al. 2000). Several authorities consider willow taxonomy, culture, and use in horticulture (Flint 1997; Cullina 2002; Newsholme 2002; US Department of Agriculture 2008).

Five subgenera comprise the willows (Argus 2010). Eight of the species we describe are members of the subgenus *Vetrix*: *S. bebbiana, S. candida, S. cordata, S. discolor, S. eriocephala, S. humilis, S. myricoides,* and *S. petiolaris.* They are shrubs or small trees, which in horticulture are known as osiers and sallows (Newsholme 2003). In subgenus *Salix,* tall shrubs or small trees, are *S. lucida* and *S. serissima.* Our characteristic river floodplain species sandbar willow, *S. interior,* is in subgenus *Longifoliae.* Our wetland *S. pedicellaris,* with transcontinental range to the Yukon, belongs to subgenus *Chamaetia,* characterized by many low, creeping, dwarf, arctic, and mountain species.

The key to 12 commonly occurring species is designed for identification during the growing season and emphasizes plant form and leaves. Keys of reproductive characteristics are available in Voss and Reznicek 2012 and Smith 2008. The definitive taxonomic treatment of the genus in the *Flora of North America* is provided by Argus (2010).

KEY TO SPECIES OF *SALIX*

1. Low or dwarf shrubs with erect branches to 1.5 m.
    2. Mature leaves glabrous or nearly so, leaves glaucous beneath; leaves relatively broad, blades <5 times longer than wide; stipules absent; capsules glabrous . . . *S. pedicellaris,* p. 282
    2. Mature leaves pubescent, densely tomentose beneath; leaves relatively narrow, blades >5 times longer than wide; stipules present; capsules pubescent . . . *S. candida,* p. 266
1. Erect shrubs more than 1.5 m high.
    3. Petiole glandular at or near its junction with the blade; leaves relatively broad, <5 times longer than wide.
        4. Leaves glaucous beneath; catkins appear long after the leaves; fruiting in late summer or autumn . . . *S. serissima,* p. 286
        4. Leaves not glaucous beneath; catkins appear with the leaves; fruiting in early summer . . . *S. lucida,* p. 278
    3. Petiole not glandular at or near its junction with the blade; leaves either relatively broad, blades <5 times longer than wide, or relatively narrow, blades >5 times longer than broad.
        5. Mature leaves pubescent, at least beneath.
            6. Leaves linear to narrowly lanceolate, blades >5 times longer than wide.
                7. Leaf blades densely tomentose beneath; young branchlets gray-white, pubescent . . . *S. candida,* p. 266
                7. Leaf blades with appressed silky hairs beneath; young branchlets glabrous or sparsely pubescent.
                    8. Margins of leaf blades remotely and often sharply toothed; petioles 0.5–5 mm long; clonal spreading shrub; river edges, lakeshores, sandbars . . . *S. interior,* p. 274

       8. Margins of leaves finely serrate, at least above the middle, sometimes entire near the base; petioles 3–10 mm long; forming clumps by basal sprouting; fens, bogs, shrub thickets ... *S. petiolaris*, p. 284

  6. Leaves broadly lanceolate to oblanceolate or oblong, elliptic or narrowly ovate; blades <5 times longer than wide.

    9. Base of leaf usually rounded to subcordate; margins regularly serrate, serrulate, or denticulate.

      10. Leaves oblong-lanceolate, the tips long-acuminate; young leaves reddish ... *S. eriocephala,* p. 272

      10. Leaves oblong-ovate to broadly lance-oblong, the tips acute or abruptly acuminate; young leaves not reddish ... *S. cordata*, p. 268

    9. Base of leaf acute or tapered; margins entire, undulate or irregularly crenate to serrate or glandular-serrulate.

      11. Branchlets divaricately spreading; catkins appearing with the leaves; bracts yellowish to straw-colored; capsules on pedicels 3–5 mm long ... *S. bebbiana*, p.264

      11. Branchlets not divaricately spreading; catkins appearing before the leaves; bracts dark brown to black; capsules on pedicels 1–2.5 mm long.

        12. Leaf blades smooth or the main veins slightly raised above; mature branchlets glabrous, lustrous, or pruinose ... *S. discolor*, p. 270

        12. Leaf blades often slightly rugose and with somewhat revolute margins, the veinlets impressed above; branchlets pubescent or glabrous but dull, not shiny ... *S. humilis*, p. 276

5. Mature leaves glabrous or the midrib and petioles sometimes pubescent.

  13. Leaves glaucous or whitened beneath.

    14. Margins entire to undulate-crenate, sometimes revolute, lacking definite teeth.

      15. Branches divaricately spreading; leaves dull green above, sometimes rugose beneath; catkins appearing with the leaves ... *S. bebbiana*, p. 264

      15. Branchlets not divaricately spreading; leaves dark green and shiny above; catkins appearing before the leaves ... *S. discolor*, p. 270

    14. Margins finely, distinctly, or regularly serrate.

      16. Leaves linear-lanceolate, elliptic-lanceolate to ovate-lanceolate, long-attenuate or tapering to an acute or acuminate tip.

        17. Leaf blades more or less equally tapered at both ends.

18. Leaves thin to membranaceous; petioles yellowish; stipules small or absent . . . *S. petiolaris*, p. 284

18. Leaves thick and leathery; petioles not yellowish; stipules ovate-lanceolate to broadly reniform, prominent on vigorous shoots . . . *S. myricoides*, p. 280

17. Leaf blades unequally tapered at the two ends, the tip acuminate to long-attenuate, the base abruptly tapered, rounded, or subcordate . . . *S. eriocephala*, p. 272

16. Leaves broadly elliptic, ovate, oblong-lanceolate or obovate, acute to short-acuminate or blunt to rounded at the tip . . . *S. myricoides*, p. 280

13. Leaves not glaucous beneath, green on both sides or only slightly paler beneath.

19. Leaves relatively narrow, blades >5 times longer than wide; margin entire or nearly so or with teeth irregularly and widely spaced; blade base cuneate; stems slender . . . *S. interior*, p. 274

19. Leaves relatively broad, blades <5 times longer than wide; margin finely and regularly spinulose-serrulate, blade base rounded or subcordate; stems stout . . . *S. cordata*, p. 268

SALICACEAE
Salix bebbiana Sargent
Beaked Willow, Bebb's Willow

Size and Form. Tall shrub with few ascending stems, 2–4 m, or small tree with single trunk to 8 m; branches divaricately spreading. Michigan Big Tree: girth 91.4 cm (36.0 in), diameter 29.1 cm (11.5 in), height 9.4 m (31 ft), Leelanau Co.

Bark. Grayish, rough and scaly, becoming furrowed with age.

Leaves. Alternate, simple, deciduous; blades 3–7 cm long, 1–3 cm wide, relatively broad, <5 times longer than wide; firm; elliptic, oblong-lanceolate, or obovate-oval; acute to short-acuminate; cuneate to slightly rounded; entire to undulate-crenate, sometimes irregularly gland-toothed, somewhat revolute; young leaves gray-pubescent to silky-hairy both sides, mature leaves dull green and finely pubescent, the lower surface rugose, with prominent veins, glaucous, gray-pubescent to tomentose or glabrate both sides when old; petioles 0.5–1 cm long, pubescent; stipules on indeterminate shoots 2–5 mm long, acute, toothed, persistent.

Stems-Twigs. Slender, flexible, round, divaricate, young twigs brownish, covered with fine gray pubescence, older twigs reddish to brown, slightly pubescent to glabrous; leaf scars small, crescent-shaped; bundle scars 3.

Winter Buds. Terminal bud absent; lateral buds small, 3–5 mm long, appressed, covered by 1 scale with color and pubescence as the twigs; petiole-bud scale scars projecting, often conspicuous on basal portion of old twigs.

Flowers. Unisexual, plants dioecious; catkins with or slightly before the leaves; April–May; unisexual; terminating short axillary shoots, whitish-pubescent, with several small leaves at the base of the shoots; male catkins 1–3 cm long, narrowly cylindrical, very hairy, stamens 2; pistillate catkins 2–4 cm long, becoming 4–8 cm in fruit, 1–2 cm wide, loosely flowered, pistils long-stalked, finely gray-pubescent, contracted above the base into a long, slender neck or beak. Flower bracts greenish-yellow to straw-colored, 1–2 mm long, lanceolate-oblong, thinly to densely hairy, with silky hairs extending 1–1.5 mm beyond bract apex. Insect and wind pollinated.

Fruit. Capsule; May–June; 0.6–1 cm long, tapering from a broad oval base into a long, slender beak about 7 mm long, pubescent; pedicels 3–5 mm long, pubescent; capsules lax in fruit; seeds tiny with cottonlike hairs attached.

Distribution. Locally common throughout the state. County occurrence (%) by ecosystem region (Fig. 19): I, 89; II, 77; III, 100; IV, 100. Entire state 87%. Transcontinental from NL to AK and widely distributed in the northern part of e. North America (s. to NJ, MD, IL) and throughout w. North America; lacking in the mid-South and the s.e. and Gulf coastal plain. Found across the Bering Straits to far e. Russia and Siberia.

Site-Habitat. Open to lightly shaded, moist to wet, circumneutral to basic sites; swamps, lakeshores, riverbanks, sedge meadows, northern and southern shrub thickets, wetland edges; in gaps as a shrub or small tree in coniferous or deciduous forests. Associates include many trees and shrubs of these diverse habitats.

Notes. Shade-intolerant, persisting under light shade. Several varieties have been recognized in different areas of its wide geographic range. In the Pacific Northwest and elsewhere, it is a main source of wood termed diamond willow, which has striking, diamond-shaped markings beneath the bark. Used medicinally for many reasons because of the presence of at least 112 different salicylates, especially salicin and its esters, flavonoids, and tannins. Salicin is closely related to ASA, the active ingredient of aspirin. See "The Willows— *SALIX*" for medical and other uses. The name *bebbiana* is for M. S. Bebb, a distinguished willow scholar. It forms natural hybrids with *Salix candida*, *S. humilis*, and *S. petiolaris* (Argus 2010).

Chromosome No. $2n = 38$, $n = 19$; $x = 19$

1. Vegetative shoot, × 1/2
2. Flowering shoot with male catkins, × 1/2
3. Male flower, × 5
4. Flowering shoot with female catkins, × 1/2
5. Female flower, × 5
6. Twig with vegetative shoots and expanded female catkins, × 1/2
7. Mature fruit, capsule, × 2 1/2
8. Seed with cottonlike hairs attached, × 5
9. Winter twig, × 1/2

KEY CHARACTERS

- tall shrub to small tree with few ascending stems
- leaf blades elliptic to obovate-oval, tip acute to short-acuminate, margin entire or sparingly crenate-serrate; mature leaves dull green and finely pubescent, the lower surface rugose, with prominent veins, gray-pubescent to tomentose
- flowers and fruit capsules terminating short leafy shoots; capsule and pedicel pubescent
- habitat open, moist to wet sites; not confined to wetlands

SALICACEAE
Salix candida Willdenow
Sage Willow, Hoary Willow

Size and Form. Low, much-branched shrub, 0.2–1.5 m; branches widely diverging. Asexual reproduction occurs primarily by layering.

Bark. Smooth or rough, gray or brown (Smith 2008).

Leaves. Alternate, simple, deciduous; blades 3–12 cm long, 0.5–2 cm wide, relatively narrow, >5 times longer than wide; firm; oblong-lanceolate to linear-lanceolate; acute to acuminate; cuneate or abruptly narrowed; entire, slightly revolute; white-tomentose above when unfolding, becoming thinly tomentose and dark green with impressed veins that resemble sage leaves; when young, dense white tomentum beneath conceals the veins; midrib prominent, yellow to reddish-brown; petioles 3–10 mm long, tomentose, not glandular near the junction with leaf blade; stipules small, lanceolate-glandular, tomentose, deciduous on determinate shoots, persistent and vigorous on indeterminate shoots, 3–10 mm long.

Stems-Twigs. Slender, flexible, many bud-scale scars, young shoots with woolly, fluffy, grayish-white tomentum, becoming yellowish to light brown, older stems reddish; leaf scars crescent-shaped; bundle scars 3, conspicuous.

Winter Buds. Terminal bud absent; lateral buds sharp-pointed, 3–6 mm long, covered by 1 scale.

Flowers. Unisexual, plants dioecious; catkins with or slightly before the leaves; May; unisexual; terminating short axillary shoots with several small leaves at the base; male catkins ovoid, 1–2.5 cm long; stamens 2; female catkins cylindrical, densely flowered, 2–5 cm long in fruit; pistils tomentose. Flower bracts obovate, 1.5–2.5 mm long, pale to dark brown, with silky hairs extending 1–1.5 mm beyond bract apex, persistent. Insect and wind pollinated.

Fruit. Capsule; June; ovoid-conic to lanceolate, densely white-woolly tomentose, 5–8 mm long, pedicel 0.5 mm long.

Distribution. Locally common in southern lower Michigan; occasional in northern lower Michigan and eastern upper Michigan; rare in western upper Michigan. County occurrence (%) by ecosystem region (Fig. 19): I, 68; II, 47; III, 100; IV, 25. Entire state 59%. Transcontinental species: NL to AK, s. to WA, e. to NE, IL, NJ.

Site-Habitat. Open, wet and wet-mesic, poorly drained, circumneutral to basic wetlands; fens, rich tamarack swamps, wet meadows, stream borders, marsh edges, marshy lakeshores. Associates include tamarack, poison-sumac, winterberry, bog birch, alder-leaved buckthorn, shrubby cinquefoil, ninebark, meadowsweet, and shrub dogwoods.

Notes. Shade-intolerant, persisting in light shade. The name *candida* refers to the white-tomentose leaves. A good indicator of basic wetlands.

Chromosome No. $2n = 38$, $n = 19$; $x = 19$

1. Branch with vegetative shoots and leaves, reduced
2. Lower side of leaf blade, × 1
3. Fruiting shoot with female catkins, × 1
4. Fruit, capsule with bract attached

KEY CHARACTERS

- low, much-branching shrub, 0.2–1.5 m; branches widely diverging
- leaf blades narrow, >5 times longer than wide; margin entire, revolute; upper surface dark green with impressed veins that resemble sage leaves, densely white-tomentose beneath; stipules present on indeterminate shoots
- fruit a capsule, densely white-tomentose
- habitat open basic wetlands, especially prairie fens and rich tamarack swamps

SALICACEAE
Salix cordata Michaux
Sand-dune Willow, Furry Willow

Size and Form. Low, coarse shrub 1–3 m, branches erect or ascending.

Bark. Brown.

Leaves. Alternate, simple, deciduous; blades relatively large, 3–10 cm long, 1–4 cm wide; thick and leathery, oblong-ovate or broadly lance-oblong; acute or abruptly acuminate; base broadly rounded to subcordate on early leaves and to cordate on late leaves of indeterminate shoots; margin finely spinulose-serrulate, teeth pointing outward, to 1 mm long and tipped with enlarged glands; silky-woolly above when young or permanently so above, glabrate beneath when mature; deep green both sides beneath woolly pubescence; veins prominent beneath, midvein heavily pubescent; petioles stout, 2–10 mm long, pubescent, somewhat clasping; stipules large, ear-like, 6–15 mm long, obliquely ovate to cordate, glandular-serrate.

Stems-Twigs. Rather short and stout, yellow or yellowish-brown to dark brown, covered with a dense white to grayish tomentum when young; older branches glabrous, leaf scars crescent-shaped; bundle scars 3.

Winter Buds. Terminal bud absent; lateral buds midsized, 3–8 mm long, acute to obtuse, covered by 1 brown to reddish-brown scale.

Flowers. Unisexual, plants dioecious; catkins before or with the leaves, May–June; terminating short axillary shoots with 3–5 leaves at base, which become 2–3 cm long; shoot and catkin rachis pubescent; male catkins 2–4.5 cm long, stamens 2; female catkins 2–4 cm long in flower, becoming 6–8 cm long in fruit. Flower bracts oblong dark brown at apex, densely villous with long hairs extending beyond the bract apex. Insect and wind pollinated.

Fruit. Capsule; mid-June through July; lance-ovoid, glabrous, 5–8 mm long, pedicels 0.5–1 mm long, glabrous.

Distribution. Occasional to frequent along the sandy shores and dunes of the Great Lakes, especially in northern lower Michigan, eastern upper Michigan, and mid- to southwestern lower Michigan. County occurrence (%) by ecosystem region (Fig. 19): I, 13; II, 47; III, 71; IV, 0. Entire state 35%. Centered in the Great Lakes Region and the Ottawa and St. Lawrence River valleys n. to Hudson Bay, s. from ON to n.e. IL, IN, e. to n. NY, ME.

Site-Habitat. Open, wet to seasonally wet sites; generally restricted to sandy lakeshores, marshes, and dunes along the Great Lakes. Associates include tamarack, balsam poplar, creeping juniper, sweet gale, red-osier, bearberry, sand cherry, and autumn and sandbar willows.

Notes. Shade-intolerant, persisting under light shade. Sometimes called heart-leaved willow; late leaves of indeterminate shoots are distinctly cordate. Leaf-blade teeth are typically more distinct and outward pointing than in *Salix eriocephala.*

In late fall, winter, or early spring, you may be surprised to find pine cones on willow stems. Not to worry; located at the apex of willow twigs is the pinecone willow gall, often forming on *S. cordata.* Emerging from an egg on a young shoot in summer, the larva of the gnat midge, *Rabdophage strobiloides,* secretes a hormone into the willow stem and causes it to proliferate and form a multilayered, conelike structure, thus halting further development of the shoot and housing the wintering larva, which metamorphoses into the gnat with warm weather.

Chromosome No. $2n = 38$, $n = 19$

1. Vegetative shoot, × 1/2
2. Flowering shoot, male catkins, × 1/2
3. Flowering shoot, female catkins, × 1/2
4. Mature fruit, capsule, × 5
5. Winter lateral bud and leaf scar, enlarged
6. Winter twig, slightly reduced

KEY CHARACTERS

- low, coarse shrub 1–3 m, branches erect or ascending
- all vegetative parts more or less pubescent
- leaf blades thick and leathery, silky-woolly above when young or permanently so, teeth tipped with prominent glands
- habitat open dunes and sandy lakeshores of the Great Lakes

SALICACEAE
Salix discolor Muhlenberg
Pussy Willow

Size and Form. Tall, few-stemmed shrub, 2–3 m or small tree up to 6 m; crown rounded, relatively compact; sprouting occasionally from the root collar. Michigan Big Tree; girth 142.2 cm (56.0 in), diameter 45.3 cm (17.8 in), height 8.9 m (32.0 ft), Shiawassee Co.

Bark. Grayish to dark brown, smooth, becoming rough, scaly.

Leaves. Alternate, simple, deciduous; blades 3–10 cm long, 1–3.5 cm wide; firm; oblong or elliptic, oblong-lanceolate or oblanceolate; acute, short-acuminate to blunt; acute to rounded; nearly entire to irregularly crenate-serrate, especially near the middle; young leaves sparsely to densely pubescent with deciduous, rusty-colored, curly hairs; mature leaves dark to bright green, glabrous above, main veins slightly raised, densely glaucous and finely hairy beneath; petioles 5–15 mm long, glabrous or minutely hairy; stipules small or rounded to ovate, toothed, prominent on vigorous shoots and sprouts.

Stems-Twigs. Stout, round, not divaricately spreading; at first pubescent (more densely so in var. *latifolia* Andersson), becoming glabrous, dark reddish-brown, smooth, lustrous; leaf scars large, crescent-shaped; bundle scars 3.

Winter Buds. Terminal bud absent; lateral buds large, 4–10 mm long, acute to obtuse, covered by 1 scale, light to dark brown, glabrous.

Flowers. Unisexual, plants dioecious; catkins much before the leaves, fully developed before leaves expand; March–April; terminating short axillary shoots with several early-deciduous inner bud scales or small, silky-haired leaves at base; male catkins 2–3 cm long, stamens 2; female catkins densely flowered, 2–6 cm long (12 cm in fruit); Flower bracts oval to oblong, dark brown to blackish at apex with silky hairs extending 1–1.5 mm beyond bract apex. Insect and wind pollinated.

Fruit. Capsule; May–June; 5–12 mm long, long-beaked, minutely silky-pubescent on pubescent pedicels 1.5–4 mm long.

Distribution. Locally common throughout the state. County occurrence (%) by ecosystem region (Fig. 19): I, 92; II, 83; III, 86; IV, 88. Entire state 88%. Transcontinental, NL to BC and NT, s. to CO, MO, IL, KY, NC.

Site-Habitat. Open to lightly shaded, moist to wet sites, rooting in circumneutral and basic soils; nutrient-rich and poor conifer swamps, fens in transition to bogs, northern and southern shrub swamps, riverbanks, lakeshores, wet meadows and abandoned fields, seasonally wet roadside ditches; occasionally upland. Associates included many tree and shrub species of diverse ecosystems and their communities.

Notes. Shade-intolerant, persisting under light shade. *S. discolor* is one of the most variable willows in morphology and hybridizes with several other species. In the variety *latifolia* Andersson, the twigs, buds, and lower leaf surfaces are pubescent. It is nearly equally common throughout the state and sometimes may resemble *Salix humilis;* see Voss and Reznicek 2012 for distinguishing floral and vegetative differences. Once the outer bud scale is shed, the strikingly white appearance of male catkins (the telescoped concentration of silky flower bract hairs) is highly prized by gardeners and florists in floral displays. Florists typically use twigs of the cultivated European species, *S. caprea* L., European pussy willow or goat willow, because of the silky softness of its catkins (Dirr 1990). However, for long-lasting spring display combined with the extreme cold hardiness of a native species, *S. discolor* is preferred. For landscaping, *S. discolor* is quite susceptible to stem canker and is considered inferior. The name *discolor* refers to the contrast between the dark green upper surface and densely glaucous lower surface.

Chromosome No. $2n = 76, 95, 114, n = 38, 57; x = 19$

Similar Species. The Eurasian pussy willow or goat willow, *Salix caprea* L., is a counterpart species.

270

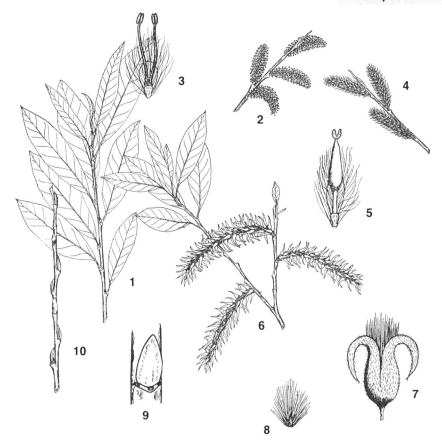

1. Vegetative shoot, × 1/2
2. Flowering shoot with male catkins, × 1/2
3. Male flower, hairs on subtending bract, × 5
4. Flowering shoot with female catkins, × 1/2
5. Female flower, hairs on subtending bract, × 5
6. Twig with vegetative shoots and expanded female catkins, × 1/2
7. Mature fruit, capsule, × 5
8. Seed with cotton attached, × 10
9. Winter lateral bud and leaf scar, enlarged
10. Winter twig, × 1/2

KEY CHARACTERS
- tall shrub to small tree
- leaf blades oblong or elliptic, oblong-lanceolate or oblanceolate and relatively broad, entire or irregularly crenate-serrate; contrasting dark green above and pale or densely glaucous beneath
- flowers in male and female axillary catkins on different plants, lacking small leaves at the base
- fruit a pubescent capsule
- habitat open, moist to wet sites, rooting in acidic to basic soils

271

SALICACEAE
Salix eriocephala Michaux

Heart-leaved Willow, Diamond Willow

Size and Form. Medium to tall shrub, 2–6 m; many erect or ascending branches; forming clump clones by basal sprouting.

Bark. Smooth to somewhat rough, gray; diamond-shaped markings beneath the bark.

Leaves. Alternate, simple, deciduous; blades relatively long and narrow, 0.5–15 cm long, 1–4 cm wide, oblong-lanceolate; acuminate to long-attenuate, rounded to subcordate on vigorous indeterminate shoots, rounded to cuneate on determinate and flowering shoots; granular-serrulate, teeth pointing upward; young blades pubescent and strongly reddish or purplish, later glabrous, dark green above, green or somewhat whitened beneath; in autumn rigid and veiny and often a deep reddish color; petioles 0.5–1.5 cm long, minutely pubescent, without glands; stipules ovate to reniform, 0.5–1.5 cm long, remotely glandular-dentate, large and persistent on indeterminate shoots and sprouts.

Stems-Twigs. Slender, flexible, round, glabrous, yellowish, gray-hairy when young, becoming reddish-brown to dark brown; leaf scars small, crescent-shaped; bundle scars 3.

Winter Buds. Terminal bud absent; lateral buds small, 2–5 mm, sharp-pointed, covered by 1 yellowish to dark brown scale.

Flowers. Unisexual, plants dioecious; catkins developing with or slightly before the leaves; April–May; terminating short axillary shoots with several small leaves at base, often many catkins in a series along ascending branches; male catkins 1–4 cm long, stamens 2; female catkins 2–8 cm long, ovary glabrous. Flower bracts densely hairy, 1–1.5 mm long, dark brown at apex with silky hairs extending 1–1.5 mm beyond the apex. Insect and wind pollinated.

Fruit. Capsule, May–June; lance-ovate, glabrous, 4–6 mm long, numerous, crowded, divergent, reddish when young, becoming brownish at maturity, pedicels 1–2 mm long.

Distribution. Locally common in southern lower Michigan, declining in occurrence and abundance northward. County occurrence (%) by ecosystem region (Fig. 19): I, 84; II, 60; III, 57; IV, 38. Entire state 69%. Widely distributed throughout e. North America, w. to SK and s. to CO, OK, LA, AL to FL.

Site-Habitat. Open to lightly shaded, wet, circumneutral to basic, poorly drained sites; swamps, swales, stream banks and river floodplains, lakeshores, ditches, dunes with seasonally high water tables. Associates include many willows, dogwoods, bog birch, shrubby cinquefoil, poison-sumac, swamp rose, highbush blueberry, winterberry.

Notes. Shade-intolerant, persisting under light to moderate shade. In older field guides it may be described as *Salix cordata* Muhl., *S. rigida* Muhl., or several other species. Subcordate or heart-shaped blade base is characteristic predominately of late leaves of indeterminate shoots. Willow pinecone galls are common on shoot tips. The name diamond willow derives from diamond-shaped markings beneath the bark.

Chromosome No. $2n = 38$, $n = 19$; $x = 19$

Similar Species. Similar to sand-dune willow, *Salix cordata* in some morphological characters but different in habit, geographic distribution, and site preferences.

272

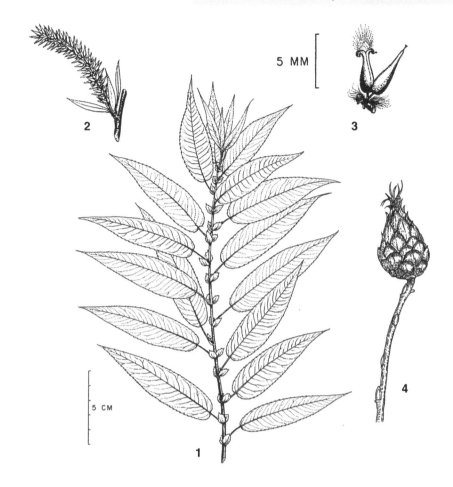

5 MM

5 CM

1

2

3

4

1. Indeterminate vegetative shoot, primarily late leaves, × 2/5
2. Fruiting shoot with female catkin, fruit a capsule, × 1
3. Capsules; one dehiscing, × 3
4. Willow pinecone gall, × 1/2

KEY CHARACTERS

- medium to tall shrub, 2–6 m; many erect or ascending branches sprouting from base
- blades long and narrow (to 15 cm long, 4 cm wide), tip acuminate to long-attenuate, young blades pubescent and reddish or purplish, later dark green above, whitened beneath, stipules conspicuous
- floral ovaries and fruit capsules glabrous
- habitat open, wet, poorly drained basic sites

Distinguished from sand-dune willow, *Salix cordata*, by its unfolding leaves very sparsely pubescent, older leaves usually somewhat whitened below, glabrous or nearly so; taller habit, more widely geographically distributed throughout the state and characteristic of wet, poorly drained sites; occurrence in swamps and swales rather than generally restricted to sandy lakeshores and dunes along the Great Lakes.

SALICACEAE
Salix interior Rowlee

Sandbar Willow

Size and Form. Tall shrub, 2–4 m, or small tree to 6 m, forming large, many-stemmed clones by stems arising from very long-spreading surface roots. Michigan Big Tree: girth 66.0 cm (26 in), diameter 21.0 cm (8.3 in), height 15.8 m (52 ft), Wayne Co.

Bark. Smooth to somewhat rough; branchlets yellowish-brown or red-brown, becoming gray on older stems.

Leaves. Alternate, simple, deciduous; blades 5–15 cm long, 0.5–1.5 cm wide, typically >10 times longer than wide; linear-lanceolate; acuminate; cuneate, tapering gradually to a very short petiole; remotely toothed, glandular-denticulate, teeth outward-pointing, irregularly and widely spaced or entire or nearly so; glabrous bright green above, distinctly veined, paler green or whitened beneath and silky-hairy when young, becoming glabrous; petioles 0.5–5 mm long; stipules absent or rudimentary on early leaves or narrowly ovate to lanceolate to 1.5 mm long and caduceus on late leaves.

Stems-Twigs. Slender, flexible, round, small, 3–6 (10) cm in diameter; glabrous, glossy, and giving a varnished appearance, reddish-brown; at first silky-hairy, becoming glabrous; epidermis peeling on branchlets; leaf scars small, half-moon-shaped; bundle scars 3, conspicuous.

Winter Buds. Terminal bud absent; lateral buds small, 2–4 mm long, appressed, covered by 1 reddish-brown scale, hairy at first, becoming glabrous.

Flowers. Unisexual, plants dioecious; catkins with the leaves; April–May; terminating axillary shoots with several narrow leaves at base from buds of the previous year or produced in midsummer of current year on long, indeterminate shoots; male catkins dense, 2–5 cm long, sometimes panicled, stamens 2; female catkins lax, 2–6 cm long in fruit, terminal on leafy shoots; pistils densely to thinly silvery-hairy, long-beaked with prominent stigmas. Flower bracts ovate to lanceolate, puberulent, pale yellowish-green, deciduous before capsules mature. Insect and wind pollinated.

Fruit. Capsule; May–July; 5–9 mm long, narrowly lanceolate, usually glabrous; pedicels 0.5–1.5 mm long.

Distribution. Common to abundant throughout the state except for the interior of northern lower Michigan (District II-8, High-plains). County occurrence (%) by ecosystem region (Fig. 19): I, 95; II, 70; III, 100; IV, 75. Entire state 84%. Transcontinental and distributed throughout e. North America except the coastal plain and inland sites of the Southeast: NC, SC, GA, FL, AL.

Site-Habitat. Open, seasonally flooded, wet, poorly drained, circumneutral to basic sites; a common floodplain species; newly formed land of rivers and stream edges, sandbars, perimeter of islands in rivers and lakes, dunes, lakeshores, and wet edges of swamps. Associated with many trees and shrubs of the front, first bottom, and backswamp of rivers and other open, wet, and poorly drained sites.

Notes. Very shade-intolerant, rarely persisting in light shade. Tolerant of sedimentation, seasonal flooding, and inundation during the initial part of growing season. On the remotely denticulate, mature early leaves, the greatest distance between teeth is generally 3–6 mm. Formerly known as *Salix exigua* Rowlee. One of the most tolerant of woody plants to lack of oxygen when waterlogged or inundated during part of the growing season. Tolerant of medium acid (pH 5.6) to calcareous (pH 8.0) soils. Commonly infested with willow pinecone galls caused by the larvae of the willow pinecone gall midge, *Rhabdophaga stobiloides.*

Chromosome No. $n = 38$, $n = 19$; $x = 19$

Similar Species. Closely related to narrow-leaf or coyote willow, *Salix exigua* Nuttall; of smaller stature, leaf margins usually entire, range overlapping *S. interior* in the Great Plains and occurring in all southwestern and western states where *S. interior* is absent (Argus 2010).

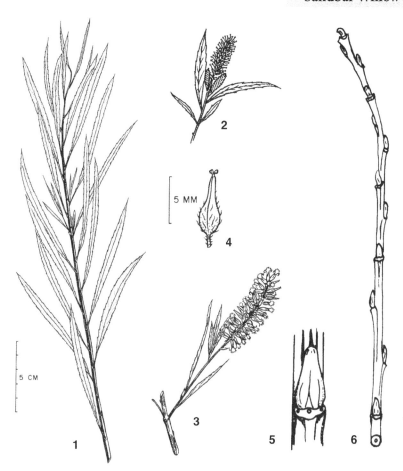

1. Indeterminate vegetative shoot
2. Flowering shoot, male catkin
3. Fruiting shoot, female catkin with capsules
4. Fruit, capsule
5. Winter lateral bud and leaf scar, enlarged
6. Winter twig, × 1

KEY CHARACTERS

- tall shrub, 2–4 m, forming many-stemmed clones
- leaf blades linear-lanceolate, typically >10 times longer than wide; both ends acuminate to acute, margin remotely serrulate-toothed with teeth irregularly and widely spaced or entire or nearly so
- stems slender, glossy, and giving a varnished appearance, reddish-brown
- habitat river and stream margins and floodplains, lakeshores, dunes

SALICACEAE
Salix humilis Marshall
Upland Willow, Prairie Willow

Size and Form. Low to medium sprangly shrub, 1–2 m, occasionally 3 m; many stems arising from the base or by layering and forming a compact spreading clone.

Bark. Smooth, becoming roughened, gray.

Leaves. Alternate, simple, deciduous; crowded toward the tip of the shoot; blades 2–11 cm long, 1–2.5 cm wide, relatively broad, <5 times longer than wide; oblanceolate or narrowly obovate; acute to abruptly short-acuminate; cuneate to acute; entire or often undulate-crenate, with minute gland-tipped teeth from midblade to apex, somewhat revolute; upper surface dull dark green, puberulent to glabrate, with impressed venation, lower surface rugose, gray-pubescent to glabrate or somewhat glaucous if lacking pubescence, veins prominent, reticulate; petioles 0.5–1.5 cm long, pubescent; stipules occasional or absent on early leaves, conspicuous and persistent on late leaves of indeterminate shoots, lanceolate, 3–10 mm long, acute-toothed, pubescent.

Stems-Twigs. Moderately slender, 3–5 mm basal diameter, round, flexible, pubescent, becoming glabrate, dull; depending on size and exposure, from greenish-yellow, yellowish-brown, reddish-brown to dark brown; leaf scars small, crescent-shaped; bundle scars 3.

Winter Buds. Terminal bud absent; lateral buds closely spaced, midsize, 4–8 mm long, elliptical, flattened with broad blunt tips, covered by 1 scale, light tan to reddish, hairy.

Flowers. Unisexual, plants dioecious; catkins much before the leaves; March–May; terminating very short axillary shoots with several very small, early deciduous, often silky-hairy leaves at base; male catkins sessile, 1–1.5 cm long, stamens 2; pistillate catkins 1.5–2 cm long, becoming 2–5 cm long in fruit; pistils tomentose. Flower bracts oblanceolate-obtuse, dark brown to blackish at apex with silky hairs extending 1–1.5 mm beyond bract apex. Insect and wind pollinated.

Fruit. Capsule; late April–June, slender, long-beaked, tomentose, 7–9 mm long; capsule splitting along two sutures to release cottony-tufted seeds, 0.6–0.8 mm long.

Distribution. Locally common throughout the state. County occurrence (%) by ecosystem region (Fig. 19): I, 74; II, 77; III, 100; IV, 88. Entire state 78%. Widely distrib-uted in e. North America, NL to MA, s. to TX, FL.

Site-Habitat. Open, dry to moist sites, typically rooting in basic substrates of otherwise acidic, nutrient-poor soils; upland prairies; dry sandy, fire-prone, outwash and lake plains; pine and oak savannas and barrens; sometimes on mesic to wet meadows, lakeshores. Associates include jack pine, northern pin oak, black cherry, New Jersey tea, prairie redroot, sand cherry, low sweet blueberry, sweetfern, hazelnuts.

Notes. Shade-intolerant, persisting in light shade. Tolerant and characteristic of dry, acidic, sandy, fire-prone plains; for example, the Highplains District (II-8) of northern lower Michigan (Fig. 19). Sprouting vigorously following fire. Generally, leaf blades become more lanceolate and proportionally longer than wide on indeterminate shoots than determinate shoots and at the apex compared to the base of determinate shoots. Plants smaller in habit, var. *tristis* (Aiton) Griggs, occur primarily in southernmost Michigan. The name *humilis* refers to the low stature of individuals.

Chromosome No. $2n = 38, 76, n = 19, 38; x = 19$

Similar Species. The pussy willow, *Salix discolor*, is most morphologically similar but typically a tall shrub or small tree with few stems.

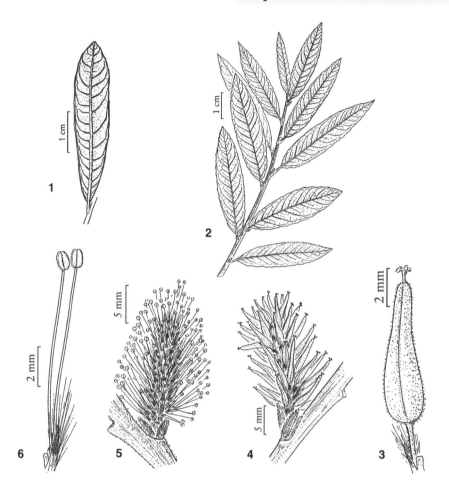

1. Leaf, × 1
2. Vegetative shoot, × 2/3
3. Fruit, capsule, 2 mm
4. Flowering shoot, female catkin, 5 mm
5. Flowering shoot, male catkin
6. Male flower

KEY CHARACTERS

- low to medium spreading shrub, typically 1–2 m
- leaf blades oblanceolate or narrowly obovate, lower surface rugose, gray-pubescent to glabrate with prominent reticulate veins, margin entire or undulate-crenate with minute gland-tipped teeth
- flowers borne in catkins terminating axillary shoots with small green leaves at base
- habitat dry to moist upland sites, often pine-oak savannas and barrens

Distinguished from pussy willow, *Salix discolor*, by low, spreading clones on dry to dry-mesic prairies and barrens; leaf blades oblanceolate or narrowly obovate, rugose and gray-pubescent beneath, not white-glaucous.

277

SALICACEAE
Salix lucida Muhlenberg
Shining Willow

Size and Form. Tall shrub, multiple ascending stems, 1–3 m or small tree 4–6 m.

Bark. Smooth, becoming roughened with age, brown.

Leaves. Alternate, simple, deciduous; blades relatively large, 4–15 (20) cm long, 1.5–4.5 (8) cm wide, relatively broad, <5 times longer than wide; leathery and coriaceous; lanceolate to ovate-lanceolate; long-acuminate or attenuate; rounded or cuneate; sharply serrate with gland-tipped teeth; shining green, with reddish-brown hairs when very young, becoming glabrous above, paler green and glabrous beneath, petioles 0.5–1.5 cm long, glabrous or sparsely pubescent with early-shedding reddish-brown hairs, glandular near the junction with leaf blade; stipules on indeterminate shoots and vigorous flowering branchlets, conspicuous, persistent, subcordate to reniform, 2–4 mm long and wide, marginal glands distinctive.

Stems-Twigs. Slender, flexible, mostly midsized, 2–5 mm in diameter, round, young shoots with early-shedding reddish-brown hairs, becoming glabrous; yellowish-brown to reddish-brown, very shiny; leaf scars half-moon-shaped; bundle scars 3, conspicuous.

Winter Buds. Terminal bud absent; lateral buds appressed, 4–10 mm long, covered by 1 scale with fused margins, shining chestnut-brown, glabrous.

Flowers. Unisexual, plants dioecious; catkins with the leaves, May-June; terminating relatively long shoots (to 8 cm) from buds with several small leaves at the base; male catkins stout, 1.5–5 cm long, stamens 3–6, anthers golden prior to anthesis; densely flowered with conspicuous yellow bracts; female catkins 1.5–5 cm long in fruit, narrower than the male. Flower bracts pale yellow, oblong, obtuse, pubescent on back with silky hairs extending 1–1.5 mm beyond bract apex. Insect and wind pollinated.

Fruit. Capsule; June–July; borne on pedicels 0.5–1.5 mm long; slender, ovoid-conical, 4–7 mm long, rounded at base, long-beaked, light brown, glabrous, bracts deciduous after flowering; seeds tiny, tufted with silky white hairs.

Distribution. Locally frequent throughout the state, increasing in abundance northward. County occurrence (%) by ecosystem region (Fig. 19): I, 66; II, 83; III, 100; IV, 100. Entire state 78%. Transcontinental, NL to AK, s. to NM, IA, e. to n. VA.

Site-Habitat. Open, wet, poorly drained, basic to slightly acidic sites; wetlands, including swamps, swales, lakeshores, streamside alder thickets. Associates include northern white-cedar, bog birch, shrubby cinquefoil, and red-osier and silky dogwood.

Notes. Shade-intolerant, persisting in light shade. A conspicuous species with beautiful shining foliage and branches. The golden anthers of male catkins are especially attractive in spring.

Chromosome No. $2n = 76$, $n = 38$; $x = 19$

Similar Species. Resembles autumn willow, *Salix serissima*, and bay-leaved or laurel willow, *S. pentandra* L., a European tree species occasionally escaped from cultivation. See detailed keys in Soper and Heimburger 1994; and Voss and Reznicek 2012.

278

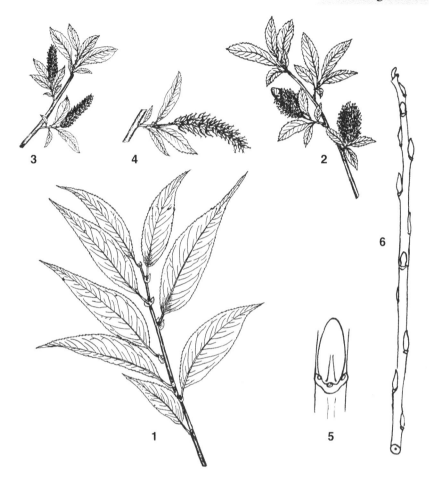

1. Vegetative shoot, × 1/2
2. Flowering shoot with male catkins, slightly enlarged
3. Flowering shoot with female catkins, × 1/2
4. Fruiting shoot with an expanded female catkin and fruits, capsule, × 1/2
5. Winter lateral bud and leaf scar, enlarged
6. Winter twig, × 1

KEY CHARACTERS

- tall shrub, 1–3 m with multiple stems, to small tree
- leaf blades large, leathery, and coriaceous, lanceolate-ovate, with long-acuminate to attenuate apex, shining green and glabrous both sides; unfolding leaves with reddish-brown hairs, stipules on indeterminate shoots conspicuous; petioles with glands near junction with blade
- twigs shining yellowish-brown and glabrous when mature, young shoots with reddish-brown hairs
- fruit a glabrous capsule, seeds shed June–July
- habitat open, wet places, rooting in neutral to calcareous soils

SALICACEAE
Salix myricoides Muhlenberg
Blueleaf Willow, Bayberry Willow

Size and Form. Low, spreading shrub, 1–2.5 m; rarely a small tree to 5 m; branches clustered, spreading or ascending, forming clones by layering.

Bark. Smooth, becoming rough with age, grayish-brown.

Leaves. Alternate, simple, deciduous; blades relatively large, 4–12 cm long, 1–6 cm wide, blades <5 times longer than wide; thick and leathery; mostly lanceolate or broadly elliptic, occasionally ovate to obovate (see diagram upper left); acute or abruptly short-acuminate; cuneate or rounded (equally tapered at apex and base); margins finely, distinctly, or regularly serrate with gland-tipped teeth; dark green above, strongly glaucous beneath, sometimes appearing bluish-white, often hairy when young, becoming glabrous; petioles stoutish, 4–12 mm long, pubescent and often extending along the blade midrib; stipules on indeterminate shoots ovate-lanceolate to broadly reniform-subcordate, margins gland-toothed, acute, 3–10 mm long, prominent and persistent.

Stems-Twigs. Stoutish, round, puberulent to tomentose when young, becoming smooth, yellowish to chestnut to dark brown; leaf scars half-moon- to crescent-shaped; bundle scars 3.

Winter Buds. Terminal bud absent; lateral buds 3–6 mm long, blunt, brown, silky-hairy, covered by 1 scale.

Flowers. Unisexual, plants dioecious; catkins before or with developing leaves; May–June; terminating axillary shoots (4–7 cm long in females) from buds of the previous year with several small leaves at base; flowers on both male and female catkins numerous and crowded at first along the rachis, becoming lax as capsules develop and mature; rachis and peduncle pilose-tomentose; male catkins 2–4 cm long, stamens 2; female catkins 6–8 (10) cm long, rachis slightly silky-hairy to glabrate. Bracts of male and female catkins obovate to oblanceolate, dark brown or blackish, 1–2 mm long, densely villous with silky hairs extending 1–1.5 mm beyond bract apex. Insect and wind pollinated.

Fruit. Capsule; May–June, narrowly lance-ovoid, 4–8 mm long, glabrous; pedicels 1–6 mm long.

Distribution. Occasional to frequent in lower Michigan and eastern upper Michigan,

especially on dunes and sites along the shores of lakes Huron and Michigan and inland lakes. County occurrence (%) by ecosystem region (Fig. 19): I, 39; II, 50; III, 71; IV, 12. Entire state 43%. Glaciated e. North America; centered in the Great Lakes region, ON, QC to Hudson Bay; New England s. to IL, IN, OH, PA.

Site-Habitat. Open, wet-mesic to wet and poorly drained sites with calcareous substrates; open dunes, sandy-gravelly lakeshores, interdunal depressions along the Great Lakes. Associates include trembling aspen, balsam poplar, creeping and dwarf junipers, bearberry, sand-dune and autumn willows.

Notes. Unfolding leaves at shoot tips often reddish, glabrous, or with copper-colored hairs. Described in older sources as *Salix glaucophylloides* Fernald. The name *myricoides* means "resembling *Myrica* in leaf shape." The variety *albovestita* (Ball) Dorn is known for its densely white-villous branchlets.

Similar Species. Unfolding young leaves similar to those of *Salix cordata*, but mature leaves are thick and leathery. Pussy willow, *S. discolor*, also has whitened or glaucous lower blade surfaces, but it has entire blade margins and is a larger shrub or small tree.

280

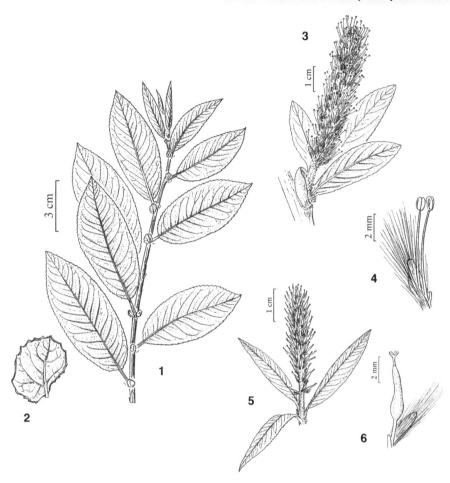

1. Vegetative shoot, × 1/2
2. Stipule, enlarged
3. Flowering shoot, male catkin, × 1/2
4. Stamens and bract, × 3
5. Flowering shoot, female catkin, enlarged
6. Fruit, capsule, × 3

KEY CHARACTERS

- low, spreading shrub, 1–2.5 m, clustered branches
- leaf blades thick and leathery, relatively large (to 12 cm long, 6 cm wide), margins finely, distinctly, or regularly serrate with gland-tipped teeth; dark green above, strongly glaucous beneath, sometimes appearing bluish-white
- fruit a glabrous capsule
- habitat open, moist to wet calcareous sites; dunes and interdunal depressions along lakes Michigan and Huron

SALICACEAE
Salix pedicellaris Pursh

Bog Willow

Size and Form. Low shrub with few erect stems, 0.3–1.0 m; often decumbent near the base; lower branches becoming buried in organic matter, developing adventitious roots and the ascending branches forming large clonal patches by layering.

Bark. Grayish or light brown on young stems, turning dark grayish-brown and roughened with age.

Leaves. Alternate, simple, deciduous; blades 1.5–6 cm long, 0.6–2 cm wide, relatively broad, usually <5 times longer than wide; firm and coriaceous; narrowly oval to narrowly or broadly elliptic, rarely ovate (see diagram on facing page); obtuse, rounded, rarely acute; rounded or tapered; margin strictly entire, slightly revolute; bright to dark green above, glaucous beneath; midrib prominent, yellow to reddish-brown; both sides glabrous with prominent reticulate venation; petioles 2–6 mm long, glandular near the junction with leaf blade; stipules absent.

Stems-Twigs. Slender, flexible, round, yellowish to olive-green, becoming reddish to purplish-brown or grayish; leaf scars small, moon-shaped; bundle scars 3.

Winter Buds. Terminal bud absent; lateral buds small, 2–5 mm long, pointed, covered by 1 brown scale.

Flowers. Unisexual, plants dioecious; catkins with the leaves; May–June; terminating axillary leafy shoots 1–4 cm long with several small leaves at base; male catkins 0.5–2 cm long, stamens 2; pistillate catkins 1–2 cm long, becoming longer in fruit, 2 cm wide, rather loosely flowered. Flower bracts oblong, 1 mm long, greenish-yellow to yellowish-brown, villous toward the apex with silky hairs extending >1 mm beyond apex. Insect and wind pollinated.

Fruit. Capsule; May–August; conspicuously stalked, 2–5.5 mm long; glabrous; oblong-conical, obtuse, 4–8 mm long, yellowish to dark reddish-brown.

Distribution. Locally occasional to common in wetlands throughout the state. County occurrence (%) by ecosystem region (Fig. 19): I, 53; II, 50; III, 86; IV, 62. Entire state 55%. Northern parts of e. North America; transcontinental to YT, s. to OR, ID; Great Lakes Region s. to IL and e. to NJ.

Site-Habitat. Open, wet, poorly drained, moderately acidic to basic wetlands; fens, bogs, muskegs, swamps, wet meadows, marsh edges. Associates include tamarack, northern white-cedar, and shrubs ranging from leatherleaf, bog-rosemary, bog-laurel, and highbush blueberry to those of more basic soils, bog birch, speckled alder, mountain holly, shrubby cinquefoil, meadowsweet, red-osier, sage, and pussy willows.

Notes. Very shade-intolerant. Distinguished among willows by complete lack of pubescence. A good indicator of fens. The name *pedicellaris* refers to the distinctive long stalks of capsules.

Chromosome No. 2n = 76, *n* = 38; *x* = 19

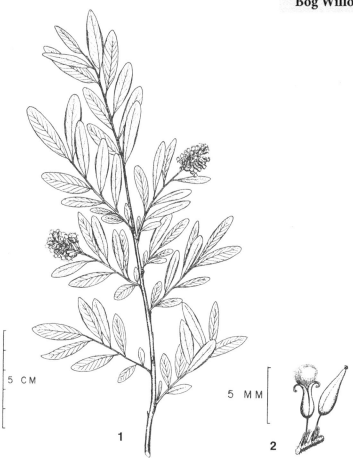

1. Branch with vegetative shoots and female flowering shoots
2. Capsules, one dehiscing

KEY CHARACTERS

- low shrub, decumbent or with ascending branches, 0.3–1.0 m; stems reddish to purplish-brown
- leaf blades elliptic-oblong to obovate-oblong; rounded, margin strictly entire, slightly revolute; glaucous lower surface; stipules absent
- fruit a glabrous capsule
- habitat open wetlands; bogs, fens, swamps, marsh edges, wet meadows
- general lack of pubescence in all vegetative parts

SALICACEAE
Salix petiolaris J. E. Smith

Slender-leaved Willow, Meadow Willow

Size and Form. Low to medium, much-branched shrub, 1–3 m; branches erect, long, arching; sprouting at root collar to form multistemmed clonal clumps. Michigan Big Tree: girth 33.0 cm (13.0 in), diameter 10.5 cm (4.1.0 in), height 10.4 m (34 ft), Leelanau Co.

Bark. Smooth, becoming roughened with age, gray.

Leaves. Alternate, simple, deciduous; blades 2–8 cm long, 0.5–1.5 cm wide, relatively narrow, >5 times longer than wide; thin to membranous, numerous and often overlapping along the stem; linear-lanceolate to lanceolate; acuminate; cuneate to acute; finely glandular serrulate, at least above the middle, entire or nearly so toward the base; usually hairy, often red-tinged, with rusty-colored hairs when young, soon glabrate or glabrous, green, lustrous above, glaucous and glabrous or thinly silky-hairy beneath; petioles 3–10 mm long, yellowish; stipules absent.

Stems-Twigs. Slender, flexible, round, leafy, at first yellowish-green to olive-brown, pubescent or glaucous, becoming glabrous and dark reddish-brown to purple, seldom shining; leaf scars small to inconspicuous, half-moon-shaped; bundle scars 3.

Winter Buds. Terminal bud absent; lateral buds small, 2–5 mm, acute to obtuse, dark brown or reddish-brown, covered by 1 scale.

Flowers. Unisexual, plants dioecious; catkins with the leaves; May–June; terminating very short axillary shoots with few small, linear leaves at base and often many in a series along indeterminate branches; male catkins 1–2 cm long, ellipsoid-obovoid, stamens 2; female catkins, 1.5–2.5 cm, rather loosely flowered and lax in fruit. Flower bracts linear-lanceolate to oblong-spatulate, yellowish or pale brown, acute, thinly pilose with silky hairs extending >1 mm beyond bract apex. Insect and wind pollinated.

Fruit. Capsule; May, 5–8 mm long, lanceolate-conic, thinly silvery-pubescent, slender-beaked; pedicels slender, 2.5–5 mm long, pubescent; seeds 500,000/lb (1,100,000/kg).

Distribution. Locally common throughout the state. County occurrence (%) by ecosystem region (Fig. 19): I, 79; II, 67; III, 100; IV, 100. Entire state 70%.Transcontinental northern and midwestern; NL to NT; MN s. to CO, e. to IL, IN, OH, PA, New England.

Site-Habitat. Open, wet, poorly drained but nonstagnant sites with basic substrate and often topsoil that may be acidic peat; moist meadows, bogs, fens, shrub thickets with alders, along the edges of pools, swamps, rivers, and lakes; ditches and ravines; also in dry upland sites such as jack pine woodlands. Associates may include a great many tree and shrub species typical of basic to acidic sites (e.g., northern fen; see Kost et al. 2007).

Notes. Shade-intolerant, persisting in light shade. Useful in basketmaking. Rooting in circumneutral or calcareous soils and associating with many species so disposed, for example, shrubby cinquefoil, bog birch, speckled alder, and northern white-cedar.

Chromosome No. $2n = 38$, $n = 19$; $x = 19$

Similar Species. Most are similar morphologically to silky willow, *Salix sericea* Marsh.

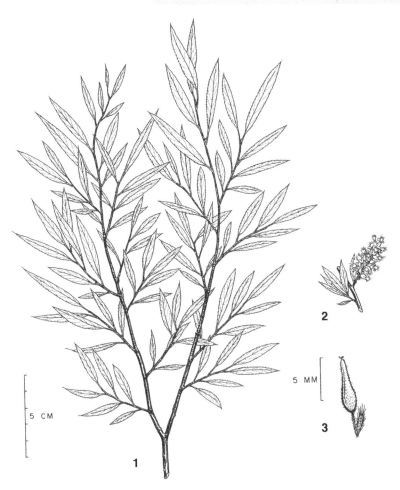

1. Branches with vegetative shoots
2. Fruiting shoot with female catkin, capsule
3. Capsule.

KEY CHARACTERS

- low to medium, much-branched shrub, 1–3 m; multistemmed clonal clumps
- leaf blades linear-lanceolate, >5 times longer than wide; acuminate, usually thinly silvery-pubescent when young, becoming glabrate, green, lustrous above, glaucous and glabrous or thinly silky-hairy beneath; petioles 3–15 mm long; stipules absent
- stems yellowish-green to olive-brown, pubescent or glaucous, becoming glabrous and dark brown to reddish-brown
- habitat open, wet, poorly drained but nonstagnant sites with basic substrate

SALICACEAE
Salix serissima (L. H. Bailey) Fernald
Autumn Willow

Size and Form. Medium to tall shrub, 1–4 m, usually multiple stems sprouting from the root collar and forming a clump clone.

Bark. Smooth, older stems grayish-brown.

Leaves. Alternate, simple, deciduous; blades 5–9 cm long, 1–3.5 cm wide, relatively broad, <5 times longer than wide; elliptic, elliptic-lanceolate, or oblong-lanceolate; acute to short-acuminate or long-tapered; acute to obtuse or rounded; finely glandular-serrulate; dark green, glabrous, highly glossy above, paler and whitened to strongly glaucous beneath, venation reticulate; petioles 0.4–1 cm long, yellowish-brown, distinctive glands near junction with leaf blade; stipules absent or minute, often reduced to a small glands at the node.

Stems-Twigs. Slender, flexible, round, yellowish- to olive-brown or reddish-brown, glabrous, shiny; leaf scars half-moon-shaped; bundle scars 3.

Winter Buds. Terminal bud absent; lateral buds blunt to sharp-pointed, 3–6 mm long, yellowish-brown, glabrous, covered by 1 scale.

Flowers. Unisexual, plants dioecious; catkins long after the leaves; May–June; terminating leafy axillary shoots; male catkins 1–3.5 cm long, short and stout, oblong to oblong-oval; stamens 5; female catkins 1.5–5 cm long and 2 cm wide in fruit. Flower bracts obovate, 2–3 mm long, pale yellow, white-pilose with silky hairs extending 1–1.5 mm beyond bract apex. Insect and wind pollinated.

Fruit. Capsule; August–October; pedicels thick, 0.5–2 mm long, narrowly conical, light brown, thick-walled, glabrous, 7–12 mm long, deciduous after dehiscence.

Distribution. Locally occasional to common in wetlands throughout the state. County occurrence (%) by ecosystem region (Fig. 19): I, 37; II, 30; III, 86; IV, 50. Entire state 40%. Locally occasional throughout the state. Transcontinental in n. parts of e. North America, NL to YT and BC., s. to CO; IL e. to NJ.

Site-Habitat. Open, wet, poorly drained, circumneutral to basic sites; shrub thickets, fens, swamps, open dunes, swales, shores, and stream sides. Associates include northern white-cedar, red maple, balsam poplar, speckled alder, bog birch, shrubby cinquefoil, winterberry, poison-sumac, swamp rose, shrub dogwoods.

Notes. Very shade-intolerant, persisting in light shade. The name *serissima* relates to late fruiting and autumn seed dispersal; flowering occurs in spring as with other willows.

Chromosome No. $2n = 76$, $n = 38$; $x = 19$

Similar Species. Shining willow, *Salix lucida*, is most similar but differs in leaf, flower, and much earlier seed-dispersal time (Voss and Reznicek 2012).

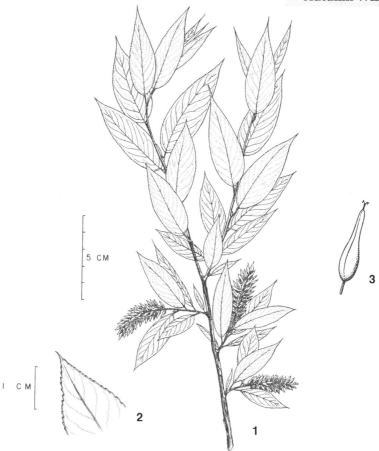

1. Branch with vegetative long-shoots and fruiting shorter shoots with catkins
2. Leaf blade, upper part
3. Fruit capsule

KEY CHARACTERS

- medium to tall shrub, 1–4 m, with multiple erect branches sprouting from base
- leaf blades elliptic-lanceolate, margin finely glandular-serrulate; dark green, glabrous, highly glossy above, paler and whitened to strongly glaucous beneath; petioles with glands near junction with leaf blade
- twigs yellowish- to olive-brown or reddish-brown, glabrous, shiny
- flowers expanding after the leaves, May–June
- fruit a glabrous capsule, late dehiscing and releasing seeds August–October
- habitat open to lightly shaded wetlands of calcareous substrate; fens, swamps

ADOXACEAE
Sambucus canadensis Linnaeus
Common Elder, Elderberry

Size and Form. Erect, few-branched shrub to 4 m; stems arise from a common base to form a broad, rounded, irregular crown of arching branches; shallow-rooted; clone-forming by basal sprouting from root collar and tip rooting. Michigan Big Tree: girth 35.6 cm (14.0 in), diameter 11.2 cm (4.4 in), height 7.9 m (26.0 ft), Leelanau Co.

Bark. Thin, gray, with conspicuous white, warty lenticels.

Leaves. Opposite, deciduous, pinnately compound, 10–30 cm long; leaflets 5–9 (usually 7), the lowermost often 3-parted, short-stalked, 5–15 cm long, 3–6 cm wide; elliptic, lanceolate, or ovate-oblong; acute or acuminate; rounded or tapered, often inequilateral; sharply serrate, lustrous, glabrous, bright green above, paler and glabrous or pubescent on veins beneath; rachis pubescent or glabrous, petiole 4–5 cm long.

Stems-Twigs. Angled in cross section, stout, swollen at the nodes, soft wooded and easily broken, yellowish-gray, warty, with conspicuous raised white lenticels, ill-smelling when crushed; leaf scars opposite, large, broadly crescent-shaped or triangular; bundle scars 5; pith soft, white, continuous, very large and conspicuous.

Winter Buds. Terminal bud lacking on indeterminate shoots, the current shoot dying back to the uppermost lateral buds (see Fig. 14B), lateral buds solitary or multiple, 2–5 mm long, conical or ovoid divergent, glabrous, light reddish-brown to green, 4–5 pairs of bud scales exposed.

Flowers. June--July; bisexual, borne in conspicuous, erect, large, terminal, flat-topped or slightly convex cymes, broader than long, 10–20 cm across, 200–400 flowers; small, 5–6 mm across; white; corolla 3–5 lobed, petals united; stamens 5. Insect pollinated.

Fruit. Drupe; September–October; dark purple to black, shiny, without bloom, 4–6 mm across, numerous on flat-topped cymes, with rose-red stalks, fruit clusters heavy, drooping; edible when ripe and cooked; used in pies, jellies, and jams and for wine; juice crimson; pits 3–5, about 3.5 mm long; seeds small, 232,000/ lb (510,400/kg).

Distribution. Common throughout the state; most abundant in the Lower Peninsula. County occurrence (%) by ecosystem region

(Fig. 19): I, 100; II, 90; III, 86; IV, 75. Entire state 93%. Widely distributed in the USA, adjacent e. Canada, and n. Mexico, lacking in the interior West and Pacific Northwest.

Site-Habitat. Open to moderately shaded, wet-mesic to mesic, moderately well to somewhat poorly drained, circumneutral, fertile sites. Common in conifer and hardwood swamps, floodplains, stream banks, forest edges, ditches; lacking on dry, sandy upland sites. Associated with American elm, red maple, black ash, yellow birch, blue-beech, bog birch, shrub dogwoods, winterberry, wild black currant, American hazelnut, red raspberry.

Notes. Moderately shade-tolerant, slow-growing in shade, relatively fast-growing in full sunlight. Typically occurring at lower latitudes and elevations than red-berried elder, *Sambucus racemosa*, where their geographic ranges overlap. Although vegetative parts may cause poisoning, the fruits may be used in jellies, wines, and pies; flowers may be used to flavor candies and jellies. Fruits are excellent wildlife food, especially for birds such as quail and pheasant. DeGraaf and Witman (1979) list 41 birds that use common elder for food, cover, or nesting. Bruised leaves, inner bark, and blossoms were used at one time to treat various ailments (Deam 1924). Severely browsed by white-tailed deer. Transplants well but is used rarely in landscaping. The elders include 20–30 species of deciduous shrubs or small trees, rarely herbs, in temperate regions of the Northern Hemisphere but extend into Africa, South America, Malesia, Australia, and New Zealand, especially in mountainous regions.

Chromosome No. $2n = 36$, $n = 18$; $x = 18$

Similar Species. Red-berried elder, *Sambucus racemosa*, is the closely related species

1. Flowering shoot with leaves, reduced
2. Fruiting shoot, drupe, reduced
3. Winter lateral bud and leaf scar, × 2
4. Winter twig, × 1/2

KEY CHARACTERS

- erect, few-branched shrub to 4 m
- leaves opposite, pinnately compound, with serrate leaflets
- twigs stout with conspicuous white warty lenticels; pith very large, white
- buds small, light reddish-brown to green
- fruit a purplish-black drupe borne in large numbers on a flat-topped cyme
- habitat wet-mesic, relatively fertile swamps and floodplains

Distinguished from red-berried elder, *Sambucus racemosa*, by its larger leaves with more leaflets; stems with white lenticels, pith white; flowers and purplish-black drupes borne in large flat-topped cymes; winter buds small, conical, light reddish-brown to green.

of eastern North America. It also occurs in western Europe and western Asia. In Europe the counterpart species is European elder, *S. nigra* L. The blue elder, *S. caerulea* Raf., occurs from British Columbia and Alberta south to California and New Mexico.

ADOXACEAE
Sambucus racemosa Linnaeus
Red-berried Elder, Red Elderberry

Size and Form. Tall coarse shrub to 4 m with arching branches; forming clones by basal sprouting and root suckers or rhizomes. Michigan Big Tree: girth 50.8 cm (20.0 in), diameter 16.2 cm (6.4 in), height 8.2 m (27 ft), Keweenaw Co.

Bark. Thin, gray, and warty with conspicuous lenticels.

Leaves. Opposite, deciduous, pinnately compound, 10–25 cm long; leaflets 5–7 (usually 5), short-stalked, 4–10 cm long, 2–4 cm wide; ovate-oblong to ovate-lanceolate; acute or acuminate; rounded to subcordate, usually inequilateral; sharply serrate; glabrous, lustrous, bright green above, pubescent and paler beneath; rachis pubescent, petioles 2.5–5 cm long.

Stems-Twigs. Angled in cross section, stout, soft wooded, easily broken, yellowish-brown, young shoots usually slightly to densely pubescent, warty, with conspicuous orange lenticels; leaf scars broadly crescent-shaped or triangular, large, bundle scars 5; pith soft, dark brown or orange-brown, continuous, very large and conspicuous.

Winter Buds. Terminal bud lacking on indeterminate shoots, solitary or multiple, large, 7–12 mm long; globular, reddish-purple, stalked; 2–3 pairs of bud scales exposed.

Flowers. April–May; bisexual, borne in pyramidal, compound cymes (or cymose panicle), 3–8 cm across, 100–200 flowers; small, 3–4 mm across, yellowish-white, corolla 3- to 5-lobed, stamens 5. Insect pollinated.

Fruit. Drupe; June–July; bright red, 5–7 mm across, borne in elongate pyramidal clusters; juice crimson-purple; raw berries inedible and considered toxic to humans (see "Notes"); pits globose, 3–5, dark yellowish-brown; seeds small, 286,000/lb (629,200/kg).

Distribution. Relatively common in the Lower Peninsula and common in the Upper Peninsula. County occurrence (%) by ecosystem region (Fig. 19): I, 89; II, 83; III, 86; IV, 100. Entire state 88%. Transcontinental in North America, s. to GA in e. North America and to CA and the Southwest in w. North America. Native throughout Europe and w. Asia.

Site-Habitat. Open to moderately shaded, mesic and wet-mesic, relatively fertile sites; mesic forests and swamp edges; open areas such as roadsides, fencerows, edges of woods, stream banks, and lakeshores. Lacking in deep shade of hemlock–northern hardwoods forests, present in gaps and forest edges. Associated with American elm, yellow birch, red maple, black ash, sugar maple, beech, basswood, white spruce, mountain holly, maple-leaved viburnum, common elder, soapberry.

Notes. Moderately shade-tolerant; slow-growing in shade, fast-growing in full sunlight. Typically occurring at higher latitudes and elevations than common elder where their geographic ranges overlap. Common throughout temperate areas of western North America from British Columbia and Alberta to California and New Mexico. Raw or unripe fruit, as well as leaves, seeds, roots, and bark, have cyanide-producing glycosides, which are poisonous to humans. *Cooked fruits* may be used in jellies, wines, and pies.

Chromosome No. $2n = 36$, $n = 18$; $x = 18$

Similar Species. Very closely related to the European red elder, *Sambucus racemosa* L., and now treated as subsp. *pubens* (Michaux) House or var. *pubens* (Michaux) Koehne of *S. racemosa* (Voss and Reznicek 2012). *Sambucus canadensis* is a closely related species in Michigan and elsewhere in North America. In China, Japan, and Korea, the counterpart species is *S. sieboldiana* Graebn. *S. kamtschatica* E. L. Wolf is reported from Japan's Hokkaido Island and northern Honshu. In northeastern China, related species include *S. buegeriana* (Nakai) Blume ex Nakai, *S. manshurica* Kitag., and *S. williamsii* Hance.

1. Leaf, reduced
2. Flowering shoot, × 1/2
3. Fruiting shoot, drupe, reduced
4. Winter lateral bud and leaf scar, × 2
5. Winter twig, × 1

KEY CHARACTERS

- erect, few-branched shrub to 4 m
- leaves opposite, pinnately compound, sharply serrate
- twigs stout with conspicuous orange lenticels; pith very large, brown to orange-brown
- buds large, reddish-purple
- fruit a bright red drupe borne in pyramidal cymes
- habitat mesic to wet-mesic, relatively fertile forests and swamps

Distinguished from common elder, *Sambucus canadensis*, by its smaller leaves and fewer leaflets; stems with orange lenticels, pith dark brown; flowers and bright red fruits borne in pyramidal cymes; winter buds large, globular, red or reddish-purple.

ELAEAGNACEAE
Shepherdia canadensis (L.) Nuttall

Soapberry, Buffaloberry

Size and Form. Low to medium, upright, spreading, unarmed shrub, 0.3–3 m.

Bark. Dark brown to dark gray, minutely hairy.

Leaves. Opposite, simple, deciduous, blades 2–4 cm long, thick, stiff; elliptic to ovate, obtuse; rounded or tapered; entire but roughened with marginal brown scales on lower surface; dark green and slightly stellata-pubescent above, densely scurfy-pubescent beneath, and silvery brown with snowflakelike scales, veins impressed above; petioles short, grooved, rusty-scurfy, 4–6 mm long.

Stems-Twigs. Slender, scurfy; current shoots green, becoming covered with reddish-brown scales, which turn gray with age; not thorny; leaf scars minute, raised, half round; bundle scar 1; pith small, round, continuous.

Winter Buds. Terminal bud present, 5–10 mm long, oblong, stalked, scurfy, 2–4 valvate scales exposed, appearing naked; lateral buds solitary or multiple, smaller, 2–5 mm long, similar in other respects.

Flowers. April–May just before the leaves; unisexual, plants dioecious; strongly aromatic; yellowish, borne in short spikes on twigs of the preceding year, inconspicuous, 3–5 mm across, 4 tiny yellowish sepals, petals lacking; either 1 pistil or 8 stamens; plants dioecious. Insect pollinated.

Fruit. Achene, enclosed in a fleshy peri-anth, the enlarged pulpy base of the calyx tube July–August; ovoid, 4–6 mm long, yellowish-red to bright red, juicy, insipid; soapy to touch.

Distribution. Hardy throughout the state, especially northward, where it typically occurs in lake-moderated coastal ecosystems. County occurrence (%) by ecosystem region (Fig. x19): I, 18; II, 50; III, 86; IV, 50. Entire state 39%. Distributed in North America from NL to AK, s. to NY, OH, MN, SD, NM.

Site-Habitat. Open, sandy, gravelly, or rocky sites with circumneutral to basic soils; lakeshores, cliffs, riverbanks, limestone outcrops; dunes, fens, calcareous marshes, lightly shaded understory and edges of forests. Associates, may include some 54 woody species in communities as diverse as boreal forest, coastal fen, and 11 other native communities of cobble and bedrock lakeshores and cliffs (Kost et al. 2007).

Notes. Shade-intolerant. Very cold and drought hardy. Difficult to transplant and rarely used in cultivation. A pioneer species invading naturally open and human-disturbed sites of all kinds. Roots bear nodules with nitrogen-fixing bacteria; soil is enriched where it grows. The fruits put in water can be whipped into a frothy, edible but bitter mixture known as Indian ice cream. The bitter compound is termed saponin; the common name soapberry refers to this soaplike froth. Still widely used by aboriginal peoples in British Columbia (Pojar and MacKinnon 1994).

Chromosome No. $2n = 22$, $n = 11$; $x = 11$

Similar Species. Thorny buffaloberry or silver buffaloberry, *Shepherdia argentea* Nutt., is centered in the northern Great Plains and adjacent Canada, with outliers as far south as Nevada and northern New Mexico and Arizona. Tolerant of drought and very cold temperatures. Blackfoot Indians believed that buffalo relished the fruits; fruits were cooked into a sauce and served with buffalo meat.

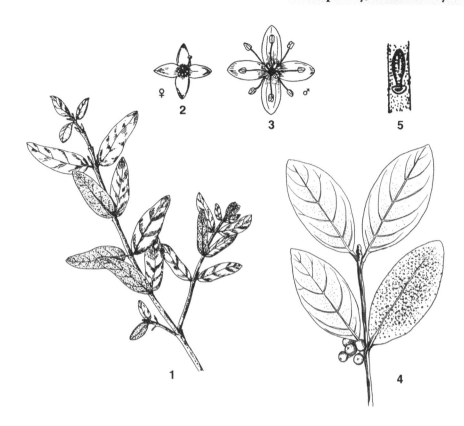

1. Vegetative shoot, × 1/2
2. Female flower, × 5
3. Male flower, × 5
4. Fruiting shoot, achene enclosed in a fleshy perianth, × 1
5. Winter lateral bud, enlarged

KEY CHARACTERS

- low to medium, upright, unarmed, spreading shrub
- leaves opposite, dark green and slightly scurfy-pubescent above, densely silvery-pubescent beneath mixed with brown scales
- twigs reddish-brown, scurfy
- fruit fleshy, yellowish to bright red, juicy, insipid or bitter

SMILACACEAE
Smilax hispida Rafinesque-Schmaltz
Bristly Greenbrier

Size and Form. Moderately stout, woody vine to 15 m long, few to many stems, high-climbing to 6 m or more; branches diffuse, climbing by tendrils; forming clones by rhizomes.

Leaves. Alternate, simple, deciduous, often retained during winter; blades 3–12 cm long, 3–9 cm wide; highly variable: oval, elliptic, elliptic-ovate, ovate-lanceolate; acute to obtuse or abruptly cuspidate; rounded or subcordate; margin entire or often fringed, with minute bristle-tipped teeth; thin, lustrous dark green, glabrous above and beneath, with bristles on the midrib; usually 2–3 pairs of primary veins from the base, arcuate, impressed; leaves breaking off above the petiole base; petioles 1–2 cm long, bearing pairs of tendrils at their base.

Tendrils. Very slender, paired, twining, attached at the base of the petioles.

Stems-Twigs. Slender, round above, slightly ridged near the base; green, glabrous, the lower stem and 1st-year sprouts (juvenile phase) densely armed with straight, bristlelike prickles of different lengths to 1.2 cm, greenish at first, becoming black; upper stem and flowering branchlets (adult phase) unarmed or nearly so; leaf scars usually not visible, the leaves breaking off above base with part of petiole remaining and covering the true leaf scar, pith lacking, vascular bundles scattered throughout.

Winter Buds. Terminal bud absent; lateral buds slender, 3-sided, pointed, divergent; 1 exposed scale; concealed by the partly clasping base of leaf petiole.

Flowers. June; unisexual, plants dioecious; borne in few- to many-flowered umbels in axils of the leaves, peduncles glabrous, to 7.5 cm long, more than twice as long as the subtending petioles, each bearing less than 25 flowers; small; greenish-yellow; sepals 3, lanceolate; petals 3. Insect pollinated.

Fruit. Berry; October–November; globose, black, without bloom, 5–8 mm across; mostly 1-seeded, seed shiny, reddish brown; peduncles at least twice as long as the subtending petioles; persistent.

Distribution. Occasional to common throughout the state except in eastern upper Michigan, where it is rare to absent. County

occurrence (%) by ecosystem region (Fig. 19): I, 79; II, 57; III, 14; IV, 75. Entire state 65%. Widely distributed in e. North America, ON, s. to FL, w. to TX, n. to SD, MN.

Site-Habitat. Open to lightly shaded, dry to wet-mesic sites, common on circumneutral surface soils and those with calcareous substrates; floodplains, oak-hickory and beech–sugar maple forests, hardwood-conifer swamps, oak-pine savannas, forest edges, dunes, roadsides, fencerows, old fields. Associates include a very diverse range of tree and shrub species of these communities and areas.

Notes. Shade-intolerant, establishing and persisting in light to moderate shade. Prickles vary greatly in size and form, always needle-like and sharp. In floodplains, tolerates short periods of inundation and sedimentation in spring. Six species are native in the state, 4 of which are herbaceous and lack armature (Voss and Reznicek 2012). Widespread in temperate to tropical regions, 310 species comprise *Smilax* (Judd et al. 2008).

Chromosome No. $2n = 32$, $n = 16$

Similar Species. Common greenbrier, *Smilax rotundifolia* L., is a similar woody climbing vine and frequent associate in Michigan and its southern range. The fringed greenbrier, *S. bona-nox* L., and cat greenbrier, *S. glauca* Walter, are also similar woody associates in the Midwest and South. Many similar woody species occur in East Asia, including *S. sieboldii* Miq., *S. riparia* A.DC. var. *ussuriensis* (Regel) Kitag., and *S. discotis* Warb.

1. Flowering shoot, reduced
2. Fruiting shoot, berry, × 1/2
3. Juvenile or sprout stem, reduced
4. Adult stem, reduced
5. Upper (distal) part of stem in winter with persistent leaf bases, reduced
6. Winter twigs with tendrils attached to persistent leaf bases, enlarged

KEY CHARACTERS

- moderately stout, woody climbing vine, tendrils borne near base of petiole
- leaf blades oval, elliptic, elliptic-ovate; margin roughened and usually fringed, with minute bristle-tipped teeth, 5–7 primary veins, arcuate
- twigs slender, round, green, the lower part of stems densely armed with straight, needlelike, blackish bristly prickles; upper stem and flowering branchlets usually unarmed
- fruit a berry, globose, black without bloom, mostly 1-seeded

Distinguished from common greenbrier, *Smilax rotundifolia*, by its slender stems, round above and only slightly angled below, prickles blackish, needlelike, not broadened at base; leaf blades elliptic to ovate, margin usually fringed with minute teeth; peduncles bearing fruit more than twice as long as subtending petioles; fruit without bloom.

SMILACACEAE
Smilax rotundifolia Linnaeus

Common Greenbrier,
Round-leaved Greenbrier

Size and Form. Stout, tough, woody climbing vine to 5 m or forming dense tangles to 3 m; branches diffuse; climbing by tendrils borne in pairs near the base of the petiole; spreading rapidly by long rhizomes and forming clones.

Leaves. Alternate, simple, deciduous; blades 4–16 cm long, 2–15 cm wide; broadly ovate to elliptic or orbicular; acute to cuspidate; rounded, narrowed, or subcordate; entire or minutely rough-margined; leathery, medium green, dull to glossy and glabrous both sides; besides midrib 2 pairs of primary veins, parallel or arcuate; leaves breaking off above the petiole base and persistent, with tendrils intact; petioles 0.5–2.0 cm long, bearing pairs of tendrils at their base.

Tendrils. Very slender, paired, twining, attached at the base of the petioles.

Stems-Twigs. Stout, tough, sharply or obscurely 4-angled, zigzag, green, glabrous; usually few, well-spaced, stout prickles to 1 cm, straight or slightly curved, green and broadened at base, blackish at tip, confined to angles of the stem, not at nodes; pith lacking, vascular bundles scattered throughout.

Winter Buds. Terminal bud absent; lateral buds slender, 3-sided, pointed, divergent; 1 exposed scale; concealed by partly clasping base of leaf petiole.

Flowers. June; unisexual, plants dioecious; borne in few- to many-flowered umbels of which only 3–8 mature; small; bronze to greenish-yellow; sepals 3, lanceolate; petals 3; peduncles flattened, 0.6–1.5 cm, about as long as the subtending petioles; Insect pollinated.

Fruit. Berry; October–November; globose, black with bloom, 5–7 mm across, in short-stalked compact clusters, persistent into winter; mostly 2- to 3-seeded; pedicels 2–8 mm long.

Distribution. Common locally in southern lower Michigan, especially the southwestern counties bordering Lake Michigan and the lake plain of the southeast. Absent in northern lower Michigan and upper Michigan. County occurrence (%) by ecosystem region (Fig. 19): I, 24; II, 3; III, 0; IV, 0. Entire state 12%. Widespread in e. North America; similar to the range of *S. hispida*.

Site-Habitat. Open to moderately shaded, wet-mesic to mesic and seasonally wet lake plain sites; tolerates acidic surface soil and moderate drought; disturbed oak forests, acidic lake plain terraces, stream and lake borders, dunes, roadsides, fencerows, old fields, edges of woods. Associates include many early successional tree and shrub species typical of forest edges and openings in oak-hickory, dune, and lake plain forests of southwestern and southeastern lower Michigan, especially black, scarlet, and white oaks, as well as sassafras, black and choke cherries, sumacs, black chokeberry, wintergreen, and shrub dogwoods.

Notes. Shade-intolerant; most vigorous growth and fruit production in full sun. Forming almost impenetrable thickets. The leaves and stems are among the most favored browse plants for deer and rabbits. Native Americans and pioneers pounded roots to a pulp and made a fine reddish powder. It was boiled to make a jellylike pudding or used in making bread or cakes and to thicken soups. Berries are an important food of wild turkeys, ruffed and sharp-tailed grouse, ring-necked pheasants, and at least 38 species of nongame birds (Gill and Healy 1974).

Chromosome No. $2n = 32$, $n = 16$

Similar Species. The bristly greenbrier, *Smilax hispida*, is an occasional associate, although it occurs typically on more nutrient-rich soils. Differences prior to European settlement have been blurred by human disturbance. It is similar enough to be confused with the southern species *Smilax bona-nox* L., fringed greenbrier or sawbrier, in parts of its range to the east and south of Michigan. The cat greenbrier, *S. glauca* Walt., also has stout prickles and is most common in its midwestern and southern range.

1. Vegetative shoot, reduced
2. Fruiting shoot, berry, × 1/4
3. Winter portion of midstem with tendrils attached to leaf bases, reduced
4. Winter twig with persistent leaf base and tendrils, × 1

KEY CHARACTERS

- stout, woody climbing vine or forming dense tangles; climbing by tendrils borne near base of petiole
- leaf blades broadly ovate to orbicular, entire or minutely rough-margined, usually 5 primary veins, parallel or arcuate
- stems moderate, usually sharply 4-angled, green, armed with a few stout, straight, or slightly curved prickles green and broadened at base, tip blackish
- fruit a berry, globose, black with bloom, 2- to 3-seeded

Distinguished from bristly greenbrier, *S. hispida*, by its stout stems, usually strongly 4-angled, prickles stout, broadened at base; leaf blade broadly ovate to circular, margin entire or minutely roughened; peduncles bearing fruit about as long as subtending petioles; fruit with bloom.

297

SOLANACEAE
Solanum dulcamara Linnaeus

European Bittersweet, Bittersweet Nightshade

Size and Form. Perennial climbing vine, reaching a length of 1–3 m, climbing by twining; also creeping; woody only near base; spreading by rhizomes, rooting at nodes.

Leaves. Alternate, simple, deciduous; blades 2.5–8 (11) cm long, 1.5–5 cm wide, lower stem early leaves (*preformed* in dormant bud), broadly ovate to elliptic, unlobed; acute or acuminate; rounded or subcordate; entire; leaves at midshoot and the indeterminate upper shoot (*neoformed* in growing season) are lobed with 1–2 (3) pairs of small but deeply cut basal lobes or sometimes inequilateral with 1 lobe at base; all leaves blade margins entire; dark green, glabrous to minutely downy; unpleasant odor when crushed; petioles 1.5–3 cm.

The distinctive sequence of forms, from base of stem to apex, include (1) unlobed early leaves, broadly ovate; (2) transition 1-lobed leaves; and (3) late leaves on the indeterminate shoot, usually 2-lobed, sometimes 3-lobed.

Stems-Twigs. Woody at and near the base, olive-green and brownish above, flexuous, branching; lenticels many, conspicuous; slightly pubescent when young, becoming glabrous; unpleasant odor when bruised or crushed; leaf scars raised at base but flattened at top, half round; bundle scar 1, relatively large but sometimes separated into 3 small scars; pith hollow except at nodes.

Winter Buds. Small, 1.5–3 mm, solitary, sessile, rounded, densely white-hairy; about 4 indistinct outer scales, ciliate.

Flowers. June–August; bisexual; corolla bluish-purple with a central column of yellow stamens borne on branched cymes with pedicels 3–4 cm long; somewhat saucer-shaped; 5 spreading petal lobes. White-flowered forms are more frequent than in many species. Insect pollinated.

Fruit. Berry; August–October; many-seeded; red with a thin translucent covering; juicy. Parts of cymes with dried berries persistent in winter. Berries reported poisonous to most people, especially when not fully ripe; *avoid experimentation.* Seeds 350,000/lb (700,000/kg).

Distribution. Common in lower Michigan, locally frequent to common in upper Michigan. County occurrence (%) by ecosystem region (Fig. 19): I, 97; II, 80; III, 71; IV, 62. En-

tire state 86%. A Eurasian species commonly naturalized throughout most of the conterminous USA and adjacent Canada.

Site-Habitat. Moderately open and disturbed areas, especially wet-mesic and wet, poorly drained sites; favors circumneutral and basic soil conditions; hardwood- and conifer-dominated swamps, edges of marshes, cutover deciduous forests, old fields, fencerows, lakeshores, stream banks, vegetable and flower gardens, disturbed places of all kinds. Because of wide dispersal by birds to distant sites and in successful competition with native plants, it may appear indigenous.

Notes. Shade-intolerant, persisting in light shade, fast-growing. Native from Europe to North Africa and East Asia; introduced to gardens in North America and has become widely naturalized. Possibly widespread in southern lower Michigan by the 1890s (Voss and Reznicek 2012). The name *dulcamara* is from Latin *dulcis*, "sweet," and *amarus*, "bitter." The common name bittersweet nightshade is reported to be derived from the roots; if chewed they first taste bitter and then sweet (Soper and Heimburger 1982). However, chewing roots is not recommended, even experimentally, because this plant belongs to the nightshade family, which includes many plants containing poisonous compounds. We have not listed bittersweet as a common name because it is preferably applied to *Celastrus scandens*. The family Solanaceae contains ca. 75 genera and >2,000 species in topical and temperate regions. Voss and Reznicek (2012) describe 6 herbaceous species, including the potato, *Solanum tuberosum* L. (see Pollan 2002), that are found in Michigan, mostly not native.

Chromosome No. $2n = 24, 48, 72$; $n = 12$; $x = 12$

1. Leaf-blade forms from base (a, unlobed), lower midshoot (b, 1-lobed), midshoot (c, two-lobed), upper shoot (d, two lobed), × 1/2
2. Flowering shoot, × 1/2
3. Fruiting shoot, berry, × 1
4. Winter lateral bud, side view, enlarged
5. Winter lateral bud and leaf scar, enlarged
6. Winter twig, reduced

KEY CHARACTERS

- perennial climbing vine to 3 m; climbs by twining
- leaf blades of the mid- and upper stem with small basal lobes, margins entire
- stem woody near base, herbaceous above, green, becoming brownish
- flowers bluish-purple with a yellow core of stamens, borne on long-stalked cymes
- fruit a red berry, persisting in winter, may be poisonous if eaten
- habitat widespread in disturbed sites, especially wet to wet-mesic swamps; a Eurasian species now widely naturalized; all parts more or less poisonous if ingested

MOUNTAIN-ASHES, ROWANS, WHITEBEAMS—*SORBUS*

Sorbus is an interesting group of shrubs and trees with distinctive geography, habitats, multiple leaf forms, showy flowers and fruits, and a worldwide distribution in temperate Europe, North America, and Asia (Chant 1986; Satake et al. 1993); 258 species are recognized. The alternatively arranged leaves of the whitebeams (subgenus *Aria*, temperate Europe and Asia) are simple. In the mountain-ashes or rowans (subgenus *Sorbus*, Europe and North America), they are pinnately compound with ovate leaflets. Hybridization occurs within and between these two distinctive groups, producing individuals of intermediate morphology and habit.

Of the pinnately compound species, one of the most attractive and widely used in horticulture is the rowan, European mountain-ash, or quickbeam, *Sorbus aucuparia*. It is native in much of Europe, western Asia, and Siberia. The service tree, *S. domestica*, occurs throughout Europe and can grow to a height of about 18 m. The large pear-shaped fruits are fermented to produce an alcoholic beverage, and the bark is used for tanning leather. Native in eastern North America are *S. americana* and *S. decora*, described in the following pages. Other species of this rowan group are the western mountain-ash, *S. scopulina*, a shrub to 1–2 m of the North American Rocky Mountains; Sitka mountain-ash, *S. sitchensis*; and *S. cascadensis*, a shrub to 2–5 m and native of British Columbia south to northern California. In mountainous Japanese habitats, three shrubs, 1–2 m tall, include *Sorbus gracilis*, and *S. sambucifolia*. A small tree, *S. commixta*, 5–10 m, also occurs in mountain forests of all the main islands. In China Vilmorin's rowan, *S. vilmorinii*, and the Chinese mountain-ash, *S. pohuashanensis*, are shrubs or small trees of the western mountains.

Of the whitebeam group, *Sorbus aria*, a tree to 12 m, typically grows on limestone substrates in central Europe; it has unlobed leaves with serrate leaf margins. *S. torminalis*, a tree to 20 m of Europe, North Africa, and western Asia, has simple, deeply lobed leaves. Other notable species are the Korean mountain-ash, *S. alnifolia*, of Korea, China, and Japan; and *S. japonica* of Japan.

The large quantities of pomes produced per tree are used in making jams and jellies. The fruits are at first unpalatable to birds, but after freeze-ferment cycles they are eaten by robins, grosbeaks, thrushes, waxwings, and others. The pulp is digested and the seeds excreted, leading to widespread dispersal and subsequent establishment in diverse sites. Once the fruits ferment on the mountain-ash trees, cedar waxwings often swarm in droves to feast on the fragrant pomes. Nature takes its course, and many become tipsy, grasp fearlessly at branches, or (disoriented) are observed gently and happily floundering unsteadily on the ground underneath. As Cullina (2002) observes,

"[T]he alcoholic content of the fruits must provide quite a little nip to warm their weary wings."

KEY TO SPECIES OF *SORBUS*

1. Winter buds, current shoots, lower surface of leaflets, and inflorescence parts all densely white-pubescent; fruit orange; introduced . . . *S. aucuparia*, p. 301
1. Winter buds, current shoots, lower surface of leaflets, and inflorescence parts not densely white-pubescent; fruit bright orange or coral red; native.
   2. Compound leaves 15–30 cm long; leaflets >3 times as long as wide; leaflet shape more lanceolate than oblong; inner bud scales nonvillous; comparatively late flowering, June–July; fruit relatively small, 4–7 mm across . . . *S. americana*, p. 302
   2. Compound leaves 10–20 cm long; leaflets <3 times as long as wide; leaflet shape more oblong than lanceolate; inner bud scales villous; comparatively early flowering, May–June; fruit relatively large, 8–10 mm across. . . . *S. decora*, p. 304

SORBUS AUCUPARIA L.—EUROPEAN MOUNTAIN-ASH

The European mountain-ash or rowan tree is the most commonly planted mountain-ash due to its fast juvenile growth, profuse crops of showy white flowers, and orange fruits. It is a small tree, 6–12 m high, with a slender, short trunk that separates 1–2 m above the ground into stout, spreading branches to form a rounded crown. The leaves are odd-pinnately compound as in the American and showy mountain-ashes, but the leaflets are blunt, rounded, or short-pointed at the apex; and densely white-pubescent beneath. Other distinguishing features are the fruit, twigs, and winter buds. The fruit is somewhat larger than that of the American mountain-ash. The twigs are pubescent; the buds are woolly and not gummy. A cultivar with orange-yellow fruit is typically planted. From ornamental plantings in urban and suburban areas birds disperse seeds locally and probably widely as well. Infrequent to rare in southern lower Michigan; rare or absent elsewhere. County occurrence (%) by ecosystem region (Fig. 19): I, 18; II, 3; III, 0; IV, 25. Entire state 12%. Chromosome number: $2n = 34$, $n = 17$; $x = 17$. The trunk of this small tree is a prime target of the yellow-bellied sapsucker, *Sphyrapicus varius*, a medium-sized woodpecker. European mountain-ash is probably the most common *Sorbus* encountered in southern lower Michigan, although it rarely escapes cultivation. It occurs in disturbed sites, forest edges, and swamps, especially near cities and towns, but it is also dispersed by birds farther afield.

ROSACEAE
Sorbus americana Marshall

American Mountain-ash

Size and Form. Tall shrub or small tree with single or multiple stems, 4–7 m high and 10–25 cm in diameter. Trunk short, branches stout, spreading, forming a narrow, rounded crown. Michigan Big Tree: girth 111.8 cm (44 in), diameter 35.6 cm (14.0 in), height 21.6 m (71 ft), Leelanau Co.

Bark. Thin, light grayish-brown, smooth, becoming slightly roughened and scaly with age; conspicuous horizontally elongated lenticels; inner bark fragrant.

Leaves. Alternate, odd-pinnately compound, 15–30 cm long; leaflets 9–17, 5–10 cm long, 1–2.5 cm wide; lateral leaflets 3.4–4.5 times longer than wide; leaflets sessile or nearly so except terminal; lanceolate to oblong-lanceolate; long-acuminate, sharp-pointed; cuneate or rounded; finely, singly, sharply serrate above the entire base, glabrous; dark yellowish-green above, paler beneath, turning clear yellow in autumn; rachis green, becoming reddish, usually with several glands and a few long hairs at the base of the leaflet; petioles slender, grooved, enlarged at base, 3–6 mm long.

Stems-Twigs. Stout, at first reddish-brown and hairy, becoming glabrous, dark brown; mild odor of bitter almond (i.e., cherrylike) when bruised or broken. Current shoots arise from the first lateral bud below the flower cluster; they may develop as a new shoot or remain a dormant bud, becoming enlarged and resembling a terminal bud; leaf scars crescent-shaped, raised on the persistent base of the petiole, which when shed reveals the true leaf scar; bundle scars 5.

Winter Buds. Terminal bud large, 1–1.8 cm long, conical to oblong, acute, pointed with curved apex, red to purplish-red, 3–4 scales exposed, lustrous, outer buds glabrous except at apex and along margins, more or less glutinous; lateral buds smaller, tightly appressed, scales rounded on the back, purplish-red, lustrous, glutinous.

Flowers. May–June, after the leaves; bisexual; white, showy, ca. 6 mm across; borne on terminal short pedicels in many-flowered dense, flat or dome-shaped corymbs; 7–20 cm across, pedicels and inflorescence branches glabrous or nearly so; calyx urn-shaped, 5-lobed, sepals glabrous; petals 5, obovate to nearly orbicular; stamens 15–20; styles 2–3. Insect pollinated.

Fruit. Pome; October but persistent in winter; subglobose, 4–7 mm across; bright orange or coral red; flesh thin, bitter; eaten readily by birds after multiple freezes. Seeds 144,900/lb (318,780/kg).

Distribution. Frequent to locally common in upper Michigan, occasional in northern lower Michigan, absent in southern lower Michigan. County occurrence (%) by ecosystem region (Fig. 19): I, 0; II, 27; III, 86; IV, 88. Entire state 25%. A boreal, northern, and mountain species reaching the southern limit of its midwestern range in east-central lower Michigan. Distributed in n.e. North America, w. Great Lakes Region s. at high elevations in the Appalachian Mountains and along the Atlantic coastal plain to GA.

Site-Habitat. Open to lightly shaded, wet to wet-mesic sites, strongly acidic to circumneutral soils; swamps, swamp borders, and edges of streams; rocky lakeshores; conifer and conifer-hardwood forests. Associates include black spruce, tamarack, balsam fir, red maple, Labrador-tea, leatherleaf, bog-rosemary, bog-laurel, blueberries, and black chokeberry.

Notes. Shade-intolerant. Tolerates extremes of cold air and soil, wet, acidic, low-nutrient and low-oxygen conditions. Fruit is high in ascorbic acid and used by Native Americans as tea to prevent or cure scurvy. Fruit high in pectin and makes good jelly but is best picked when overripe after multiple freezes.

Chromosome No. $2n = 34$, $n = 17$; $x = 17$

Similar Species. Closely related to the showy mountain-ash, *Sorbus decora* (Sarg.) C. K. Schneider. See "Mountain-Ashes, Rowans, Whitebeams–SORBUS" for the related and widely planted European mountain-ash, *S. aucuparia* L.

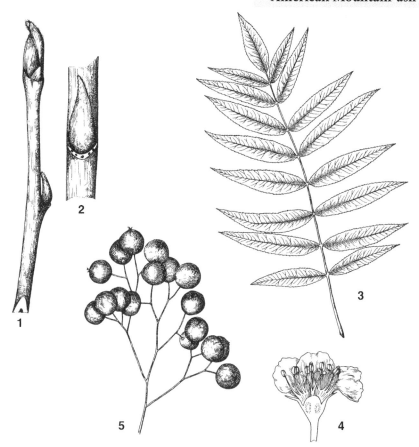

1. Winter twig, × 1
2. Portion of twig, enlarged
3. Leaf, × 1/3
4. Vertical section of flower, enlarged
5. Portion of a fruiting cyme, × 1

KEY CHARACTERS

- tall shrub or small tree, 4–7 m
- leaves pinnately compound; leaflets lanceolate, sharply serrate, 5–10 cm long, laterals 3.4–4.5-times longer than wide; blade tips long-acuminate
- terminal bud present, large, conical to oblong, purplish-red, outer scales glutinous, inner scales glabrous
- fruit a bright orange-red pome, spherical, borne in dense clusters, persistent in winter
- habitat northern, wet-mesic to wet, acidic to circumneutral sites with light to moderate shade
- absent in southern lower Michigan

Distinguished from showy mountain-ash, *Sorbus decora,* by its longer leaves, longer lateral leaflets (>3 times as long as wide vs. <3 times), leaflets more lanceolate (in contrast to oblong), nonvillous inner bud scales, flowers at least a week earlier, smaller fruit.

Distinguished from European mountain-ash, *Sorbus aucuparia* L., by its longer leaflets, glabrous or nearly so, tip long-acuminate, sharp-pointed; winter buds with outer bud scales glutinous; smaller fruit; a native species.

ROSACEAE
Sorbus decora (Sarg.)
C. K. Schneider

Showy Mountain-ash

Size and Form. Tall shrub or small tree, single or multiple stems to 10 m; branches initially erect and then spreading, forming a narrow, rounded crown. Michigan Big Tree: girth 144.8 cm (57 in), diameter 46.1 cm (18.1 in), height 17.7 m (58 ft), Mackinac Co.

Bark. Thin, smooth, with horizontally elongated lenticels, light grayish-green, becoming roughened or scaly on old trees, dark gray inner bark fragrant.

Leaves. Alternate, deciduous, odd-pinnately compound, 10–20 cm long; leaflets 11–15, 3–8 cm long, 1–2.5 cm wide; laterals 2.4–3 times as long as wide, the lowermost pair often smaller; leaflets sessile or nearly so except the terminal, oblong to narrowly oval, terminal leaflet broadly elliptic to obovate; leaflet tips abruptly short-acuminate to blunt; inequilateral, rounded, or somewhat narrowed; margin sharply and mostly singly and sharply serrate above the entire base, teeth more or less spreading; upper surface dark green to bluish-green and glabrous, lower surface much paler, minutely papillose and more or less persistently pubescent, especially along the midrib, rachis slightly pubescent with resinous glands and long hairs at the base of the leaflets; petioles sparsely hairy to glabrate, 3–6 cm long.

Stems-Twigs. Grayish-tomentose at first, soon becoming glabrate and reddish-brown with scattered elliptical lenticels and slightly scaly with age; short-shoots common; mild odor of bitter almond (i.e., cherrylike) when bruised or broken; leaf scars crescent-shaped to broadly U-shaped, raised on the persistent base of the petiole, which when shed reveals the true leaf scar; bundle scars 5, arranged in a single curved line.

Winter Buds. Terminal bud present, relatively large, narrowly conical, sharp-tipped, often with a curve, scales shiny dark reddish-brown, sticky, the 2–3 outer ones glabrous, the inner ones conspicuously brownish-tomentose; lateral buds smaller, tightly appressed.

Flowers. June–July; after the leaves; bisexual; white, showy, 8 mm across; borne on short, stout hairy pedicels in many-flowered open clusters 6–15 cm across, more open than in *S. americana*; calyx urn-shaped, 5-lobed, sepals glabrous or very sparsely pilose; petals

orbicular, 4–5 mm across; stamens numerous; styles 2–3. Insect pollinated.

Fruit. Pome; August–September, persistent in winter; subglobose, 8–10 mm across; bright red; flesh thick, eaten readily by birds. Seeds 127,200/lb (279,840/kg).

Distribution. Occasional in southern lower Michigan, frequent along lakeshores in northern lower Michigan and upper Michigan. County occurrence (%) by ecosystem region (Fig. 19): I, 24; II, 33; III, 86; IV, 88. Entire state 39%. Widely distributed across n. Canada and adjacent n. states to central MB with more southerly disjunct occurrences. A northern and boreal species, it is more common than *Sorbus americana* in northern Michigan; reported from 9 counties in southern lower Michigan, where *S. americana* is absent.

Site-Habitat. Open to moderately shaded wet to wet-mesic sites, acidic to mildly basic soils; moist sites of conifer, conifer-hardwood, and mixed hardwood forests; rocky lakeshores and lakes; edges of forests along Lake Superior and the north shore of Lake Huron; wooded dunes. Associates include most of those for *Sorbus americana* and mountain alder.

Notes. Shade-intolerant, persisting in light shade. Flowers develop a week or so later than in *S. americana*. Repeated freezing reduces the astringent or bitter taste of ripening fruit.

Chromosome No. $2n = 68$, $n = 34$; $x = 17$

Similar Species. In the upper Great Lakes Region, it is similar to mountain-ash, *Sorbus americana* Marsh., and occasionally associated with it. More easily distinguished from European mountain-ash, *S. aucuparia* L., which is often planted in lower Michigan and has winter buds densely covered with white woolly hairs.

1. Flower, enlarged
2. Fruiting shoot with leaves, pome, × 1/2
3. Winter lateral bud and leaf scar, × 2
4. Winter twig, × 1

KEY CHARACTERS
- tall shrub or small tree to 10 m with narrow, rounded crown
- leaves pinnately compound; leaflets oblong to narrowly oval, sharply serrate, 3–8 cm long, 2.4–3 times as long as wide, leaflet tips abruptly short-acuminate to blunt
- terminal bud relatively large, purplish-red, outer scales glabrous, sticky, inner scales conspicuously brownish-tomentose
- fruit a shiny red pome, spherical, borne in open clusters, persisting in winter
- habitat moist sites of dunes, rivers, and rocky shores of lakes Michigan, Superior, and Huron

Distinguished from the American mountain-ash, *Sorbus americana*, by its shorter leaves, shorter lateral leaflets (<3 times as long as wide vs. >3 times), leaflets more oblong (in contrast to lanceolate); villous inner bud scales, flowers at least a week later; larger fruit.

ROSACEAE
Spiraea alba Du Roi

Meadowsweet

Size and Form. Low to medium, erect shrub, 0.3–2 m, with stiff, wandlike stems; forming spreading clones by rhizomes.

Bark. Gray to brown, at first smooth, outer bark exfoliating in narrow strips with age.

Leaves. Alternate, deciduous, simple, blades 3–7 cm long, 1–2 cm wide, 3–4 times as long as wide; oblong to narrowly oblanceolate, acute; cuneate; sharply and finely serrate; firm; green; essentially glabrous both sides; petioles nearly sessile, short, 1–5 mm long.

Stems-Twigs. Stems often branched, slender, narrow, tough; current shoots brown, minutely pubescent, marked with vertical lines, becoming grayish-brown, glabrous; angled near nodes; pubescent when young; leaf scars minute, much raised, half round or crescent-shaped, bundle scar 1; pith small, round, continuous.

Winter Buds. Terminal bud absent, buds small, sessile, ovoid, about 1 mm long, sparingly hairy near tip; about 6 scales exposed; a lateral bud becomes the end bud when the shoot tip dies back.

Flowers. July–August; bisexual; borne in dense terminal panicles, elongate, conical to cylindrical, narrower than wide, glabrous or nearly so; persistent through the winter; peduncles and pedicels more or less finely pubescent; white; 6–8 mm in diameter; calyx short-campanulate, 5-lobed; petals 5; stamens 15 to many. Insect pollinated.

Fruit. Follicle, borne in an aggregate, usually 5; September; glabrous, about 3 mm long, containing 2–5 seeds; seeds linear, 1–2 mm long; panicles persistent in winter.

Distribution. Common throughout the state. County occurrence (%) by ecosystem region (Fig. 19): I, 97; II, 100; III, 100; IV, 100. Entire state 99%. Widely distributed in e. North America, NL, NS w. to AB; New England and mid-Atlantic regions s. to GA; w. Great Lakes Region w. to NE, ND.

Site-Habitat. Open, wet to wet-mesic sites, often with fluctuating water tables and circumneutral to strongly basic, sandy soil; acidic topsoil as well; open wetlands, wet meadows and fens, shrub thickets, muskeg, lakeshores, stream and beaver dam margins, interdunal swales, roadside ditches. Associates include a great number and range of trees and shrubs, from tamarack, black spruce, leatherleaf, and bog-rosemary to northern white-cedar, balsam poplar, willows, speckled alder, shrub dogwoods, bog birch, and shrubby cinquefoil.

Notes. Shade-intolerant. Prior to European settlement, periodic fire was a major factor in maintaining wet meadows, fens, and shrub thickets in early successional status without trees. Also known as wild spiraea, pipestem, queen-of-the-meadow, Quaker lady, and willowleaf spiraea. The name *Spiraea* is from the Greek *speira* ("wreath" or "band"). Worldwide there are ca. 100 species of *Spiraea*, which occur mostly in northern temperate regions and at moderately high elevations in Mexico and the Himalayas. They are a group of many similar species and free in their hybridization habits such that many taxa have been described. In 1940 Rehder listed 226 species, varieties, cultivars, and hybrids that had been given names, largely due to their inherent variation and wide use and experimentation in horticulture and gardens. A patron of Robinson's English Flower Garden complained that, although no spiraea was worthless, there were too many of them, they were too similar, and too many flowered at the same time (Coats 1992).

Chromosome No. $2n = 36$, $n = 12$; $x = 12$

Similar Species. Closely related is broad-leaved meadowsweet, *Spiraea latifolia* (Aiton) Borkh., an eastern species (or variety. *S. alba* var. *latifolia*) that intergrades with *S. alba* in Michigan (see Voss and Reznicek 2012 for a discussion). The pink-flowered *S. tomentosa* L., steeplebush or hardhack, is common in southern lower Michigan and widely distributed in eastern North America. Also related is the pink-flowered *S. salicifolia* L., of southeastern Europe to northeastern Asia and Japan; in Michigan it rarely escapes cultivation.

1. Leaf, × 1/2
2. Flowering shoot, × 1/2
3. Fruiting shoot, follicle, enlarged
4. Winter lateral bud, side view, enlarged
5. Winter lateral bud and leaf scar, front view, enlarged

KEY CHARACTERS

- Low to medium, erect shrub, 0.3–2 m; stems slender, stiff, wandlike, often branched
- Leaf blades narrowly oblanceolate, sharp and finely serrate, 3–4 times as long as wide
- Flowers white, borne on elongated narrow panicles
- fruit a follicle, borne in an aggregate, glabrous; panicles persistent in winter
- habitat open, wet, usually circumneutral to basic, sandy sites

Distinguished from steeplebush, *Spiraea tomentosa*, by its stems often branched, leaf blades narrowly oblanceolate and essentially glabrous below, sharp and finely serrate; flowers white.

ROSACEAE
Spiraea tomentosa Linnaeus
Hardhack, Steeplebush

Size and Form. Low, erect shrub, 0.3–1.2 m; with few branches; forming dense and spreading clones by rhizomes.

Bark. Grayish-brown, smooth, outer bark loosening and peeling with age.

Leaves. Alternate, deciduous, simple, blades 2–7 cm long, 1–3 cm wide, ovate to oblong, acute to blunt; cuneate to rounded; coarsely crenate-serrate, often unequally doubly serrate; dark green and usually glabrous above, somewhat wrinkled, yellowish or tawny-rusty below, with dense tomentum, prominently veined; petioles tomentose, sessile or 1–4 mm long.

Stems-Twigs. Stems single, occasionally branched, slender, angled, covered with tawny or rusty tomentum, becoming glabrate and reddish-brown; leaf scars very small, much raised, half round, often appearing partially torn; bundle scar 1; pith small, round, continuous.

Winter Buds. Terminal bud absent; buds very small, sessile, ovoid, 1–2 mm long, ovoid, so woolly that several scales are indistinct; a lateral bud becomes the end bud when the shoot tip dies back.

Flowers. July–August; bisexual; borne in dense terminal panicles, elongate, conical to cylindrical, narrower than wide; densely tomentose; 10–20 cm long terminating in a long, slender point, 11–20 flowers per cm of axis, somewhat branched below, tomentose throughout, persistent through the winter; flowers 1.5–3 mm across; peduncles and pedicels densely pubescent; pink to deep rose; calyx short-campanulate, 5-lobed; petals 5; stamens 15 to many. Insect pollinated.

Fruit. Follicle, borne in an aggregate of follicles, usually 5; September; 2–3 mm long, pubescent, light tawny beneath, soon glabrate, usually oblong-ovate, containing 2–5 seeds; seeds linear, 1–2 mm long; panicles persistent in winter.

Distribution. Common in southern lower Michigan; increasingly infrequent northward; absent in western upper Michigan. County occurrence (%) by ecosystem region (Fig. 19): I, 63; II, 17; III, 43; IV, 12. Entire state 40%. Present throughout e. North America except AL, FL; w. of Mississippi River from TX n. to MN; OR, WA.

Site-Habitat. Open, wet to wet-mesic, primarily acidic sites; wet meadows, borders of marshes, bogs, muskeg, sandy-peaty lakeshores. Associates include many species of trees and shrubs on acidic topsoils, bogs and muskegs, as well as others characteristic of circumneutral soils. Often associated with *Spiraea alba*.

Notes. Shade-intolerant. Tolerant of soils of low nutrient and oxygen availability. The name *Spiraea* is from the Greek *speira* ("wreath" or "band"). The name hardhack, applied particularly to the pink-flowering species, is thought to be derived from the difficulty encountered by early settlers in hacking their way through dense colonies, especially in the West.

Chromosome No. $2n = 24$, $n = 12$; $x = 12$

Similar Species. Primarily in eastern North America is the similar white-flowered, *S. alba* L., meadowsweet. Closely related in the West is the deep-rose-flowered hardhack, *S. douglasii* Hooker, with recognized varieties along the Pacific coast from Alaska to California and in the interior West. In coastal British Columbia, in most sites and swamps, it forms dense thickets and reaches a height of 2 m. Also related is the pink-flowered Eurasian species *S. salicifolia* L., which in Michigan rarely escapes cultivation.

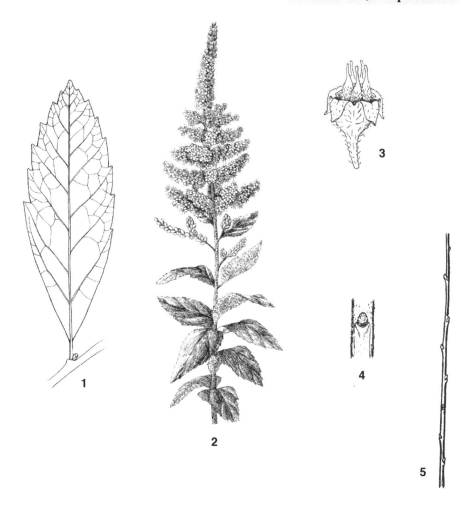

1. Leaf, × 1
2. Flowering shoot, × 1/2
3. Young fruit, follicle, × 10
4. Winter lateral bud and leaf scar, enlarged
5. Winter twig, reduced

KEY CHARACTERS

- low, erect shrub, 0.3–1.2 m; stems single, occasionally branched
- leaf blades ovate to oblong, narrowly oblanceolate, coarsely crenate-serrate, usually glabrous above, white or tawny-rusty below, with dense tomentum
- flowers borne on elongate panicles, densely tomentose; pink or deep rose
- fruit aggregate of follicles, borne on narrow panicles, persistent in winter
- habitat open, wet and wet-mesic sites, usually acidic

Distinguished from meadowsweet, *Spiraea alba*, by its leaf blades ovate to oblong, coarsely crenate-serrate, yellowish or tawny-rusty below, with dense tomentum; inflorescences densely tomentose; flowers pink to deep rose.

Spiraea × *vanhouttei* (Briot) Carrière

Bridal-Wreath

Size and Form. Medium, mound-shaped shrub, 1–2.5 m; branches gracefully arching and spreading to 3 m; fountain- or vase-shaped, round-topped, with branches flowing to the ground.

Bark. Smooth and yellowish-brown on young branches, gray and somewhat shreddy with age.

Leaves. Alternate, simple, deciduous; blades 1.5–4 cm long, 1.2–3 cm wide; rhombic-ovate or somewhat obovate; acute; cuneate or rounded; usually slightly 3-lobed and 3-veined; margin incised-serrate, teeth irregular, obtuse, mucronate; dark green above, pale bluish-green below, glabrous; shedding in late autumn without color change or occasionally turning a delicate, dull reddish-orange or orange-pink shade; petioles 5–12 mm long, glabrous.

Stems-Twigs. Slender, circular in cross section, young shoots pink on exposed side, turning reddish and then brown, glabrous; epidermis starts peeling on stems >2 mm; sylleptic branching common.

Winter Buds. Terminal bud absent, buds very small, pointed to blunt; 4–8 scales exposed, sharp-pointed, ciliate and sparsely pubescent, light brown; pointing outward.

Flowers. May–June; bisexual; borne on many-flowered, umbellike racemes; 3–5 cm across, showy, pure white, 5–8 mm across, sepals upright or spreading in fruit, petals orbicular, twice as long as the partly sterile stamens. Insect pollinated.

Fruit. Follicle, sterile, not produced.

Distribution. County occurrence (%) by ecosystem region (Fig. 19): I, 8; II, 7; III, 0; IV, 25. Entire state 8%. Cultivated occurrences in Michigan not included; rarely found outside cultivation. Widely reported in USDA Plants (http://plants.usda.gov) in e. North America.

Site-Habitat. Extensively cultivated as an ornamental shrub. Tolerates a wide range dry-mesic to mesic soil-site conditions. Collections outside cultivation include lakeshores, along railroads and roadsides, and an abandon quarry (Voss and Reznicek 2012).

Notes. Shade-intolerant, persisting but less vigorous flowering in moderate shade. It is a hybrid between East Asian species *Spiraea cantoniensis* Lour. and *S. trilobata* L. Undoubt-edly the most popular of all spiraeas due to its arching, fountainlike form combined with the profusion of white flowers. Hardy in urban environments and northern winters (to -34°C [-30°F]) without injury. Few insect or disease pests. Present in most towns in the Midwest, the western Great Lakes Region, New England, the mid-Atlantic region, and the northern parts of eastern North America.

Similar Species. Not surprisingly, it is similar in many respects to the parent species Reeves' meadowsweet, *S. cantoniensis* Lour., which is native in Japan and China, and to the Asian meadowsweet, *S. trilobata* L., of northern China and Siberia, which has generally smaller features.

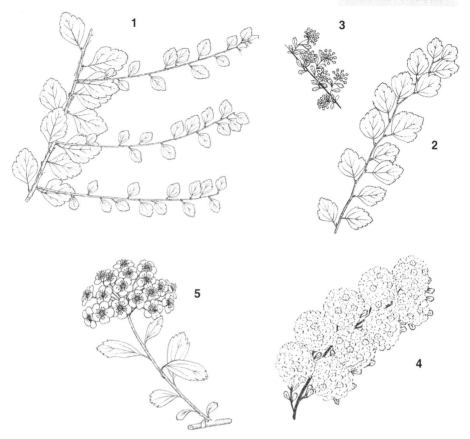

1. Branch with current vegetative shoots, reduced
2. End part of an indeterminate vegetative shoot, reduced
3. Stem with flowering shoots, flowers in bud, reduced
4. Stem with many flowering shoots, reduced
5. Flowering shoot, reduced

KEY CHARACTERS

- medium, mound-shaped shrub, 1–2.5 m; branches gracefully arching, spreading, and flowing to the ground
- leaf blades rhombic-ovate, usually slightly 3-lobed; margin irregularly serrate-glabrous, retaining leaves to early winter
- flowers borne on many-flowered, umbellike racemes, 3–5 cm across, showy, pure white, flowering mid- to late spring
- widely cultivated, rarely found in nature

STAPHYLEACEAE
Staphylea trifolia Linnaeus

Bladdernut

Size and Form. Tall, erect, rather stiffly and heavily branched shrub to 4 m; sprouting from roots and forming large clonal patches; lower branches occasionally layering or tip rooting.

Bark. Smooth and green at first, becoming slightly roughened by lenticels; greenish-brown to reddish-brown or grayish, striped or flecked with white lenticels.

Leaves. Opposite, trifoliate, deciduous; leaves to 25 cm long; leaflets 4–10 cm long, 2–6 cm wide, the terminal long-stalked and slightly larger than the laterals; laterals sessile or short-stalked; ovate to obovate; short-acuminate; rounded to cuneate; finely serrate; light green at first, turning dark green, glabrous above, pubescent, at least along veins beneath; turning dull pale yellow in autumn; petioles 2.5–12 cm long.

Stems-Twigs. Slender, rounded, glabrous, lustrous, greenish-gray at first with white fissures (i.e., lenticels), turning reddish-brown, slightly ridged or warty; leaf scars opposite, sometimes slightly separated, half round; bundle scars 3 or divided into 5, 7, or more; stipule scars rounded or elongated; pith large, continuous, white.

Winter Buds. Terminal bud absent; buds small, 3–5 mm long, end buds larger than laterals, sessile, ovoid, pointed, reddish-brown or brown, glabrous, usually 4 blunt scales visible; uppermost lateral buds sometimes paired at twig tip.

Flowers. April–May; bisexual; borne in terminal, drooping panicles, 3–6 cm long, opening when leaves are 1/4 to 1/2 grown; white to cream-colored, cylindrical, about 1 cm long; calyx greenish-white, sometimes pinkish; corolla white, pedicels about 1 cm long; ovary pubescent. Insect pollinated.

Fruit. An inflated capsule, bladderlike; September–October; prominently 3-angled, 3-celled, ellipsoidal and usually obovate, veiny, greenish and pubescent at first, becoming yellowish-brown and glabrous or nearly so, to 8 cm long, 5 cm wide; each cell containing 1–4 pale brown seeds, becoming loose and rattling around inside; capsules persistent over winter; capsule buoyant, adapted for water dispersal.

Distribution. Locally common in southern lower Michigan; rare in northern lower Michigan; absent in upper Michigan. County occurrence (%) by ecosystem region (Fig. 19): I, 84; II, 7; III, 0; IV, 0. Entire state 41%. Distributed widely in e. North America with its core in the Midwest and mid-South; parts of New England s. to VA; w. of the Mississippi River in MO and adjacent states.

Site-Habitat. Moderately open, nutrient-rich sites; alluvial floodplains—levee and second bottom, stream banks, nutrient-rich deciduous woods; also rocky woods, and wooded sand dunes and ridges in southern Ontario. Associates include silver maple, American elm, red ash, blue-beech, willows, common elder, nannyberry, pawpaw, wild black current, running strawberry-bush, Virginia creeper, riverbank grape, and poison-ivy.

Notes. Moderately shade-tolerant. Adapted to brief inundation by floodwaters in the growing season, as well as light sedimentation. The inflated bladderlike capsule favors seed dispersal by streams and rivers. Distinguished from wafer-ash, *Ptelea trifoliata* L., by its opposite leaves and branching and bladderlike fruiting capsule. The name *Staphylea* is from Greek *staphyle*, "bunch of grapes," referring to the drooping cluster of flowers. Powdered bark was used medicinally by Native Americans. The seeds were considered sacred and used in the rattles of the medicine dance (Smith 1928).

Chromosome No. $2n = 78$, $n = 39$

Similar Species. The Sierra bladdernut, *Staphylea bolanderi* A. Gray, is endemic to California, uncommon in the Sierra Nevada, Cascade Range, and eastern Klamath Mountains. In Japan *Staphylea bumalda* DC. is similar in morphology and habitat. Related shrubs of southern Europe are *S. pinnata* L. and *S. colchica* Steven.

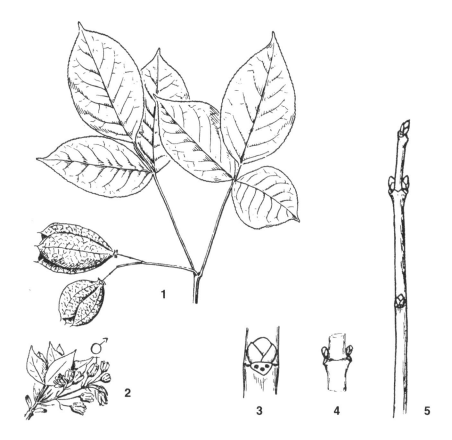

1. Fruiting shoot with leaves, × 1/2
2. Flowering shoot, reduced
3. Winter lateral bud and leaf scar, enlarged
4. Winter twig with paired lateral buds, enlarged
5. Winter twig, reduced

KEY CHARACTERS

- tall, erect shrub to 4 m, forming clonal patches by root suckering
- leaves trifoliate, leaflets ovate with short-acuminate tip, petioles long
- branches and bark with white lines or stripes of lenticels
- fruit a bladderlike capsule, yellowish-brown, persisting in winter
- habitat moist, nutrient-rich sites, especially floodplains of southern lower Michigan

CAPRIFOLIACEAE
Symphoricarpos albus (L.)
S. F. Blake

Snowberry

Size and Form. Low, erect, much-branched, spreading shrub to 1 m; forming clones by rhizomes.

Bark. Gray, older bark shreddy, fibrous, blackish with age; inner bark purplish.

Leaves. Opposite, simple, deciduous, often sessile; blades 2–5 cm long, 1–3 cm wide; ovate to elliptic-oblong; acute or rounded; rounded or cuneate; margin entire or wavy-toothed; dark green above, glabrous or with few appressed hairs, paler but not white, hairy beneath, at least along veins; leaves sessile or petioles short, pubescent, 1–3 mm long.

Stems-Twigs. Slender, opposite, round, young shoots puberulent, becoming glabrous; light yellowish-brown, becoming purplish-brown to gray and blackish; leaf scars raised, half round, partly connected by ridges, bundle scar 1; pith small, round, brownish at nodes, stems hollow between nodes.

Winter Buds. Terminal bud lacking; laterals small, sessile, solitary or multiple, ovoid-oblong, somewhat flattened, 2–3 mm long; 3 pairs of keeled, pointed outer scales; bud scales of end buds persistent.

Flowers. June–July; bisexual; short-pediceled (0.5–2 mm long), mostly solitary in leaf axils, in pairs, or few-flowered in terminal, interrupted, spikelike racemes, pink to white; sepals 5-toothed; corolla bell-shaped, 4–5 lobed, 5–8 mm long; stamens 4–5. Insect pollinated.

Fruit. Drupe; September–October; bright white, waxy, round, appearing berrylike, spongy, 5–10 mm across, 2 pits; withering but persisting in winter; bitter, may be toxic, not edible; seeds 76,000/lb (167,200/kg).

Distribution. Infrequent to locally common throughout the state. County occurrence (%) by ecosystem region (Fig. 19): I, 34; II, 60; III, 57; IV, 88. Entire state 51%. Transcontinental in USA and Canada, s. to WV and VA in the East and NM and Mexico in the West.

Site-Habitat. Wide tolerance of upland topography and soils with dry to mesic soil-water availability throughout the state as long as high light irradiance is present; floodplains, pine barrens, disturbed forests, dunes, alvar and other limestone bedrock sites. Associated with a wide range of plants, especially where natural or human-caused disturbances have provided well-lighted landscapes, edges, and open woodlands.

Notes. Shade-intolerant, persisting in light shade. Fifteen species of *Symphoricarpos* are reported. All are in North America and Mexico except one in East Asia. Snowberries were once widely used in wildlife plantings to provide cover and food. Fruits are toxic when eaten in quantity. Stems, leaves, and roots are considered poisonous, causing vomiting and diarrhea. Snowberries were called "corpse-berries" by some native peoples (Kershaw, MacKinnon, and Pojar 1998). For medicinal uses, see Marles et al. 2000. Sometimes cultivated for the fruits but not recommended. In Michigan we recognize *S. albus* var. *albus*. More often planted and sometimes escaped is the larger-leaved *S. albus* var. *lavigatus* (Fernald) S. F. Blake, a native of the Pacific Northwest and adjacent northern states (Voss and Reznicek 2012).

Chromosome No. $2n = 36, 54$; $n = 18, 27$; $x = 9$

Similar Species. Very similar is the western snowberry or wolfberry, *Symphoricarpos occidentalis* Hook., which occurs widely in the Rocky Mountains to Alaska and coastal British Columbia. It is known in Michigan only from disturbed places (Voss and Reznicek 2012). Other commonly occurring Rocky Mountain species include longflower snowberry, *S. longiflorus* A. Gray; mountain snowberry, *S. oreophilus* A. Gray; and roundleaf snowberry, *S. rotundifolius* A. Gray. Widely distributed along the Atlantic coast, southeastern states, southern Appalachian Mountains, and mid-South is the coralberry or Indian currant, *S. orbiculatus* Moench. The counterpart in East Asia is Chinese snowberry, *S. sinensis* Rehder, which has blue-black fruit with bloom.

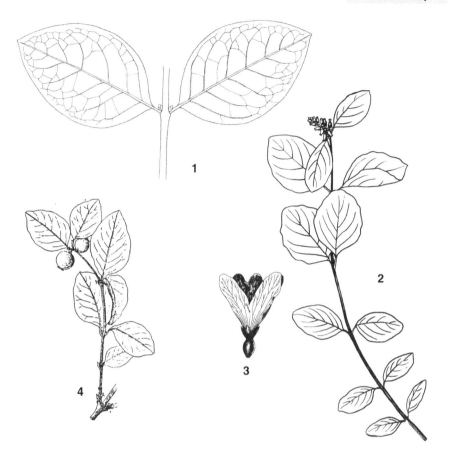

1. Leaves, × 1
2. Flowering shoot, × 1/2
3. Flower, × 4
4. Fruiting shoot, drupe, × 1/2

KEY CHARACTERS

- low, erect, much-branched shrub to 1 m
- leaves opposite, sessile or nearly so, blades ovate to elliptic-oblong
- stems slender, opposite, pith lacking between nodes on older stems
- fruit a bright white, waxy drupe; persistent in winter; bitter, inedible

OLEACEAE
Syringa vulgaris Linnaeus

Common Lilac

Size and Form. Tall, upright shrub or small tree to 7 m; typically growing in dense, spreading clumps of many stems. Michigan Big Shrub: girth 165.1 cm (65 in), diameter 52.6 cm (20.7 in), height 9.1 m (30 ft), Mackinac Co.

Bark. Relatively thin, gray to brownish, splitting into long, narrow, scaly ridges, scales gradually flaking off.

Leaves. Opposite, simple, deciduous; blades 5–12 cm long, 3–8 cm wide; heart-shaped (cordate-ovate); acuminate; truncate to cordate; entire; dark green to bluish-green above (or dusty white due to powdery mildew), paler beneath; glabrous; petioles 1.5–3 cm long.

Stems-Twigs. Moderately stout, 6 mm or more thick, stiff, olive green, forking repeatedly because inflorescences are terminal such that adult shoot growth ends in a pair of lateral buds (i.e., end buds), young shoots glandular-pubescent, becoming glabrous, lenticels prominent; leaf scars crescent- or shield-shaped, raised, bundle scar 1; pith moderate, continuous, round, pale.

Winter Buds. Terminal bud often abortive in adult growth; cessation of growth ends in a pair of lateral buds from which emerge diverging shoots in spring; end and lateral buds ovoid, somewhat pointed, outer scales fleshy, about 4 pairs.

Flowers. May, about the time leaves reach full growth; bisexual; borne in compact terminal panicles 10–20 cm long, panicles in pairs from paired end buds; showy and very fragrant; white to violet, purple, and deep reddish-purple; pedicels and calyx glandular-puberulent, sepals about equal, corolla tubular, slender, about 1 cm long, lobes ovate, spreading, rounded, about 5 mm long. Insect pollinated.

Fruit. Capsule, oblong, 2-celled, bearing 4 seeds, flattened, smooth, leathery, brown; 1–1.5 cm long and 5 mm wide, glabrous, acute, persistent throughout the summer. Seeds 86,000/lb (189,200/kg).

Distribution. Native in southeastern Europe and Asia. Common and hardy throughout Michigan and the upper Great Lakes Region. County occurrence (%) by ecosystem region (Fig. 19): I, 34; II, 33; III, 57; IV, 62. Entire state 39%.

Site-Habitat. Grows well in most upland sites except drought-prone, sandy soil.

Notes. Shade-intolerant, persisting in light shade. The 25–30 species of the genus are native in Asia and southeastern Europe. Many cultivars, over 400 (possibly 800–900) offer a variety of colors (white to reddish purple), double flowers, and variegated leaves (Dirr 1990). One of the most widely planted shrubs in North America, it often persists indefinitely in abandoned homesteads, cemeteries, and hedgerows. A classic example of shrub form due to multiple forking, in turn from terminal inflorescence and aborting terminal buds. Powdery mildew, a fungal disease, appears as a dusty white coating on leaf surfaces. It is caused by many different species of fungi, which infect almost all ornamental plants. For best flowering, the inflorescences should be removed after flowering ceases.

The Michigan Big Tree is from Mackinac Island. In 1947 Michigan botany professor Carl D. La Rue studied and measured many of the largest individuals. Most had been "trained" to a single stem and were the common lilac, *Syringa vulgaris* L., but white lilacs, *S. vulgaris* var. *alba* Weston, were common. Diameter at the base of the 10 largest individuals ranged from 15.7 to 23.6 inches. From ring counts he estimated the age of the oldest tree to be 118 years. The 64th annual Mackinac Island Lilac Festival was held in June 2013— 64 years after the publication of his article (La Rue 1948).

Chromosome No. $2n = 46$ (38, 44), $n = 23$

Similar Species. Of the many species native in Europe and Asia, Rehder (1940) describes 20. Major ones include *Syringa persica* L., probably native in Afghanistan and

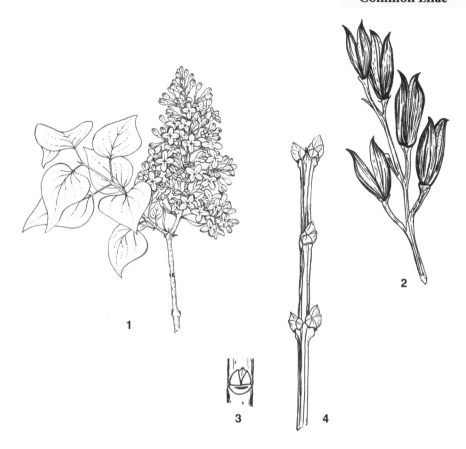

1. Stem with vegetative shoot and flowering shoot, × 1/4
2. Persistent capsules on fruiting shoot, × 1
3. Winter lateral bud and leaf scar, enlarged
4. Winter twig, reduced

KEY CHARACTERS

- tall shrub, typically growing in clonal clumps of multiple stems
- stems forking repeatedly, adult growth ending in a pair of lateral buds (i.e., end buds)
- leaves opposite, heart-shaped, margin entire, glabrous
- flowers in paired, long, compact terminal panicles, showy, white to dark purple, very fragrant
- widely planted and characteristic of urban and suburban settings and old farmsteads

Kashmir and widely planted in Iran (former-ly Persia); the bushy *S. villosa* Vahl., native in China and Korea; the Hungarian lilac, *S. josikaea* Jacq.; the Amur lilac, *S. reticulate* subsp. *amurensis* (Rupr.) P. S. Green & M. C. Chang, native in northeastern China and North Korea; and the Japanese lilac, *S. reticulata* (Blume) H. Hara. Most species are shrubs, occasionally small trees to 10 m, such as the Amur and Japanese lilacs.

TAXACEAE
Taxus canadensis Marshall

Ground-hemlock, Yew

Size and Form. Low, straggling, multistemmed shrub from a single base, sometimes prostrate or trailing, ascending or rarely erect branches to 2 m; often forming clones by layering.

Bark. Very thin, scaly, brown to reddish.

Leaves. Evergreen; spirally arranged but twisted at the base to appear opposite and 2-ranked; linear, abruptly narrowed to a sharp point; 1–2 cm long, 1.3–2 mm wide, a keel in the center of both sides extends the entire length; dark green above, yellow-green beneath with no white or yellowish stomatal lines; petioles very short.

Stems-Twigs. Slender to moderate, glabrous, green at first, turning brownish, ends of branches generally upturned.

Winter Buds. Terminal bud present; buds minute, rounded, keeled; bud scales more or less lanceolate.

Strobili. Unisexual, plants usually monoecious. Male buds are present in winter; each small cluster of pollen sacs projects from a basal, cuplike group of tiny scales, and the stalk elongates and discharges pollen in April–May. Female strobili, individual minute ovules, occur on short stalks bearing 3 pairs of scales and are borne in the axils of the leaves of last year's shoots; receptive to pollination April–May and mature in August of the 2nd summer following pollination. Wind pollinated.

Seeds. Seed surrounded by a cup-shaped, scarlet, fleshy aril about 5–10 mm long, nearly covering the dark-colored seed; aril juicy, sweet, resembling the fruit of angiosperms, not poisonous; seeds are reported poisonous; seeds 21,000/lb (46,200/kg).

Distribution. Occasional to locally rare; occurrence increasing northward, most likely in cool sites adjacent to the Great Lakes in northern lower Michigan and upper Michigan. County occurrence (%) by ecosystem region (Fig. 19): I, 45; II, 77; III, 100; IV, 88. Entire state, 65%. Ranging in n. and mtn. sites from NL to MB, New England s. to NC, w. Great Lakes region s. to TN.

Site-Habitat. Shaded, mesic and wet-mesic; circumneutral to basic sites. It was common in poorly drained, wet-mesic, northern white-cedar swamps, hardwood-conifer swamps, and mesic northern hardwoods forests prior to European settlement; now

it is very rare in these northern forests and swamps, as well as dune and swale complexes and limestone cliffs. Voss and Reznicek (2012) observe that it is "favored by moist winds from Lake Michigan and often luxuriant on wooded dunes and in coniferous woods near the shore." On islands in the Great Lakes, especially those lacking deer, *Taxus* can be astounding, forming impenetrably dense, thick growth to 2 m tall over large areas (personal communication, A. Reznicek). Associated with northern white-cedar, balsam fir, eastern hemlock, black and white spruces, balsam poplar, red maple, striped and mountain maples, poison-sumac, *Ribes* spp., creeping-snowberry, and other ericaceous species that grow on acidic soil surfaces of swamps.

Notes. Very shade-tolerant, slow-growing; long-lived; very hardy. It is the only conifer of our region lacking resinous odor or taste. *Taxus* species are trees throughout the temperate world, except in eastern North America. When present in large numbers in the northern forest, it is reported to inhibit the development of the typical dense sugar maple seedling layer in the ground cover. However, ground-hemlock has been severely browsed by deer and moose for decades and is rare in many places where it once was frequent or abundant. Curtis (1959) notes, "This plant is preferred above all others by deer and is sought out and utilized by them to the point of virtual extinction in many [Wisconsin] counties." Wood is fine-grained, durable, hard. Ground-hemlock is not a true hemlock (*Tsuga*). The seed and green, especially wilted, foliage contain the poisonous alkaloids referred to as taxine. The scarlet aril surrounding the seed is not poisonous; the "hemlock" Socrates is said to have

318

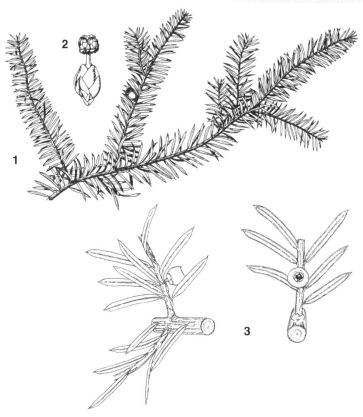

1. Vegetative shoot with seed covered by a fleshy aril, × 1/2
2. Male cone, × 5
3. Underside of leaves and arillate seed shown from two angles

KEY CHARACTERS

- straggling shrub to 2 m
- leaves sharp-pointed, dark green above, yellowish-green beneath, appearing 2-ranked
- seed black, surrounded by a scarlet, fleshy, cuplike aril

Distinguished from the Japanese yew, *Taxus cuspidata*, by its rare occurrence only in native forest ecosystems, foliage appearing 2-ranked, needles smaller (1–2 cm long, 1.3–2 mm wide), needles not keeled, somewhat darker green above and beneath, and lacking two yellowish-green stomatal bands.

taken for suicide is from the herb *Conium maclatum* L., in the Apiaceae. Sensitive to hot, dry conditions; limited in landscaping to ground cover in cool, shaded places.

Chromosome No. $2n = 24$, $n = 12$; $x = 12$

Similar species. The western or American yew, *Taxus brevifolia*, Nutt., of western North America, is a tree 5–15 m. The English yew, *T. baccata* L., of Europe, North Africa, and western Asia reaches 12–20 m; many cultivars are used in landscaping. The Chinese yew, *T. chin-*ensis, (Pilger) Rehder, of China and Taiwan is a shrub or tree to 12 m. The Japanese yew, *Taxus cuspidata* Siebold & Zucc., of Japan, Korea, and northeastern China, is widely planted and hardy in our region. Many cultivars are used in landscaping, especially because of their dark green foliage, which makes a dense mass and a very effective screen. To date cultivars of this species have tended to be hardier and outperform ground-hemlock in landscape use (Dirr 1990; Flint 1997).

ANACARDIACEAE
Toxicodendron radicans (L.) Kuntze
Poison-Ivy

Size and Form. Often a climbing, creeping, or trailing vine, attaching to bark of tree trunks by aerial rootlets, and branching outward over 1 m. Also it may be a low shrub to <1 m, with erect, ascending branches, forming clonal patches by rooting stems and rhizomes.

Bark. Gray, slightly roughened.

Leaves. Alternate, deciduous, pinnately compound, 15–30 cm long; 3 leaflets, leaflets 5–15 cm long, 3–10 cm wide; terminal leaflet long-stalked, lateral leaflets nearly without stalk; ovate to elliptical; short-acuminate; rounded; entire or sparingly or coarsely dentate or sinuate; usually more or less wrinkled; glabrous, lustrous yellowish-green above, more or less pubescent beneath; not aromatic; oil on leaf surfaces is allergenic and causes acute dermatitis; petioles pubescent, 5–20 cm long.

Stems-Twigs. Slender to stout, circular in cross section, tan, sparingly pubescent at first, becoming glabrate, lenticels raised and usually conspicuous; leaf scars large, broadly V-shaped or crescent-shaped, located below the buds, not surrounding them; bundle scars 6–8, scattered or in a line; pith whitish, continuous.

Winter Buds. Terminal bud erect, stalked, fingerlike, 4–6 mm long, naked, flattened tan, scurfy-pubescent; lateral buds much smaller, appressed.

Flowers. May–June; usually unisexual, plants dioecious; borne in loose, ascending paniculate clusters; small; greenish-white. Insect pollinated.

Fruit. Drupe; August–September; borne in open clusters, whitish or yellowish; shiny, globose, 4–6 mm across, glabrous, persisting in winter.

Distribution. Common to abundant in southern lower Michigan, locally common in northern lower Michigan, absent in upper Michigan. County occurrence (%) by ecosystem region (Fig. 19): I, 92; II, 33; III, 0; IV, 0. Entire state, 54%. Widely distributed in e. North America and from BC and AK to NM in the Rocky Mountains.

Site-Habitat. Open areas and the understory of open woodlands, often climbing tree trunks to >10 m; present on virtually any kind of soil except extremely acidic, wet peat; stream banks, lake margins, dunes, fencerows, and where birds are likely to disperse the seeds. Especially vigorous on calcareous soils, in floodplains, and on mounds in swamps, where it climbs swamp trees. Associates include a great number of trees and shrubs in dry to dry-mesic and wet-mesic forests and open disturbed areas of nearly all Michigan landforms, especially river floodplains.

Notes. Shade-intolerant, persisting in light shade; rapid-growing; forming long-lived clones. *Toxicodendron* is from the Greek and means "poison tree." The oil secreted by all plant parts is highly allergenic to the touch, producing acute skin irritation accompanied by itching, swelling, and formation of watery blisters. Persons who have come in contact with the plant should wash the affected parts with rubbing alcohol as soon after exposure as possible. Vigorous scrubbing with strong soap and water is an alternative treatment. Such treatments may not prevent dermatitis but may reduce the spread of excess sap that has not immediately reacted with the skin. For details see Voss and Reznicek 2012; and Smith 2008.

Chromosome No. $2n = 30$, $n = 15$; $x = 15$

Similar Species. The similar western poison-ivy, *Toxicodendron rydbergii* (Rydb.) Greene, is widely distributed in the West—Rocky Mountains, Pacific Northwest, and Great Plains—as well as Minnesota and Michigan, and in the far north from Nova Scotia to the Yukon. It is a low, nonclimbing, erect shrub of sandy to rocky soil, especially on shores and sand dunes in upper Michigan and near lakes Michigan and Huron in northern lower Michigan (Voss and Reznicek 2012). County occurrence (%) by ecosystem region (Fig. 19): I, 21; II, 81; III, 100; IV, 100. Entire state, 54%.

320

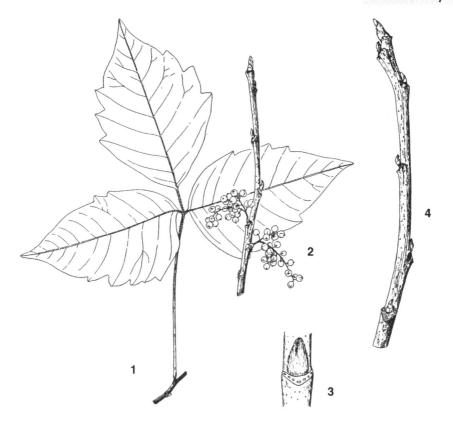

1. Leaf, × 1/2
2. Fruiting twig in fall-winter, drupe, × 1/2
3. Winter lateral bud and leaf scar, enlarged
4. Winter twig, × 1

KEY CHARACTERS

*   creeping vine, low, decumbent or erect shrub, or climbing and holding fast to tree bark with aerial roots
*   leaves trifoliate with terminal leaflet longer-stalked than laterals
*   terminal buds erect, fingerlike, tan, naked
*   drupes white, persisting in winter
*   oil of any plant part usually producing acute dermatitis

The related *Toxicodendron diversilobum* (Torr. & A. Gray) Greene, Pacific poison-oak (not a true oak species), is a shrub and climbing vine of dry interior sites in California and coastal sites from California to British Columbia. The leaves are trifoliate, but with lobes that resemble oak leaves more than those of *T. radicans*. Also the Atlantic poison-oak (not a true oak), *Toxicodendron pubescens* Mill., distributed along the Atlantic and Gulf coasts to Texas and in the mid-South, has oaklike leaves and is a low-branching and climbing vine. Poison-oak produces a violent allergic skin reaction in many people. A closely related species in Japan is *Toxicodendron radicans* subsp. *orientale* (Greene) Gillis, which closely resembles *T. radicans* in foliage and habit. Especially abundant on northern Honshu and Hokkaido islands, it grows in open to shaded sites and upland and strongly acidic wetland soils. It also produces acute dermatitis.

ANACARDIACEAE
Toxicodendron vernix (Linnaeus) Kuntze

Poison-Sumac

Size and Form. Tall, upright shrub or small tree with several stems, to 6 m and stems 6–8 cm in diameter; crown open, rounded, or flat-topped with stout, spreading, sparsely foliated branches. Michigan Big Tree: girth 38.1 cm (15 in), diameter 12.1 cm (4.8 in), height 9.4 m (31 ft), Oakland Co.

Bark. Thin, smooth, slightly streaked, light to dark gray.

Leaves. Alternate, deciduous, pinnately compound, 15–35 cm long, rachis glabrous, unwinged; leaflets 7–13, subsessile except for the stalked terminal; 5–12 cm long, 1.5–6 cm wide; elliptic to obovate; acuminate; acute or rounded; entire; thick and firm; glabrous, lustrous, dark green with scarlet midribs above, slightly pubescent beneath, turning orange to scarlet in autumn; oil of leaf surfaces highly allergenic, causing skin irritation; petioles glabrous, stout, reddish, 2–10 cm long.

Stems-Twigs. Stout, round, greenish to brownish-yellow, glabrous, lenticels raised, numerous; leaf scars broad, shield-shaped, entirely below the buds, not surrounding them; bundle scars numerous, in a line or scattered; pith large, continuous.

Winter Buds. Terminal bud present, naked, often stalked glabrate to hairy, brown, moderate, to 5 mm long, lateral buds smaller, sessile.

Flowers. June; unisexual, plants dioecious; borne in erect, loose paniculate clusters; very small, about 4 mm across, yellowish-green. Insect pollinated.

Fruit. Drupe; September; borne in large, loose, open, drooping panicles, persistent long into winter; white to light yellowish-gray, glabrous, lustrous, subglobose, 4–6 mm across.

Distribution. Infrequent to locally abundant in the southern half of the Lower Peninsula, north to Leelanau and Antrim Co.; absent in the Upper Peninsula. County occurrence (%) by ecosystem region (Fig. 19): I, 66; II, 27; III, 0; IV, 0. Entire state 40%. Widely distributed in e. North America, QC, ON s. to FL, w. to TX.

Site-Habitat. Wet to wet-mesic sites; deciduous swamps, wetlands of all landforms in southern lower Michigan; circumneutral organic soils (pH 6.8–7.2). Associates include tamarack, American elm, yellow birch, black ash, red maple, silver maple, spicebush, red-osier, silky dogwood, wild black currant, highbush blueberry, wild-raisin, *Salix* spp.

Notes. Shade-intolerant, persisting in light to moderate shade; fast-growing; short-lived. Often clone-forming by sprouting from the root collar, giving rise to a clump of genetically identical ramets. All parts of the plant are allergenic, often causing severe dermatitis; use preventive measures discussed under poison-ivy, *T. radicans*. Dermatitis is often more serious than with poison-ivy because the victim has a greater likelihood of facial and neck contact with the allergenic parts. The sap was once used to make a high-grade varnish. Species name *vernix* means "varnish," erroneously referring to the Japanese lacquer tree, *Toxicodendron vernicifluum* (DC.) E. A. Barkley & F. A. Barkley. In older literature it was recognized as a member of the genus *Rhus*.

Chromosome No. $2n = 30$, $n = 15$; $x = 15$

Similar Species. In China and Japan, *Toxicodendron trichocarpum* (Miq.) Kuntze is the counterpart species. Though occurring in wetlands, it is common in upland sites, including oak forests, and is usually a single upright stem; moderately shade-tolerant. It may reach >22 cm in diameter breast height.

1. Leaf, × 1/3
2. Fruiting shoot, drupe, × 1/2
3. Winter lateral bud and leaf scar, enlarged
4. Winter twig, × 1

KEY CHARACTERS

- tall shrub or small tree to 6 m
- leaves pinnately compound, leaflets sessile, shiny, margins entire, turning brilliant orange to scarlet in autumn
- twigs stout with shield-shaped leaf scars
- drupes white, persisting in winter
- habitat open to lightly shaded wetland sites in southern Michigan only
- oil of any plant part usually producing acute dermatitis

THE BLUEBERRIES AND CRANBERRIES–
VACCINIUM

The genus *Vaccinium* is characterized by species with markedly diverse geographic occurrence, ecological habitat, and plant habit. Including 740 species worldwide, they occur on all continents except Australia. Many are circumboreal, arctic-subarctic species. A common denominator of northern temperate species is that they are adapted, in part by ericoid mycorrhizae, to grow and thrive on acidic, nutrient-poor soils. In addition, some preferred habitats are droughty and deficient in soil-water and others wet and deficient in oxygen. Such ecosystems range from very hot and dry sites in the upper Great Lakes Region to the ever-wet forests of the coastal Pacific Northwest. In these temperate rain forests, they may form dense understories and reach >5 m, whereas in the pine barrens of Michigan they are typically <0.5 m. Of the species we describe, 5 are in the subgenus *Cyanococcus,* true blueberries; they are erect deciduous shrubs. Two species, *Vaccinium macrocarpon* and *V. oxycoccos*, belong to the subgenus *Oxycoccus;* they are creeping or decumbent evergreen shrubs.

Fire promotes sprouting and maintains clones of *Vaccinium,* which spread widely by rhizomes. They are among the first plants to rise from the ashes and blackened forest floor. Usually growing in Michigan in open sites with high irradiance, they fruit abundantly unless spring frosts kill the flowers. Universally, blueberries are noted for their edible sweet or sour fruit, which is regarded as one of the most important fruits of aboriginal peoples of Canada's boreal northwestern forest. Marles et al. (2000) summarize uses for food and medicine.

Hybridization is common among blueberries, and this often makes identification of species uncertain. Nevertheless, the 6 species of Michigan described herein are readily identified by a combination of geographic location; upland or wetland site; plant habit; and morphology of leaves, stems, and fruits. Voss and Reznicek (2012) describe additional species in Michigan

Vaccinium species are deciduous or evergreen shrubs, erect or occasionally creeping or trailing; in tropical forests they include both shrubs and epiphytes. The leaves are alternate, simple, small, entire, or very finely toothed, with teeth sometimes ending in minute hairs; leaves are not dotted with resin droplets as in *Gaylussacia*. Stems are slender, typically green or red, and often speckled, with many raised, whitish lenticels. Leaf scars are small, raised, half-circular, triangular, or crescent-shaped with 1 bundle scar. The buds are small, valvate or with 2 to several or many visible scales. The terminal bud is absent; tips of the current shoot freeze and die back at the end of the growing season to a minute stub at the side of the topmost lateral bud.

The flowers are solitary or in terminal or lateral racemes in the axils of the leaves. The fruit is an edible, many-seeded, blue, bluish-black, black, or red berry, usually with bloom. Cultivars of the highbush blueberry, *Vaccinium corymbosum*, are favored in commercial production. Seeds are very small and thus easily and widely dispersed by birds and animals. The average number of cleaned seed for *V. angustifolium* (low sweet blueberry), for example, is 1,450,000/lb (3,196,700/kg) (US Department of Agriculture 2008).

KEY TO SPECIES OF *VACCINIUM*

1. Stems upright, leaves deciduous; fruit blue to bluish-black when ripe.
    2. Medium to tall shrub, to 4 m.
        3. Tall shrub, to 4 m; leaves 2–6 cm long, ovate to elliptic-lanceolate, flowers (fruits) numerous in terminal or lateral clusters; predominant occurrence in southern lower Michigan . . . *V. corymbosum*, p. 328
        3. Medium shrub, to 1.5 m; leaves 0.8–4 cm long, broadly oval to oblong, blade apex acute; flowers (fruits) solitary in the lower leaf axils; occurrence only in upper Michigan . . . *V. ovalifolium*, p. 334
    2. Low shrub, usually less than 0.5 m.
        4. Leaves downy-pubescent beneath; stems densely pubescent . . . *V. myrtilloides,* p. 332
        4. Leaves glabrous or nearly so; stems glabrous or sparsely pubescent.
            5. Mature leaf blades acute to obtuse at apex but with visibly pointed apex; flowers (fruits) numerous in terminal or lateral clusters.
                6. Low, erect shrub with spreading branches; leaf margins distinctly and very finely serrulate . . . *V. angustifolium*, p. 326
                6. Upright shrub with stiff branches; leaf margins entire or seldom serrulate . . . *V. pallidum*, p. 338
            5. Mature leaf blades obtuse to rounded at apex, usually lacking a visible sharp point; flowers (fruits) solitary in the lower leaf axils . . . *V. ovalifolium*, p. 334
1. Stems creeping or trailing; leaves evergreen; fruit red when ripe.
    7. Leaves very small, 0.3–1 cm long; fruit borne on a pedicel with small, paired reddish bracts . . . *V. oxycoccos*, p. 336
    7. Leaves larger, 0.6–1.8 cm; fruit borne on a pedicel with small, paired green bracts . . . *V. macrocarpon*, p. 330

ERICACEAE
Vaccinium angustifolium Aiton

Low Sweet Blueberry

Size and Form. Low, upright shrub, to 50 cm; straggly, open-growing with spreading branches; strongly clone-forming by rhizomes.

Leaves. Alternate, simple, deciduous, blades small, 1.5–4 cm long, 0.5–2.0 cm wide; narrowly elliptic to oblong-lanceolate; acute; acute to cuneate, rarely rounded, thus pointed at both apex and base; margin regularly and finely serrulate their entire length, teeth bristle-tipped; thin; bright green above, green or pale green beneath; glabrous or nearly so; petioles very short, 0.5–1 mm long.

Stems-Twigs. Very slender, green to yellowish-green, becoming reddish-brown on upper surface with exposure to sun; new shoots finely pubescent, becoming glabrous or with minutely hairy lines below the nodes, minutely warty, grooved above the buds or angular; leaf scars half round, bundle scar 1; pith small, continuous.

Winter Buds. Terminal bud absent; lateral buds solitary, sessile, small, ovoid or oblong, glabrous, bud scales imbricate.

Flowers. May–June; bisexual; borne on few-flowered (usually not more than 5) racemes; calyx 5-toothed, greenish bordered with red; corolla cylindric, urn-shaped, 6–7 mm long; white and pink tinged, 5-toothed; stamens 10. Insect pollinated.

Fruit. Berry; July–August; depressed-globose, blue to bluish-black, surface bloom may be light or heavy; 0.9–1.5 cm across; many-seeded; very sweet, edible; seeds ovoid, very small, 1,450,000/lb (3,196,000/kg).

Distribution. Common to locally abundant throughout the state. County occurrence (%) by ecosystem region (Fig. 19): I, 76; II, 100; III, 100; IV, 100. Entire state 89%. Primarily a northern and boreal species ranging from NL to MN, s. to NC, TN, IL, IA.

Site-Habitat. Open to lightly shaded, dry to wet acidic sites, usually sandy-gravelly soils; pine and oak barrens, dry and dry-mesic pine and oak forests, sand prairie, bedrock glades, shorelines, and cliffs; less common in muskeg, bogs, and swamps. Reported from 16 native Michigan communities (Kost et al. 2007), its tree and shrub associates are very diverse, especially in early successional, dry and dry-mesic ecosystems.

Notes. Shade-intolerant, persisting in light shade. Earliest flowering of the blueberries.

Different clones exhibit slight differences in various characteristics, for example, leaf and berry shape, color, and presence of bloom. Important commercial wild blueberry. Following the massive pine logging of the 1870s in northern lower Michigan, blueberry clones became luxuriant and abundant in the open lands. According to firsthand accounts from 1900–1915 recorded at the University of Michigan Biological Station (near Pellston), "Blueberries like to grow in the sun, and when the big trees had been cut down, the blueberries just took over." "The big fires had come through some time after 1900. . . . We had nothing to stop fire with in those days. The blueberries were thick then; we used to go up behind camp and pick a bushel in no time" (Scholtens and Williams 2008). In contrast, at the same sites today under light shade of big-tooth aspen and pine cover, blueberry clones, likely the same ones as 100 years ago, still persist but are weak, few-branched, and patchy in occurrence; the berries are uncommon or rare and very small.

Chromosome No. $2n = 48$, $n = 24$; $x = 12$

Similar Species. Similar morphologically is the hillside blueberry, *Vaccinium pallidum*. In East Asia, *V. smallii* A. Gray of Japan and *V. koreanum* Nakai of Korea are the low blueberries of acidic sites.

326

1. Leaf, × 2
2. Vegetative shoot, × 1/2
3. Branch with flowering shoots, × 1/2
4. Fruiting shoot, berry, × 1
5. Winter lateral bud and leaf scar, enlarged
6. Winter twig, reduced

KEY CHARACTERS

- low, erect shrub <0.5 m, branches spreading
- leaves small, typically <2 cm long and 1 cm wide, regularly and finely serrulate their entire length, teeth bristle-tipped, glabrous or nearly so
- stems very slender, smooth, yellowish-green to green
- fruit a blue to blue-black berry with many seeds
- habitat usually open, dry, fire-prone, acidic, nutrient-poor sites

Distinguished from *Vaccinium pallidum*, hillside or dryland blueberry, by its occurrence throughout the state, smooth stems, leaf blades usually 2–3 times longer than wide, narrowly elliptic, regularly serrulate their entire length, and teeth bristle-tipped.

Distinguished from *Vaccinium myrtilloides*, velvetleaf blueberry, by its erect habit, glabrous stems and leaves, and blades with distinctly serrulate margins.

ERICACEAE
Vaccinium corymbosum Linnaeus
Highbush Blueberry

Size and Form. Erect, tall shrub, 1–4 m; multiple sprouts at or near the base form an open or compact clone of arching branches.

Bark. Roughened or exfoliating, reddish-brown to grayish, mottled on older branches.

Leaves. Alternate, simple, deciduous, 2–6 cm long and about half as wide; ovate to elliptic-lanceolate, acute, often bristle-tipped; cuneate; entire or obscurely serrulate; dark green above, slightly pubescent and paler beneath, becoming orange or reddish in autumn; petioles very short, 1–2 mm long.

Stems-Twigs. Slender, angled, warty, young shoots yellowish-green with warty lenticels, becoming reddish, especially on exposed sides, glabrous or hairy in long grooves above and below the nodes; older stems exfoliating in strips; leaf scars half round, bundle scar 1; pith small, continuous.

Winter Buds. Terminal bud absent; buds red to reddish-brown, glabrous, lower pair of scales prolonged to slender points; flower buds plump, 6–9 outer scales; vegetative buds solitary, sessile, slender, ovoid or oblong, glabrous, bud scales imbricate, 3–5 exposed.

Flowers. May–June; bisexual; borne in terminal or lateral racemes of 3–8 flowers; calyx 5-lobed, glaucous; corolla cylindric to urn-shaped, 0.6–1 cm long, white to pinkish, 5-toothed; stamens 10. Insect pollinated.

Fruit. Berry; July–August; bluish-black, with more or less bloom, 0.7–1.2 cm across; sweet, edible; seeds many, very small, 1,000,000/lb (2,200,000/kg).

Distribution. Common to locally abundant in southern lower Michigan; occasional or absent northward. County occurrence (%) by ecosystem region (Fig. x19): I, 82; II, 10; III, 0; IV, 0. Entire state 41%. Widely distributed in e. North America, QC, ON s. to the southeastern and Gulf coastal plain states (except FL), w. to TX; also coastal WA and BC.

Site-Habitat. Open to lightly shaded, primarily wet sites, acidic soil in the rooting zone, occasionally seasonally dry microsites adjacent to wetlands; bogs, acidic microsite hummocks in circumneutral to basic swamps, edges of lakes and marshes, shrub carr, interdunal swales. A very wide range of associates in ecosystems with both acidic and circumneutral to basic substrates, for example, in acidic, nutrient-poor bogs, tamarack, black spruce, leatherleaf, Labrador-tea; in mildly alkaline shrub carr, willows, meadowsweet; in circumneutral to basic, nutrient-rich southern deciduous swamps, American elm, red maple, blue-beech, spicebush, common elder.

Notes. Shade-intolerant, persistent in moderate shade. Excellent example of a shrub requiring acid soil substrate for establishment and growth (e.g., acidic bogs) and also common on well-aerated, acidic microsite mounds in mildly basic deciduous swamps associated with trees and shrubs characteristic of and adapted to rooting in the circumneutral or basic and oxygen-poor organic and mineral substrate. Shoots of the current year at first greenish, turning reddish-brown on sides exposed to high irradiance. Individuals produce annual fruit crops, an important food source for birds and mammals, thus the small seeds are widely distributed. Cultivars of this species furnish a large share of the blueberries of commerce. It is cultivated extensively in southwestern Michigan. Sixteen cultivars are described by Dirr (1990).

Chromosome No. $2n = 24, 48, 72$ (mostly 48); $n = 12, 24, 36$; $x = 12$

Similar Species. Most similar is likely the diploid species *V. elliottii* Chapman, the southern highbush blueberry, an upland species of the southeastern coastal plain and piedmont.

1. Leaf, × 2
2. Branch with vegetative shoots, reduced
3. Flowering shoot, reduced
4. Flowers, × 1
5. Branch with vegetative and fruiting shoots, berry, reduced
6. Winter lateral bud and leaf scar, enlarged
7. Winter twig, reduced

KEY CHARACTERS

- tall shrub to 4 m; multiple sprouts at or near the base form an open or compact clone of arching branches
- leaves entire or variously serrulate, dark green above, only slightly pubescent beneath, turning red or orange in autumn
- stems at first green, becoming reddish with age, exfoliating in strips
- fruit a many-seeded berry, bluish-black, edible
- habitat open to lightly shaded acidic wetlands (bogs) and acidic microsite mounds in fens, swamps, and other wetlands with circumneutral to mildly basic substrates

ERICACEAE
Vaccinium macrocarpon Aiton
Large Cranberry

Size and Form. Creeping evergreen shrub to 1 m; stems slender, leathery, 10–25 cm long, some erect or ascending at ends; leaf-bearing shoots extending well beyond the peduncles bearing flowers and fruit; rooting at the nodes and forming clones in bog mats.

Leaves. Alternate, simple, evergreen, blades 6–18 mm long, 2.5–6 mm wide; oblong-elliptic, blunt or rounded; cuneate or rounded; entire, flat or only slightly revolute, leathery, dark green above, glaucous beneath; glabrous; petioles very short, about 1 mm long.

Stems-Twigs. Slender, threadlike, more or less pubescent, light brown, becoming reddish-brown in autumn, extending well beyond the flowers or fruit; older stems with exfoliating epidermis and smooth, dark brown inner bark; rooting at nodes and sending up fruiting stems 10–20 cm high; leaf scars small, half round or crescent-shaped, bundle scar 1; pith small, continuous.

Winter Buds. Terminal bud absent; lateral buds very small, solitary, sessile, raised; 2 or more scales exposed.

Flowers. Mid-June to July; bisexual; 1- to 8-flowered in terminal racemes, pedicels usually 1–3 cm long, erect, pubescent, with 2 green lanceolate bracts, 3–5 mm long, 1–3 mm wide, positioned above the middle; calyx 4-cleft; corolla pinkish, deeply 4-cleft, lobes strongly reflexed, 0.5–1 cm long; stamens 8, exserted. Insect pollinated.

Fruit. Berry, September–October; borne on a pedicel 1–3 cm long with small, paired green bracts; persistent over winter; globose, red to dark purple, 1–2 cm across, sometimes persisting into winter, sour, edible; seeds many, tiny, 495,000/lb (1,089,000/kg).

Distribution. Locally common throughout the state. County occurrence (%) by ecosystem region (Fig. 19): I, 61; II, 63; III, 86; IV, 75. Entire state 65%. Primarily a boreal, northern, and mountain species. In e. North America, from NL to ON, s. to NC, w. to TN, MN; in w. North America, NT, AK, s. to BC, s.e. WA, coastal OR, and mtns. of n. CA.

Site-Habitat. Wet, open to lightly shaded, acidic, nutrient-poor sites and microsites; bogs, especially bog mats, muskeg, fens, inter-dunal swales, peatlands adjacent to lakes and streams. Associates include tamarack, black spruce, leatherleaf, Labrador-tea, mountain

holly, bog-rosemary, bog-laurel, willows, wild-raisin, bog birch, and shrubby cinquefoil.

Notes. Very shade-intolerant. This species is the cranberry of commerce, cultivated in New England, the Pacific Northwest, and other areas for large-scale production. Though shallow-rooted and growing in acidic, nutrient-poor substrates, it has many associates that root deeper in circumneutral and mildly basic substrates. Voss (1996) notes that gathering wild cranberries is "a pleasant occupation for those who do not mind risking wet feet (or more)."

Chromosome No. $2n = 24$, n = 12; $x = 12$

Similar Species. The small cranberry, *Vaccinium oxycoccos* L., is closely related.

330

1. Leaf, enlarged
2. Vegetative and flowering shoots, reduced
3. Flowering shoot, × 1/2
4. Flower, × 2
5. Fruiting shoot, berry, × 1/2

KEY CHARACTERS

- creeping, with threadlike stems, some shoots ascending at ends
- stems extremely slender, threadlike, more or less pubescent, light brown, becoming reddish-brown
- leaves evergreen, blades very small, tip rounded or blunt, flat, dark green above, glaucous beneath
- fruit a many-seeded berry borne on a pedicel with green bracts; persistent in winter, small, red to dark purple
- habitat open, wet acidic sites in bogs and microsites in fens

Distinguished from the small cranberry, *Vaccinium oxycoccos*, by its larger and more elliptic-oblong leaf blades, flat or only slightly revolute blade margins, tip rounded to blunt; leaf-bearing shoots extending well beyond the peduncles bearing flowers and fruits; later flowering; scalelike green bracts on the flowering pedicel 3–5 mm across; and larger fruits.

Distinguished from creeping-snowberry, *Gaultheria hispidula*, by its smaller, oblong-elliptic leaf blades with rounded or blunt tips, less hairy stems; a 1–3 cm flower- and fruit-bearing pedicel; red to purple berries; shade-intolerant; and habitat of open, acidic sites.

ERICACEAE
Vaccinium myrtilloides Michaux

Velvetleaf Blueberry, Canada Blueberry

Size and Form. Low, upright shrub, 0.02–0.5 m, much-branched, straggly; clone-forming by rhizomes.

Leaves. Alternate, simple, deciduous, blades 1.5–4 cm long, 0.5–1.6 cm wide; narrow-elliptic to oblong-lanceolate; acute; acute to rounded; entire; thin; pubescent, green above, downy or velvety-pubescent beneath; petioles very short, about 1 mm long.

Stems-Twigs. Very slender, nearly round, densely pubescent, young shoots greenish-brown, becoming reddish-brown, covered with warty lenticels, epidermis exfoliating with age; leaf scars half round, bundle scar 1; pith small, continuous.

Winter Buds. Terminal bud absent; lateral buds small, ovoid or oblong, scales sharply acute, more or less pubescent.

Flowers. May–June; with the leaves; bisexual; borne in small racemose clusters (usually not more than 5–6 per cluster), pedicels 1–3 mm long; calyx 5-toothed, glabrous; corolla broadly cylindric or bell-shaped, 4–6 mm long, greenish-white and pink tinged, 5-toothed; stamens 10, filaments hairy. Insect pollinated.

Fruit. Berry; July–August; depressed globose, bluish-black, with much bloom, 6–8 mm across, many-seeded; sour to sweet, edible; seeds very small, 1,730,000/lb (3,806,000/kg).

Distribution. Occasional in the southern half of lower Michigan; common to locally abundant in northern lower Michigan and upper Michigan. County occurrence (%) by ecosystem region (Fig. 19): I, 45; II, 83; III, 100; IV, 88. Entire state 67%. Widely distributed in n. and boreal North America, NL to NT, s. to NC, the Midwest, MN, IA; also MT, WA.

Site-Habitat. Wet to dry, acidic, nutrient-poor, oxygen deficient or well-aerated, open to moderately shaded sites; more common on moist to wet-mesic than on drier sites; bogs, muskeg, northern conifer and boreal forests, interdunal swales, bedrock woodlands; often sandy to gravelly soils of outwash and lake plains. Associates include a great number of trees and shrubs from shallow-rooted species of acidic, nutrient-poor sites to deeper-rooting species of circumneutral or mildly basic sites.

Notes. Moderately shade-tolerant. Indicator of wet-mesic, acidic sites. Fruit often sour and not as palatable as that of the low sweet blueberry. Also called sourtop.

Chromosome No. $2n = 24$, $n = 12$; $x = 12$

1. Leaf, × 2
2. Flowering shoot, reduced
3. Vegetative and fruiting shoots, berry, reduced
4. Winter lateral bud and leaf scar, enlarged
5. Winter twig, reduced

KEY CHARACTERS

- low, upright, clone-forming shrub to 0.5 m
- leaves small, entire, pubescent green above, downy-pubescent beneath
- stems slender, pubescent, with warty lenticels
- fruit a many-seeded berry, bluish-black with much bloom
- habitat acidic wetlands and wet-mesic sites, spreading to adjacent drier sites

Distinguished from low sweet blueberry, *Vaccinium angustifolium*, by its densely pubescent leaves, especially leaf blade undersurface; very pubescent greenish stems; leaf blades with entire margins; more common in wet to wet-mesic sites.

ERICACEAE
Vaccinium ovalifolium J. E. Smith
Tall Bilberry, Oval-leaved Bilberry

Size and Form. Erect, spreading, stiff-bushy or straggling shrub 0.4–1.5 m, occasionally clone-forming by rhizomes.

Bark. Greenish, becoming gray; strongly exfoliating or flaking on main stems with age.

Leaves. Alternate, simple, deciduous, blades thin, 0.8–4 cm long, 0.6–3.5 cm wide; broadly oval to oblong or elliptic; obtuse to rounded, often mucronate; rounded or broadly cuneate; entire or with few widely spaced gland-tipped teeth near the base; glabrous both sides; pale green above, light blue-green, glaucous with whitish bloom beneath; petioles 1–2 mm long.

Stems-Twigs. Slender, sharply angled or ridged, conspicuously grooved, reddish in grooves, greenish on ridges; glabrous, brownish to yellowish branchlets, becoming gray with age; leaf scars small, half round; bundle scar 1; pith small, continuous.

Winter Buds. Terminal bud absent; lateral buds solitary, sessile, small, ovoid or oblong, glabrous, 2 valvate bud scales.

Flowers. May–June; before the leaves; bisexual; nodding, solitary on pedicels 4–12 mm long, arising in the lower leaf axils on current shoots; calyx very short, 10-toothed; corolla broadly urn-shaped, pinkish, 7–10 mm long; stamens 10. Insect pollinated.

Fruit. Berry; July–August; globose, lacking a persistent calyx, bluish-black with bloom, large, 4–8 mm across; many seeded; edible, tart.

Distribution. Infrequent. Upper Peninsula only. County occurrence (%) by ecosystem region (Fig. 19): I, 0; II, 0; III, 57; IV, 25. Entire state 7%. Widely distributed from AK to WA, OR, MT, SD, and disjunct to the Upper Great Lakes Region and QC, NL.

Site-Habitat. Open to moderately shaded, wet to mesic, acidic, nutrient-poor sites; bogs, moist to wet conifer forests, mixed conifer-hardwood forests, wooded dunes, rocky sites, clearings, edges of lakeshores. In Ontario, rocky mixed woods and along shores of lakes (Soper and Heimburger 1982). Associates include black spruce, balsam fir, white pine, white birch, red maple, American mountain-ash, showy mountain-ash, velvetleaf blueberry, low sweet blueberry, leatherleaf, Labrador-tea, creeping-snowberry.

Notes. Moderately shade-tolerant, occa-sionally persisting in deep shade in forests of hemlock, beech, and maple or mixed conifers. Primarily a species of the Pacific Northwest and from Alberta and British Columbia to Alaska. It is Alaska's most common woodland *Vaccinium* species.

Chromosome No. $2n = 24$, $n = 12$; $x = 12$

Similar Species. Often occurring with and closely related to mountain blueberry, *Vaccinium membranaceum* Torrey, an understory shrub to 2 m with grayish shredding bark on old branches, leaf blades toothed throughout their length, and purple-black fruit. Also closely related to the European *V. myrtillis* L., Heidelberre.

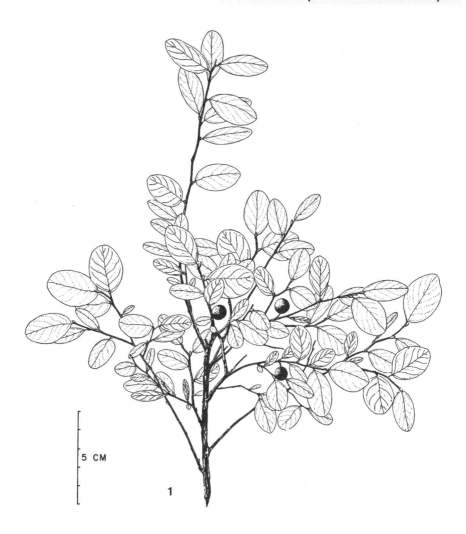

1. Individual with vegetative and fruiting shoots, berry, × 2/5

KEY CHARACTERS
- erect, spreading, stiff-bushy, or straggling shrub to 1.5 m
- leaf blades relatively large, to 4 cm; broadly oval to elliptic, entire or with few widely spaced teeth near base, glabrous both sides, glaucous or with whitish bloom beneath
- stems conspicuously grooved and ridged, strongly exfoliating outer bark on main stems
- fruit a many-seeded berry, globose, lacking persistent calyx, bluish-black with bloom
- occurrence only in upper Michigan

ERICACEAE
Vaccinium oxycoccos Linnaeus
Small Cranberry

Size and Form. Trailing or creeping dwarf shrub, evergreen, vinelike but does not climb, branchlets very slender, 10–15 cm long, some ascending; rooting at the nodes; clone-forming in bog mats.

Leaves. Alternate, simple, evergreen, blades 3–10 mm long, 1–3 mm wide; ovate-oblong to triangular, acute; rounded; entire, strongly revolute, especially toward the tip; leathery; dark green above, glaucous beneath; glabrous; petioles very short, 0.5–1 mm long.

Stems-Twigs. Very slender, threadlike, almost capillary, typically muchbranched and intertwining, more or less pubescent, light brown; older stems with exfoliating epidermis and smooth, dark brown inner bark; leaf scars small, half round, bundle scar 1; pith small, continuous.

Winter Buds. Terminal bud absent; lateral buds very small, solitary, sessile, raised; 2 or more scales exposed.

Flowers. May–June; bisexual; 1–6 flowers borne in axils of small leaves at tip of stem in terminal racemes, pedicels 1–4 cm long, erect, pubescent, with 2 red, lanceolate bracts, 1–2.5 mm long and <0.8 mm wide, positioned below or slightly above the middle; calyx 4-cleft; corolla pinkish, deeply 4-cleft, lobes strongly reflexed, 5–6 mm long; stamens 8, distinctly exserted. Insect pollinated.

Fruit. Berry; August–September; borne on a pedicel 1–4 cm long with small, paired reddish bracts, persistent in winter; globose, pale pink to dark red, 6–8 mm across, many-seeded; sour, edible; seeds very small, 662,000/lb (1,456,400/kg).

Distribution. Locally common throughout the state. County occurrence (%) by ecosystem region (Fig. x19): I, 53; II, 70; III, 86; IV, 100. Entire state 66%. A circumpolar arctic-subarctic species ranging from Greenland and NL to AK, s. to VA, w. to MN, Occurring in a variety of acidic, open sites in the coastal Pacific Northwest, including boreal montane sites, also mtns. of ID, BC, WA, OR.

Site-Habitat. Open, wet, acidic, nutrient-poor, oxygen-deficient sites, including acidic ecosystems of bogs, especially bog mats; muskeg; nutrient-poor conifer swamps; and raised, extremely acid microsites of fens; often partially buried in sphagnum mats. Associated with many trees and shrubs typical of the above sites, as well as species rooting in circumneutral to mildly basic northern fen soils such as shrubby cinquefoil, bog birch, meadowsweet, sweet gale, and Kalm's St. John's-wort.

Notes. Very shade-intolerant, thriving in full sun, persisting in very light shade.

Chromosome No. $2n = 24, 48, 72$; $n = 12, 24, 36$; $x = 12$

Similar Species. The large cranberry, *V. macrocarpon* Aiton, is closely related.

1. Vegetative and flowering shoots, reduced
2. Flowering shoot, × 1/2
3. Flowers, × 1
4. Fruiting shoot, berry, reduced

KEY CHARACTERS

- creeping, with threadlike stems, some ascending
- leaves evergreen, blades very small, tip acute, strongly revolute, dark green above, glaucous beneath
- stems threadlike, much branched and intertwining, more or less pubescent
- fruit a many-seeded berry, borne on a pedicel with red bracts, persistent in winter, very small, red
- habitat open, wet, acidic, nutrient-poor sites in bogs and microsites in fens

Distinguished from the large cranberry, *Vaccinium macrocarpon*, by its smaller leaves and fruits, leaf blade with acute tip and strongly revolute margin, flowers and fruits terminal or nearly so, earlier flowering, scalelike red bracts on the flowering-fruiting pedicel <0.8 mm across, and smaller fruits.

Distinguished from creeping-snowberry, *Gaultheria hispidula*, by its growth in open bogs; lack of wintergreen odor and flavor in all parts; stems lacking appressed hairs; leaves more ovate, glaucous beneath; and fruit red, borne on a long pedicel (1–4 cm).

ERICACEAE
Vaccinium pallidum Aiton

Hillside Blueberry,
Dryland Blueberry

Size and Form. Low, erect, stiffly branched, 20–80 cm high, clone-forming by rhizomes.

Leaves. Alternate, simple, deciduous, blades 1.5–4.5 cm long, 1.2–2.5 cm wide; ovate to broadly elliptic; acute, often mucronate; acute to rounded; entire to serrulate and gland-tipped; moderately thick; dull green above, pale beneath; minutely hairy at first, becoming deciduous; petioles short, 1–2 mm long.

Stems-Twigs. Slender, yellowish-green, distinctly warty-granular, glabrous or minutely hairy with lines descending below nodes; leaf scars half round, bundle scar 1; pith small, continuous.

Winter Buds. Terminal bud absent; lateral buds solitary, sessile, small, ovoid, glabrous, bud scales imbricate, rounded above.

Flowers. May–June; before the leaves; bisexual; borne on terminal or axillary, few-flowered racemes; calyx lobes acute, 5-toothed; corolla short-cylindric to urn-shaped, 5–8 mm long; white, often tinged red, 5-toothed; stamens 10. Insect pollinated.

Fruit. Berry; July–August; globose, dark blue or rarely black, with slight bloom, 0.6–1 cm across; many-seeded; very sweet, edible.

Distribution. Locally frequent in southern lower Michigan. County occurrence (%) by ecosystem region (Fig. 19): I, 53; II, 3; III, 14; IV, 12. Entire state 28%. Widely distributed in e. North America, ON s. to New England, w. Great Lakes Region s. to GA, AL, MS, w. to OK, KS.

Site-Habitat. Dry to dry-mesic, acidic sites, usually sandy-gravelly soils; open patches and edges of oak and oak-pine forests; old fields. Associates include black and white oaks, pignut hickory, black cherry, flowering dogwood, sassafras, low sweet blueberry, black huckleberry, American hazelnut, gray dogwood, witch-hazel, choke cherry.

Notes. Shade-intolerant, persisting in light shade.

Chromosome No. $2n = 24$ (seldom 48), $n = 12$ (24); $x = 12$

Similar Species. Most similar in Michigan is the low sweet blueberry, *Vaccinium angustifolium*; see key to species of *Vaccinium* for morphological similarities in the genus.

1. Twig with flowering shoots, reduced
2. Branch with vegetative shoots and fruiting shoots, berry, reduced

KEY CHARACTERS

- low, erect, stiffly branched clonal shrub to 80 cm
- stems slender, yellowish-green to green, distinctly warty-granular
- leaves ovate to broadly elliptic, about twice as long as wide; apex sharp-pointed; entire or minutely toothed
- fruit a blue to blue-black berry
- habitat dry to dry-mesic, acidic, nutrient-poor, open sites; restricted primarily to southern lower Michigan

Distinguished from low sweet blueberry, *Vaccinium angustifolium*, by its primary occurrence in southern lower Michigan; leaf blades ovate or broadly elliptic, entire or minutely toothed; and distinctly warty stems.

THE VIBURNUMS—*VIBURNUMS*

The large genus *Viburnum* is among the most notable for luxuriant flowering-fruiting of shrubs and small trees. Widespread in northern temperate regions and mountains elsewhere, the 220 species occur in North and South America; Europe; North Africa; and Asia south to Indonesia. It is unusual as a temperate zone genus, having apparently originated in the tropics (Winkworth and Donoghue 2005). Eight species are native in Michigan, including *V. edule,* which reaches Isle Royale from the boreal forest to the north and west, and the southern *V. prunifolium,* ranging north only into the southernmost tier of counties.

Temperate zone viburnums are primarily deciduous. Their habit is erect to straggling, medium to tall shrubs or small trees (e.g., *V. lentago*). Birds disperse their seeds widely, and they reproduce asexually by root sprouting, root-collar sprouting, and layering. Leaves are opposite, simple, either palmately or pinnately veined, and mostly lobed or toothed or both. Twigs are slender to stout and dull-colored; they have less conspicuous lenticels than those of the closely related elders (*Sambucus*). The buds are naked or covered by a single pair of valvate scales or with 2–3 pairs of overlapping scales. Petiole bases or opposing leaf scars meet around the stem or are connected by a transverse ridge. The fruit is a 1-seeded drupe borne in terminal clusters. Cullina (2002) observes that most viburnum fruits are more "flamboyant than the flowers—flat-topped or domed clusters of small, frilly off-white blossoms conglomerated into a three-to-six-inch wide lace-doily landing pad for bees." The fruits go through marked color transformations on the way from green to flamingo, red, blue, and finally almost black by fall. The genus *Viburnum* is especially distinctive from other members of the Adoxaceae (e.g., *Sambucus*) by unusual development of the gynoecium, in which 2 carpels abort and the single functional ovule is displaced, developing in 1 of the sterile cavities within the ovary (Judd et al. 2008).

Viburnum is the ancient Latin name for the wayfaring tree. Two species, long cultivated in North America, illustrate the history of the genus. A native of Europe and Britain is *Viburnum lantana* L., the wayfaring tree, thought to be the original "Viburnum" mentioned by Virgil (Coats 1992). This European hedgerow bush has white, woolly backed leaves such that the Elizabethan herbalist William Turner likened it to a wayfarer come from a dusty road. Another European (also North African and western Asian) native, *V. opulus,* the European highbush-cranberry, has been cultivated for centuries. Among the popular cultivars are the dwarf form (*nanum*) and the European snowball (*V. opulus* L var. *roseum*), which has all sterile flowers in a large globose head.

European highbush-cranberry was likely introduced to North America in colonial times. It has been widely cultivated and through widespread seed dispersal by birds has become naturalized in many places in the western Great Lakes Region (e.g., Minnesota [Smith 2008; Voss and Reznicek 2012]). It is similar to the native species in many respects and typically recognized by its primary occurrence in parks and arboretums, along streets, and in other urban environments. The most distinctive morphological difference is the size and shape of the glands on the petiole near the blade base. The European species has 2 to 8 large glands, ca. 1–1.5 mm across, which are concave and elliptical with a distinctive rim around the margin. In contrast, our species typically has 2 small stalked glands, flat-topped and round in cross section.

KEY TO SPECIES OF *VIBURNUM*

1. Leaves 3-lobed and palmately veined.
    2. Low, erect shrub to 2 m; leaves densely pubescent beneath with tiny red to brown resinous glands; petioles without glands; winter buds with 4 scales exposed, not valvate; drupes purplish-black . . . *V. acerifolium*, p. 342
    2. Tall, erect shrub to 5 m; leaves glabrous or nearly so beneath, without resinous glands; petioles with conspicuous glands; winter buds with 2 scales exposed, valvate; drupes red . . . *V. trilobum*, p. 352
1. Leaves unlobed and pinnately veined.
    3. Leaves coarsely toothed, each prominent lateral vein ending in a marginal tooth; winter buds with 4–6 imbricate scales.
        4. Leaves more or less densely downy-pubescent beneath, leaf-blade margin ciliate, petioles pubescent and short (0.2–0.8 mm), stipules present . . . *V. rafinesquianum*, p. 350
        4. Leaves glabrous beneath except in vein axils, leaf-blade margin lacking cilia, petioles long (1–2.5 cm), stipules lacking . . . *V. dentatum*, p. 346
    3. Leaves finely or irregularly toothed or entire but not coarsely dentate, lateral veins becoming indistinct near the blade margin; winter buds with 2 scales exposed.
        5. Tall shrub or small tree, 5–9 m; leaves finely and sharply serrate; winter buds dark gray; cymes sessile or nearly so; petioles irregularly undulate, wrinkled, or winged, 2–3 cm long . . . *V. lentago*, p. 348
        5. Tall, clonal shrub to 4 m; leaves irregularly crenate to entire; winter buds golden brown; cymes on short terminal peduncles; petioles not as above, 0.5–1.5 cm long . . . *V. cassinoides*, p. 344

ADOXACEAE
Viburnum acerifolium Linnaeus
Maple-leaved Viburnum

Size and Form. Low, upright shrub to 2 m; branches usually short; usually multistemmed clones formed by basal sprouting and root suckering.

Bark. Grayish-brown, smooth, becoming roughened at base, lenticels conspicuous.

Leaves. Opposite, simple, deciduous, blades 6–10 cm long and equally \ wide; suborbicular to ovate, early leaves 3-lobed, late leaves (upper leaves of indeterminate shoots) not lobed, lobes acute to acuminate; base sub-cordate to rounded, sinuses shallow; coarsely and irregularly dentate; palmately veined; thin; upper surface pubescent, becoming almost smooth at maturity, dull green above, densely stellate-pubescent, paler beneath and typically covered with tiny, red to brown resinous dots (glands); petioles round, pubescent, 1–3 cm long.

Stems-Twigs. Slender, current shoots finely pubescent, tan to reddish-brown; twigs glabrous gray; larger branches becoming ridged and 4–6 sided; lenticels conspicuous, brown; leaf scars opposite, raised, narrow, crescent-shaped, bundle scars 3; pith moderate, white, continuous.

Winter Buds. Terminal buds 5–10 mm long, green to greenish-purple, usually glabrous, 2 pairs of bud scales exposed, the lower scales very short, two-toned, if a vegetative bud if slender and pointed or a flower bud if swollen near the base; lateral buds 4–7 mm long, short-stalked, appressed, slender and pointed with 4 scales exposed.

Flowers. May–June; bisexual; borne in long-stalked cymes, usually terminal, 3–8 cm across; yellowish-white to faintly tinged with pink. Insect pollinated.

Fruit. Drupe; September; purplish-black without a bloom, ellipsoid to globose, 6–8 mm long, 8–9 mm wide, in open clusters, persistent; 1 seed, thin; seeds 13,100/lb (28,820/kg).

Distribution. Common in the Lower Peninsula; occasional to rare in the Upper Peninsula. County occurrence (%) by eco-system region (Fig. 19): I, 97; II, 93; III, 43; IV, 38. Entire state 86%. Widely distributed throughout e. North America, QC, ON s. to FL, w. to TX.

Site-Habitat. Shaded, dry-mesic to mesic, well-drained to moderately well-drained, nutrient-rich sites; oak-hickory and oak forests, mesic beech–sugar maple forests. Associates include trees of oak-hickory and beech–sugar maple communities (Barnes and Wagner 2004, 391); shrubs include choke cherry, witch-hazel, prickly gooseberry, Virginia creeper, leatherwood, alternate-leaved dogwood, and Canadian fly honeysuckle.

Notes. Shade-tolerant; slow-growing. Common and scientific names refer to maplelike foliage; a good indicator species of moderate to high nutrient availability. Although organic and upper mineral horizons may test acidic, lower horizons of calcareous substrate typically determine overall nutrient status of the site. Roots of this species penetrate into these favorable horizons, whereas surface-rooting species may suggest an acidic, nutrient-poor ecosystem. Sprouts vigorously from the root collar if stems are killed by fire. Using a hand lens one can see red or black dots (i.e., glands) on the undersides of the leaves.

Chromosome No. $2n = 18$, $n = 9$; $x = 9$

Similar Species. See key to *Viburnum* species. The 3-lobed, red-fruited, low-bush cranberry or squashberry of western North America, *Viburnum edule* (Michx.) Raf., is a counterpart species of *V. trilobum* Marsh. not *V. acerifolium*. *V. edule*, a boreal forest species, occurs on Isle Royale and Manitou Island of Keweenaw Co. (Voss and Reznicek 2012).

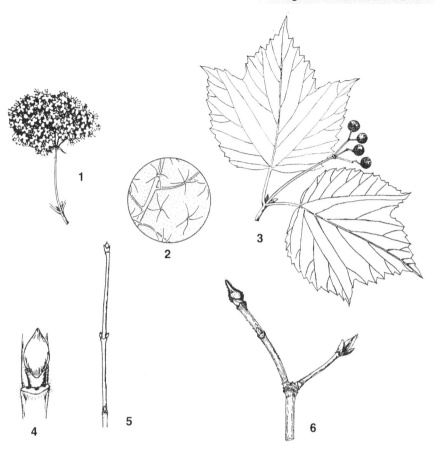

1. Flowering shoot, × 1/2
2. Enlarged lower leaf-blade surface with stellate hairs, × 6
3. Fruiting shoot, drupe, × 1/2
4. Winter lateral bud and leaf scar, enlarged
5. Winter twig, reduced
6. Winter twig, flower bud (*left*), vegetative bud (*right*), × 1

KEY CHARACTERS

- low, upright shrub to 2 m
- early leaves characteristically 3-lobed, unlobed late leaves at apex of indeterminate shoots; leaf blades stellate-pubescent beneath
- twigs finely pubescent
- fruit a purplish-black drupe borne in terminal clusters
- habitat shaded, dry-mesic to mesic, well-drained to moderately well-drained sites

Distinguished from all other *Viburnum* species except American highbush-cranberry, *V. trilobum*, and European highbush-cranberry, *V. opulus*, by its lobed, maplelike early leaves.

Distinguished from highbush-cranberry, *Viburnum trilobum*, by its low-shrub stature, pubescent branchlets and petioles, glandless petioles, and purplish-black fruit.

Distinguished from the nonnative European highbush-cranberry, *Viburnum opulus*, by its glandless petioles, purplish-black fruit, lack of bristle-tipped stipules, and common occurrence in shaded understories of native forests outside urban areas.

ADOXACEAE
Viburnum cassinoides Linnaeus
Wild-raisin, Withe-rod

Size and Form. Erect, tall, stiff-branched shrub to 4 m.

Bark. Smooth, gray.

Leaves. Opposite, simple, deciduous, blades 3–10 cm long, 2–6 cm wide, highly variable in size depending on shoot size and position in crown; elliptic to oval, oblong, the terminal, often rhombic; acute, short-acuminate or rounded; rounded or cuneate; distinctly toothed with blunt teeth, irregularly crenulate, or entire; lateral veins anastomosing before reaching the margin, scurfy-punctate upon unfolding, becoming glabrous or nearly so; dull dark green above, paler and somewhat scurfy below; petioles flat and grooved above, scurfy-punctate, stipules lacking, 0.5–1.5 cm long.

Stems-Twigs. Slender, angled below the nodes; branchlets rusty-scurfy, becoming glabrous or remaining scurfy near the tip, older branches dull, somewhat warty and ridged; leaf scars opposite, crescent-shaped, narrow; bundle scars 3; pith moderate, continuous.

Winter Buds. Golden brown, scurfy, finely pitted, valvate, with 2 scales, short-stalked. Terminal bud present either flower or vegetative; flower buds 1.5–2.5 cm long, swollen near base; vegetative buds slender, 1–2 cm long; lateral buds slender, appressed, 0.6–1.5 cm long.

Flowers. June; bisexual, borne in terminal cymes with a single branched peduncle, flat, broad, 1–4 cm across; 1–2 cm long, 0.5–1 cm wide; white, ill-scented, corolla with 5 spreading lobes; stamens 5, exserted. Insect pollinated.

Fruit. Drupe; September; globose or ellipsoid, 6–9 mm long; bluish-black to black with bloom; stone thin, elliptic or oblong-elliptic; sweet flesh, edible; 1 seeded; seeds 27,603/lb (60,727/kg).

Distribution. Common but site specific in lower Michigan; decreasing in occurrence toward the west in upper Michigan except in areas of high precipitation. County occurrence (%) by ecosystem region (Fig. 19): I, 55; II, 87; III, 86; IV, 25. Entire state 66%. Widely distributed in e. North America, NL to ON, s. to GA, LA.

Site-Habitat. Strongly favoring relatively open, wet-mesic to poorly drained wetlands; swamps, bog edges, fens, even those embedded in dry upland sites; often occurring on acidic organic matter but also on circumneutral to basic wetlands; tolerates cold microsites and northern latitudes to 49° in eastern Ontario. Associates include northern white-cedar, balsam fir, eastern hemlock, eastern white pine, white spruce, trembling aspen, balsam poplar, black ash, American elm, red maple, Canadian fly honeysuckle, mountain holly, velvetleaf blueberry, low sweet blueberry.

Notes. Moderately shade-tolerant. Indicator plant of wet-mesic to wet and typically cold microsites. Latest to bloom of all *Viburnum* species. The marked decline in occurrence in the western Upper Peninsula is likely due in part to declining precipitation toward the west, similar to the decline in occurrence of eastern hemlock and American beech. The common name withe-rod refers to its tough, flexible twigs (withes), which are shaped like rods.

Chromosome No. $2n = 18$, $n = 9$; $x = 9$

Similar Species. Our species is regarded by some authors as the northern geographic variety (i.e., var. *cassinoides* [L.] Torr. & A. Gray.) of the smooth withe-rod or possumhaw, *Viburnum nudum* L., and distinguished from it by firmer leaves with a dull upper surface, indistinct veins that anastamose before reaching the leaf margin, and sweet flesh of fruit. Smooth withe-rod ranges from the Atlantic and Gulf coastal plains from CT to FL and TX, inland at low elevations from VA and KY southward.

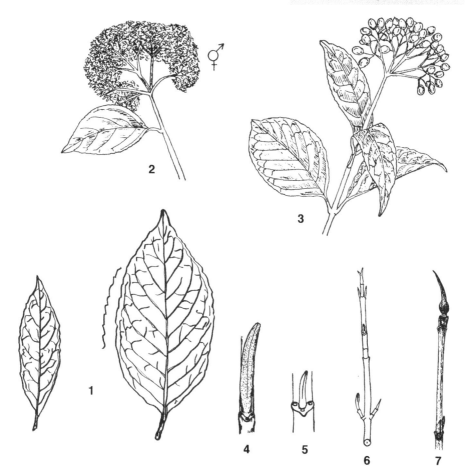

1. Leaves, reduced
2. Flowering shoot, reduced
3. Fruiting shoot, drupe, reduced
4. Winter valvate lateral bud and leaf scar, enlarged
5. Winter lateral bud with leaf scar, × 1
6. Winter twig with stalked lateral buds, reduced
7. Winter twig with terminal flower bud, × 1

KEY CHARACTERS

- erect, tall shrub to 4 m
- leaves opposite, blades dull green above, margin distinctly toothed with blunt teeth or irregularly crenulate, petiole grooved
- winter buds valvate, golden or rusty brown, pitted on the surface
- fruit a bluish-black to black drupe with bloom
- habitat wet-mesic to wet, poorly drained wetlands

Distinguished from nannyberry, *Viburnum lentago,* by its leaf blades with blunt teeth or irregularly crenulate margins; winter buds scurfy, golden brown; stalked inflorescences; and smaller fruits.

ADOXACEAE
Viburnum dentatum Linnaeus
Arrow-wood

Size and Form. Medium to tall, erect shrub to 3 m; arching stems with age form a moderately dense, irregular crown; clone-forming by root-collar sprouting.

Bark. Gray to grayish-brown, becoming roughened toward the base.

Leaves. Opposite, simple, deciduous, blades 3–9 cm long and 3–7 cm wide; ovate to orbicular, acute to acuminate; rounded to subcordate; coarse and evenly dentate, with 9–21 teeth per side, margins usually not ciliate, lateral veins simple or 1–2 forked, each extending to a marginal tooth; early leaves ovate to suborbicular with large, distinct, sharp teeth per side; late leaves lance-ovate with fewer, less distinct, and sharp teeth per side; bright green and glabrous above, paler and glabrous below except for pubescent tufts in axils of veins and a few hairs scattered along veins; turning yellow in autumn; petioles 1–2.5 cm long, glabrous or nearly so, lacking stipules.

Stems-Twigs. Young twigs slender and straight, stems smooth, 4-angled, glabrous and pale grayish-brown, becoming purplish-gray; leaf scars crescent- or widely V-shaped, bundle scars 3; pith white, continuous.

Winter Buds. Terminal bud ovoid, 4–7 mm long, sharply acute; lateral buds short-stalked, appressed, narrowly ovoid, slender, acute; buds reddish to reddish-brown, shiny, 4–6 scales; flower buds swollen near base.

Flowers. May–June; bisexual; borne in dense, stalked, flat-topped, terminal cymes 5–9 cm across; small, creamy-white; floral tube glabrous, wheel-shaped, with 5 spreading lobes; stamens 5, style short, 3-lobed, exserted, enlarged base with hairs directed upward. Insect pollinated.

Fruit. Drupe; August–September; bluish-black, ellipsoid, flattened, 5–7 mm wide, base of style persistent; pit 5–7 mm long, plump with a broad shallow groove on one side; 1-seeded; seeds 20,400/lb (44,880/kg).

Distribution. Infrequent in southern lower Michigan, rare elsewhere. County occurrence (%) by ecosystem region (Fig. 19): I, 34; II, 7; III, 14; IV, 12. Entire state 20%. NB to MI, s. to SC, GA: the northern var. *lucidulum*.

Site-Habitat. Open to lightly shaded, wet-mesic to mesic sites such as swales, swamp and marsh edges, and wet meadows; also colonizing adjacent uplands and disturbed areas such as fencerows and road edges. Wetland sites have basic soil or substrate; also grows on acidic wetland edges and disturbed uplands. Associated species of wet-mesic ecosystems and their edges include highbush blueberry, winterberry, common elderberry, spicebush, and gray dogwood.

Notes. Moderately shade-tolerant. *Viburnum dentatum* L. is a complex of several taxa, the Michigan and northern plants belonging to var. *lucidulum* Aiton. It is often recognized as the species *V. recognitum* Fernald, the northern or smooth arrow-wood. Some individuals of southern Michigan origin and of northern Lower and Upper Michigan are certainly escapes from cultivation (Voss and Reznicek 2012). The common name reportedly comes from use of straight stems by Indians at least for small game.

Chromosome No. $2n = 36, 54, 72$

Similar species. In Michigan and northern regions, the species with similar coarsely dentate blade margins is *Viburnum rafinesquianum*, downy arrow-wood. In more southerly regions, there are several varieties of the *Viburnum dentatum* L. complex, which, in contrast to our variety, *lucidulum,* have pubescent morphological parts.

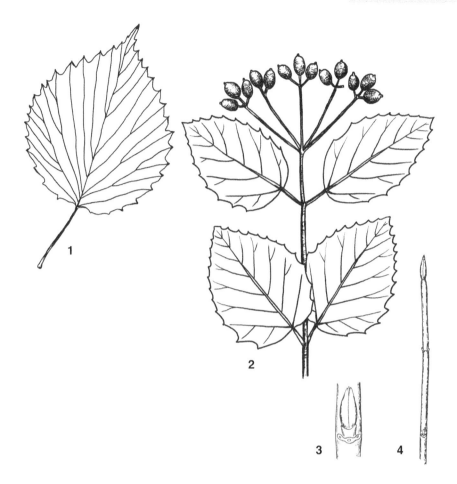

1. Leaf, × 1/2
2. Fruiting shoot, drupe, reduced
3. Winter lateral bud and leaf scar, enlarged
4. Winter twig, reduced

KEY CHARACTERS

- medium to tall shrub to 3 m
- leaves opposite, ovate to orbicular with sharply coarse, even, dentate margins, glabrous beneath except in vein axils; petioles glabrous; stipules lacking
- twigs glabrous, young stems arrow straight
- fruit a bluish-black drupe
- habitat usually wet-mesic sites with calcareous substrates

Distinguished from *Viburnum rafinesquianum*, downy arrow-wood, by its leaf blades glabrous beneath except in vein axils, glabrous and longer petioles, bud scales and leaf-blade margins lacking cilia, and stipules lacking.

ADOXACEAE
Viburnum lentago Linnaeus
Nannyberry

Size and Form. Tall shrub with bushy top or small tree, 2–8 m high and 5–20 cm in diameter. Trunks usually several, slender, often crooked; numerous tortuous branches form a wide, irregular, open, rounded crown. Clone-forming by basal sprouting and suckering from roots. Shallow root system giving rise to root suckers. Michigan Big Tree: girth 86.4 cm (34 in), diameter 27.5 cm (10.8 in), height 15.2 m (50 ft), Oakland Co.

Bark. Thin, grayish-brown, becoming reddish-brown on old trunks and broken into small, thick plates.

Leaves. Opposite, simple, deciduous, blades 5–10 cm long, half as wide; ovate to suborbicular; acuminate to acute; rounded, rarely cuneate; margins finely and sharply serrate, 6–10 teeth per cm of margin, often incurved and callous-tipped; thin and firm; lustrous bright green above, pale beneath and more or less covered with dark reddish glands; petioles broad, grooved, more or less irregularly undulate, wrinkled, or winged, 2–3 cm long.

Stems-Twigs. Slender, at first light green, rusty-pubescent, becoming glabrous and dark reddish-brown; lenticels round; leaf scars opposite, crescent-shaped, opposing scars meeting around the stem or connected by a transverse ridge; bundle scars 3; pith medium, white, continuous.

Winter Buds. Terminal bud present, either vegetative or flower; vegetative buds slender, flattened, tip elongated, 1–1.5 cm long; bud scales 2, closely valvate, brown with waxy or scurfy surface; flower buds swollen at the base, apex spirelike to acute; grayish with scurfy pubescence, 1.5–2 cm long; lateral buds small, 1–5 mm long, reddish-brown, elongated.

Flowers. May–June, after the leaves; bisexual; borne in stout-branched, scurfy, round-topped terminal cymes 7–12 cm across; small, creamy-white; calyx tubular, 5-toothed; corolla 5-lobed, cream or white, 6 mm across; stamens 5 with yellow anthers; ovary 1-celled with short, thick, green style and broad stigma. Insect pollinated.

Fruit. Drupe; September–October; fleshy, 1–1.5 cm long, ellipsoid to subglobose, flattened, bluish-black, borne in few-fruited, reddish-stemmed clusters; pit oval, flat, rough; flesh sweet, edible; seeds 5,900/lb (12,980/kg).

Distribution. Frequent to locally abundant in lower Michigan; locally frequent in upper Michigan. County occurrence (%) by ecosystem region (Fig. 19): I, 95; II, 70; III, 57; IV, 88. Entire state 82%. Widely distributed in e. North America, NS to SK, and adjacent USA, Midwest, n. Great Plains, NC, GA, AL.

Site-Habitat. Relatively open, mesic to wet-mesic sites with nutrient-rich neutral to basic soils; floodplains, fens, sedge meadows, shrub swamps, hardwood and tamarack swamps, lake margins, disturbed areas, especially sites where American elm and red ash have been killed by disease or insect attack. Associates include a great many tree and shrub species of these diverse communities throughout the state.

Notes. Moderately shade-tolerant, moderately fast-growing, stems short-lived. Stems killed by fire, but rapidly sprouting from the base or roots. Fruits are food for many birds and mammals; often planted to improve wildlife habitat. Sometimes used as an ornamental for its attractive flowers; the tendency to form sprouts in lawns and gardens makes it somewhat undesirable for landscape use.

Chromosome No. $2n = 18$, $n = 9$; $x = 9$

Similar Species. Eight native *Viburnum* species occur in Michigan and the western Great Lakes Region. The most closely related species is the blackhaw, *V. prunifolium* L., a relatively rare plant found only sporadically and primarily in the southernmost tier of counties of lower Michigan. Its leaves are only about 1/3 the size of those of *V. lentago*, the teeth are very fine, and the petiole is wingless or nearly so. The wild-raisin *V. cassinoides* L., is frequent northward, widely distributed, and occurs in wetlands and wet-mesic sites. The southern counterpart is the rusty blackhaw or southern blackhaw, *V. rufidulum* Raf.

1. Winter twig, with leaf buds, × 1 1/2
2. Winter twig, with flower bud, × 1
3. Leaf, × 3/4
4. Flower, enlarged
5. Fruiting shoot with drupes, × 1/2

KEY CHARACTERS

- tall shrub with bushy top or small tree, 2–8 m, of wet-mesic to mesic habitats
- leaves opposite; blades ovate, apex long-acuminate; margins finely and sharply serrate; tiny dark reddish dots on the undersurface; petioles broad, grooved, irregularly wrinkled or winged
- terminal vegetative buds long, slender; bud scales 2, valvate, scurfy-pubescent; flower buds swollen at the base, apex spirelike or acute
- fruit a bluish-black drupe borne on a red pedicel

ADOXACEAE
Viburnum rafinesquianum Schultes
Downy Arrow-wood

Size and Form. Low to medium, upright shrub to 2.5 m; many arching stems form a moderately dense, rounded crown; clone-forming by root-collar sprouting and root suckers.

Bark. Gray to grayish-brown, becoming roughened toward the base.

Leaves. Opposite, simple, deciduous, blades 3–9 cm long and about half as wide; ovate to elliptic, acute to acuminate; rounded to subcordate; coarsely dentate, margins ciliate, lateral veins simple or 1–2 forked, each extending to a tooth; late leaves on indeterminate shoots acuminate, entire or rarely with tiny teeth (see Fig. 6); light green, slightly pubescent above, more or less densely downy-pubescent beneath, turning yellow in autumn with a purple fungus dotting the leaves; petioles short, 0.2–0.8 cm long, pubescent; paired stipules at base bristlelike, usually longer than the petiole.

Stems-Twigs. Slender, round in cross section, tan to gray, young shoots yellowish-brown, puberulent, becoming glabrous and purplish to brownish-gray, lenticels prominent; leaf scars opposite, small, narrow, crescent- or V-shaped, bundle scars 3; pith creamy-white to tan on large stems, continuous.

Winter Buds. Terminal bud ovoid, 3–5 mm long, brown to reddish-brown, shiny, 4–6 ciliate scales exposed, keeled; flower buds swollen near base; laterals ovoid to oblong, plump, divergent, 4 scales exposed.

Flowers. May–June; bisexual; borne in dense, stalked, flat-topped, terminal cymes 5–8 cm across; small, white; floral tube pubescent, wheel-shaped, with 5 spreading lobes; stamens 5; style short, 3-lobed, exserted, enlarged base glabrous. Insect pollinated.

Fruit. Drupe; August–September; bluish-black, ellipsoid, flattened, 5–6 mm wide; pit thin, slightly 2-grooved on each side; 1-seeded, thin.

Distribution. Common to locally abundant in lower Michigan, infrequent to rare elsewhere. County occurrence (%) by ecosystem region (Fig. 19): I, 74; II, 37; III, 14; IV, 38. Entire state 52%. Widely distributed in e. North America, QC to MB, s. to GA, w. to OK.

Site-Habitat. Open to lightly shaded, dry-mesic or dry, nutrient-rich sites, especially oak-hickory and oak forests. Characteristic of calcareous, droughty, sandy-gravelly soils. Associates include black, white, and northern red oaks, pignut and shagbark hickories, hop-hornbeam, black cherry, choke cherry, maple-leaved viburnum, prickly gooseberry, and Virginia creeper.

Notes. Moderately shade-tolerant; drought-tolerant. Vigorous, few-branched sprouts from the root collar can reach 1.8 m in 2 growing seasons. The variety *affine* (B. F. Bush) House, with leaves glabrate or only pubescent along the main veins, is known mainly from southeastern lower Michigan (Voss and Reznicek 2012).

Chromosome No. $2n = 36$, $n = 18$; $x = 9$

Similar Species. In Michigan and northern regions, the species with similar coarsely dentate blade margins is *Viburnum dentatum* var. *lucidulum* Aiton, arrow-wood, or, at specific rank, *V. recognitum* Fernald (see description of *V. dentatum*).

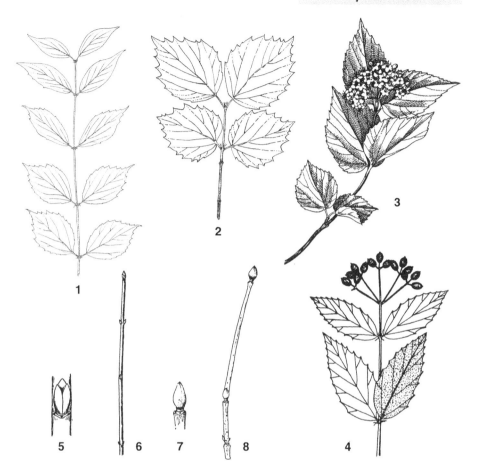

1. Late leaves on indeterminate shoot, × 1/2
2. Early leaves on determinate shoot, × 1/2
3. Flowering shoot, × 1/2
4. Fruiting shoot, drupe, × 1
5. Winter lateral bud and leaf scar, enlarged
6. Winter twig with vegetative buds, reduced
7. Winter flower bud, × 2
8. Winter twig with terminal flower bud, × 1

KEY CHARACTERS

- low, upright shrub to 2.5 m with arching branches
- leaves opposite, ovate with coarse dentate margins, lateral veins distinct, each extending to a tooth, blades pubescent beneath, petioles pubescent, stipules bristlelike
- twigs puberulent when young, becoming glabrous
- fruit a bluish-black drupe
- habitat usually dry-mesic to dry sites, often with calcareous subsoil

Distinguished from *Viburnum dentatum*, arrowwood, by its leaf-blades downy-pubescent beneath, petioles pubescent, bud scales and leaf blade margins ciliate, shorter petioles, and stipules present.

351

ADOXACEAE
Viburnum trilobum Marshall
American Highbush-cranberry

Size and Form. Tall, spreading shrub, to 5 m; many arching stems form a dense, round-topped crown; sprouts from the root collar form clones of few to many stems. Michigan Big Tree: girth 25.4 cm (10.0 in), diameter 8.1 cm (3.2 in), height 9.8 m (32.0 ft), Oakland Co. Shrub with larger diameter, 45.7 cm, in Wayne Co.

Bark. Thin, glabrous, smooth, becoming roughened and fissured, grayish-brown to gray.

Leaves. Opposite, simple, deciduous, blades 4–10 cm long and equally wide; broadly ovate to wedge-shaped, strongly 3-lobed, lobes long, longer than broad; acuminate; rounded; early leaves coarsely dentate, late leaves entire or nearly so; green, lustrous above, with sparse scattered hairs, paler and pubescent on veins beneath or becoming glabrous; leaves turning brilliant scarlet in autumn; petiole 1–3 cm long, with 2 (rarely several) small, stalked glands near the blade base, about 0.4 mm across, flat-topped and round in cross section; a pair of slender, thick-tipped stipules at base, rarely persistent.

Stems-Twigs. Stout, glabrous, light brown to grayish; lenticels small, round; vigorous sprouts (>2 m high) orange-brown, smooth and shiny; leaf scars narrow, crescent-shaped; bundle scars 3; pith large, white, continuous, encircled by a thin light brown ring of cells; stipule scars distinct.

Winter Buds. Terminal bud present except on twigs having a terminal fruit scar or where the stem has died back to lateral buds; usually flower buds stalked, 1–1.5 cm long, globose and swollen near the base; glabrous, 2 connate scales as long as the bud, splitting across the apex; vegetative buds slender, appressed, 0.5–1 cm long, green to red, glabrous, lustrous.

Flowers. May–June; bisexual; borne in flat-topped cymes 5–10 cm long; flowers white, outer flowers sterile, 1–2 cm broad, nearly wheel-shaped, deeply 5-lobed, somewhat irregular; inner flowers perfect, fertile, much smaller, 2–3 mm, broad; calyx attached to the ovary, border 5-toothed; stamens 5, elongate. Insect pollinated.

Fruit. Drupe; September; globose or short-ellipsoid, 0.8–1 cm across, orange to scarlet, translucent, persistent in winter, juicy, acidic when fresh, palatable (to some) when cooked, 1-seeded, thin.

Distribution. Locally frequent throughout the state. County occurrence (%) by ecosystem region (Fig. 19): I, 55; II, 50; III, 57; IV, 62. Entire state 54%. Transcontinental, in Canada NL to BC, in USA New England, Midwest, n. Great Plains, n. Rocky Mountains to WA.

Site-Habitat. Wet-mesic and seasonally wet sites in floodplains, along streams, in open or lightly shaded swamps. Associates include American elm, red maple, silver maple, red ash, black ash, blue ash, blue-beech, nanny-berry, common elder, red-berried elder, wild black currant, and red raspberry.

Notes. Moderately shade-tolerant. Stems grow taller in light to moderate shade than in full sunlight. Fruits used as substitute for cranberries (*Vaccinium*) in making jelly but lose their tastiness during late fall. A potentially important ornamental because of its handsome clusters of white flowers and scarlet fruits. Seldom available in nurseries, although it transplants easily. Despite its similarity to the cultivated European highbush-cranberry, *Viburnum opulus* L., our native has form and leaf differences (i.e., petiolar glands stalked and flat-topped, not concave or indented; upper blade surface with sparse scattered hairs) and a distinctive site preference for wet-mesic sites and lightly shaded wetlands and floodplains. Site requirements for the European species in nature are unknown.

Chromosome No. $2n = 18$, $n = 9$; $x = 9$

Similar Species. Closely related is the European highbush-cranberry, *Viburnum opulus* L. Our native species is often considered *V. opulus* L. var. *americanum* Aiton or *V. opulus* subsp. *trilobum* (Marshall) R. T. Clausen. Also having lobed leaves is the native low-bush

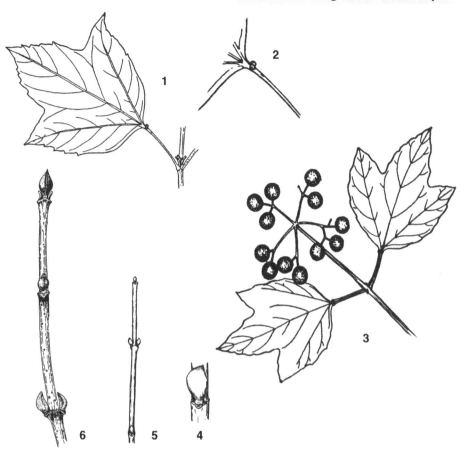

1. Leaf, × 1/2
2. Petiole with stalked glands, enlarged
3. Leaf and fruiting shoot, drupe, reduced
4. Winter lateral bud and leaf scar, enlarged
5. Winter twig with vegetative buds, reduced
6. Winter twig with lateral vegetative buds and terminal flower bud, × 1

KEY CHARACTERS

- tall shrub to 5 m with many arching stems
- leaves opposite, strongly 3-lobed, early leaves toothed, late leaves usually entire; upper blade surface with scattered hairs; petiole with usually 2 small, stalked, elliptical, flat-topped glands near base of blade
- twigs with globose, bright green, valvate buds
- fruit a red drupe borne in terminal flat-topped cymes, persistent in winter
- habitat wet-mesic and seasonally wet floodplain sites

cranberry or squashberry, *V. edule* (Michx.) Raf., which occurs on Isle Royale and in boreal forests from Newfoundland and Labrador to Alaska, also in moist sites from British Columbia and Alberta to Colorado.

VITACEAE
Vitis aestivalis Michaux
Summer Grape

Size and Form. Large, sprawling or vigorous high-climbing vine to 5–10 m with a stout trunk to 20 cm in diameter; climbing by twining tendrils.

Bark. Thin and smooth at first, becoming finely ridged, loose, stringy; reddish-brown.

Leaves. Alternate, simple, deciduous; 8–20 cm long and equally wide; blades broadly cordate-ovate to subrotund; blades of *early leaves* shallowly 3-lobed with acute apex and sinuses, not ciliate, whereas blades of *late leaves* deeply 3- to 5-lobed with acuminate tips and conspicuously rounded lateral sinuses, base cordate and deeply or narrowly V- or U-shaped, margins ciliate; *all leaves:* margins shallowly and irregularly toothed, teeth obtuse to acute, <3 mm long and mucro-tipped; when blades young and unfolding, the surface of both sides covered with reddish or rusty tomentum; mature leaves dull to bright green and glabrescent above, the lower surface *strongly glaucous*, with rusty pubescence persistent, at least along the main veins; petioles generally pubescent and often as long as or longer than the blades; stipules linear, 2–4 mm long.

Tendrils. Opposite the leaves or leaf scars, absent from every 3rd node, 10–20 cm long, slender, forked, branched, or coiled; twining, not ending in adhesive disks.

Stems-Twigs. Slender to stout, striated, round, with reddish-brown tomentose at first, becoming glabrous except at the swollen nodes; internodes medium to short; nodal diaphragms firm, relatively thick, 2–6 mm; green or dull red when young, becoming brownish; leaf scars half round or crescent-shaped; bundle scars several in a curved series; stipule scars long and narrow; pith brownish, interrupted by diaphragms at nodes.

Winter Buds. Terminal bud absent; lateral buds subglobose, deep reddish-brown, glabrous; 2 bud scales exposed.

Flowers. May–June; unisexual, plants functionally dioecious; borne only on determinate shoots or preformed parts of indeterminate shoots; inflorescence slender, elongate, short-branched cylindrical or conical panicles, 5–15 cm long, 1–2 cm wide, arising opposite the leaves; tiny, greenish-yellow to whitish, fragrant; calyx very short; petals 5, 2.5–5 mm long, coherent at apex, early deciduous, stamens 5–6. Insect pollinated.

Fruit. Berry; September–October; borne onelongate and relatively compact clusters, 1–3 cm wide; persistent; bluish-black with thin bloom, 0.7–1.1 cm in diameter; edible but sour; seeds 2–4, usually 5–6 mm long, large for the size of the fruit. Panicle axes and withered fruits persist through winter.

Distribution. Frequent in southern lower Michigan, rare in northern lower Michigan; absent in upper Michigan. County occurrence (%) by ecosystem region (Fig. 19): I, 68; II, 10; III, 0; IV, 0. Entire state 35%. Distributed widely in e. North America from ON s. to the Gulf coastal plain; central Great Plains, CA.

Site-Habitat. Open to lightly shaded, mainly dry-mesic to dry upland sites; hillside savannas, oak-hickory forests and forest edges, wooded dunes, roadsides, fencerows. Associates include many tree and shrub species of upland, fire- and drought-prone ecosystems, from northern pin, black, and white oaks; pignut hickory; sumacs; sassafras; New Jersey tea; witch-hazel; and American hazelnut to ericaceous species of acidic surface soils.

Notes. Very shade-intolerant, establishing and persisting in light to moderate shade. The species name *aestivalis* refers to the summer blooming. The diagram on the facing page illustrates part of an indeterminate shoot with late leaves; at the right is a single early leaf. The fruit is known to be eaten by many species of birds and mammals. Grape thickets and tangles provide excellent nesting cover for songbirds. Birds and squirrels use the stringy back in nest construction. Fine jellies and jams are made from the fruit.

Chromosome No. $2n = 38$, $n = 19$; $x = 19$

Similar Species. The fox grape, *Vitis labrusca*. L., the source of the Concord cultivar

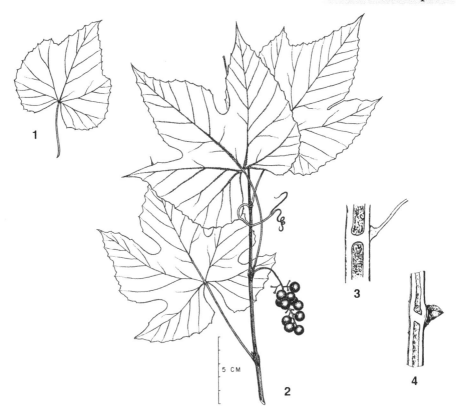

1. Early leaf, × 1/3
2. Indeterminate shoot with late leaves and fruit, berry, × 1/2
3. Stem section with nodal diaphragm and petiole base, × 1
4. Winter stem section with nodal diaphragm and lateral bud, × 1

KEY CHARACTERS

- large, vigorous, climbing vine to 10 m; climbing by twining
- blades of early leaves shallowly 3-lobed; blades of late leaves deeply 5–7 lobed; margins shallowly toothed, teeth obtuse; when young both sides covered with reddish or reddish-brown tomentum, becoming glabrous above but persisting beneath; lower surface strongly glaucous
- stems round, with reddish-brown tomentose at first, becoming glabrous except at the nodes, nodal diaphragms thick and firm, 2–6 mm
- fruit a bluish-black berry with thin bloom, persistent in winter

Distinguished from river-bank grape, *Vitis riparia,* by its stems with thick nodal diaphragms; lateral lobes of early leaves not noticeably forward pointing; margins of leaves shallowly and irregularly toothed, teeth obtuse to acute, not ciliate; adult blades whitened or glaucous beneath with reddish-brown tomentum; and predominantly dry habitats in southern lower Michigan.

of grapes, is infrequent in southern lower Michigan (Voss and Reznicek 2012). It differs from all Michigan grape species in having lower leaf blades densely and evenly covered with rust-colored hairs. Widely distributed in the Midwest, mid-South, and South is the similar graybark or winter grape, *Vitis cinerea* (Engelm.) Millardet.

VITACEAE
Vitis riparia Michaux
River-bank Grape

Size and Form. Large, vigorous high-climbing vine often to 10 m or more, attaching by tendrils, the stem to 10 cm in diameter, rarely more; also trailing or sprawling over understory trees and shrubs.

Bark. Dark brown, loose, fissured, shredding in strips.

Leaves. Alternate, simple, deciduous; blades 6–20 cm long and nearly equally wide; broadly ovate or ovate; blades of *early leaves* unlobed or shallowly 3-lobed with 2 lateral acute lobes pointing forward; many serrate teeth; acute apex and sinuses; blades of *late leaves:* deeply 3- to 5-lobed with acuminate tips, few teeth, and conspicuously rounded sinuses; base cordate to deep and broadly U-shaped; *all leaves:* margin unequally coarsely and distinctly toothed with triangular, acute, minutely ciliate teeth; glabrous, lustrous bright green above and beneath; more or less pubescent beneath, becoming glabrous except for tufts of hair in the lower vein axils, palmately veined; petioles more or less pubescent, 5–10 cm long; stipules 4–6 mm long, often persistent until fruit is formed.

Tendrils. Opposite the leaves or leaf scars, absent from every 3rd node, <5 to >20 cm long, relatively stout, 1–3 mm across, greenish to brown (exposed), branched, coiling; not ending in adhesive disks.

Stems-Twigs. Slender to stout, striated, round, green or dull red when young, becoming light brown and peeling in long, narrow, dark brown to black strips; sparsely silky-pubescent when young, soon glabrous; nodal diaphragms very thin, 0.8–2 mm thick; leaf scars half round or crescent-shaped; bundle scars several in a curved line; stipule scars long and narrow; pith small, brown, continuous, interrupted by diaphragms at nodes.

Winter Buds. Terminal bud absent; laterals subglobose to conical, acute tip, 2–3 cm long, 1–2 mm across, brown, 2–4 scales exposed.

Flowers. May–June; unisexual, plants subdioecious; borne in panicles, 8–18 cm long, 1–3 cm wide, opposite the leaves, fragrant; small; greenish; calyx minute; petals 5, coherent at apex, falling without expanding; stamens 5. Insect pollinated.

Fruit. Berry; September–October; clusters compact to loose, 4–14 cm long, 2–4 cm wide; purplish-black, with heavy bloom, 0.6–1.2 cm in diameter, very juicy and tart to sour when ripe, sweet after frost; seeds 1–4, about 5 mm long. Panicle axes and withered fruits persist through winter.

Distribution. Frequent to abundant in lower Michigan; occasional in the Upper Peninsula. County occurrence (%) by ecosystem region (Fig. 19): I, 95; II, 67; III, 43; IV, 50. Entire state 76%. Ranging widely in North America, NS to MT, e. USA s. to VA, TN, LA; Great Plains, central and n. Rocky Mountains; OR, WA.

Site-Habitat. Open to lightly shaded, mesic to dry sites with usually circumneutral to basic soils, especially alluvial soils of floodplains and stream banks; upland forests; also establishing in drier microsites in swamps; fencerows, lakeshores, dunes. Associates include a great number of tree and shrub species in floodplains and the other sites noted.

Notes. Shade-intolerant, establishing and persisting in light shade. Most common in floodplains and stream banks. Moist, litter-free, fertile alluvial soils of floodplains provide excellent establishment conditions, open to lightly shaded growth conditions, and trees for climbing. The diagram on the facing page illustrates part of a determinate shoot with early leaves and a single late leaf from an indeterminate shoot. Most common of the several grape species in Michigan. Fruits make excellent juice, jelly, and jam.

Chromosome No. $2n = 38$, $n = 19$; $x = 19$

Similar Species. Most closely related to frost or winter grape, *Vitis vulpina* L., which rarely occurs in Michigan. Also similar is the summer grape, *V. aestivalis*, which is frequent to abundant in southern lower Michigan and regions farther south.

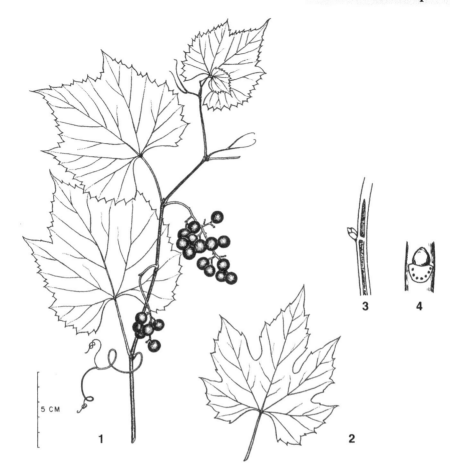

1. Long shoot with early leaves and fruit, berry, × 2/5
2. Late leaf, × 1/3
3. Stem section with nodal diaphragm and lateral bud, × 1
4. Winter lateral bud and leaf scar, enlarged

KEY CHARACTERS

- large, vigorous, high-climbing or trailing vine to 10 m or more, climbing by twining
- leaf blades of early leaves shallowly 3-lobed, with 2 lateral acute lobes pointing forward, acute apex and sinuses; margin unequally coarsely and distinctly toothed, with acute, ciliolate teeth; glabrous, lustrous bright green above and beneath, with tufts of hair in axils of lower veins
- stems with dark brown, loose bark, shredding in strips; nodal diaphragms thin
- fruit a purplish-black berry with heavy bloom; panicles and dried fruit persistent

Distinguished from summer grape, *Vitis aestivalis*, by its stems with very thin nodal diaphragms; blades of early leaves 3-lobed, with 2 lateral acute lobes pointing forward; margin unequally coarsely and distinctly toothed, with acute, minutely ciliate teeth; green glabrous beneath; and wide distribution throughout the state, especially on floodplains.

RUTACEAE
Zanthoxylum americanum Miller
Prickly-ash, Toothache-tree

Size and Form. Tall, erect, armed, aromatic shrub 2–5 m; sprouting profusely from the roots and forming dense, multistemmed clones. Michigan Big Tree: girth 38.1 cm (15 in), diameter 11.3 cm (4.8 in), height 8.5 m (28 ft), Oakland Co.

Bark. Smooth and reddish to purplish at first, becoming slightly ridged and gray to brown.

Leaves. Alternate, deciduous, pinnately compound; young leaves 0.5–1 cm long, densely pubescent, with short hairs and leaflets tipped with a rusty red gland; mature leaves 15–25 cm long; leaflets 5–11 with 2–5 pairs and a terminal leaflet, sessile or nearly so, 3–6 cm long and 3–4 cm wide; ovate-oblong to elliptic, acute or rounded, slightly notched; rounded to cuneate; entire to crenulate; glabrous, dark green above, veins impressed; paler and slightly pubescent beneath; dotted with translucent glands; aromatic when crushed; rachis and petioles prickly; petioles 2–4 cm long.

Stems-Twigs. Long-shoots stout, rigid, zigzag, dark brown; smooth with persistent, stout, broad-based prickles paired at the nodes, about 1 cm long; leaf scars semicircular; bundle scars 3; short-shoots on larger stems, distinctly ridged with annual bud-scale scars to 5 cm long; pith creamy-white, continuous. Strong odor, like lemon peel, when crushed or bruised.

Winter Buds. Terminal bud present, often partially obscured by adjacent lateral buds; all buds small, 1–3 mm long, globose, naked, woolly or matted with rusty-red hairs; bud scales indistinct; lateral buds vegetative only or with collateral flower buds surrounding the vegetative bud.

Flowers. April–May; before the leaves; unisexual, plants dioecious; borne in axillary cymes on slender pubescent pedicels; calyx lacking, 5 small petals green to yellowish-green, each petal tipped with rusty-red resinous hairs. Insect pollinated.

Fruit. Follicle, firm-walled or somewhat fleshy; August-September; reddish-brown, globose to ellipsoid, with pitted surface, 4–6 mm long, splitting into 2 cells with 1–2 round, black, lustrous seeds; strongly aromatic with a lemony odor, bitter and numbing to taste; seeds 25,600/lb (56,320/kg).

Distribution. Common throughout most of lower Michigan. County occurrence (%) by ecosystem region (Fig. 19): I, 89; II, 43; III, 29; IV, 12. Entire state 60%. Widely distributed in e. North America and w. of the Mississippi River from ND to TX.

Site-Habitat. Open to moderately shaded, seasonally wet to dry-mesic, circumneutral to basic sites; alluvial soils of levees and second bottoms of floodplains, stream banks, wet-mesic microsites in swamps. With human disturbance it has spread widely in cutover forests and their edges, roadsides and ditches, and abandoned fields. Associates include many trees, shrubs, and vines associated especially with floodplains but also wet-mesic to dry-mesic forests, including American basswood, shagbark hickory, blue-beech, redbud, American bladdernut, moonseed, river-bank grape, and running strawberry-bush.

Notes. Moderately shade-tolerant. Wood yellowish, hard. Typically clone-forming in nature; dense clones with stout prickles covering branches are extremely difficult to penetrate. Occasionally short-shoots develop into long-shoots (see, e.g., Fig. 8). Foliage resistant to insects and diseases. The genus name is derived from Greek *xanthos,* "yellow," and *xylon,* "wood." Species of the genus have been used medicinally. Powdered bark and leaves are reported to be used as a condiment and also to produce a yellow dye. Reported to have acrid sap with a numbing effect, used by Native Americans to relieve toothache. Seldom planted except to make an effective barrier. Genus with about 200 species, mostly in tropical and subtropical regions, with about 5 species in the conterminous USA.

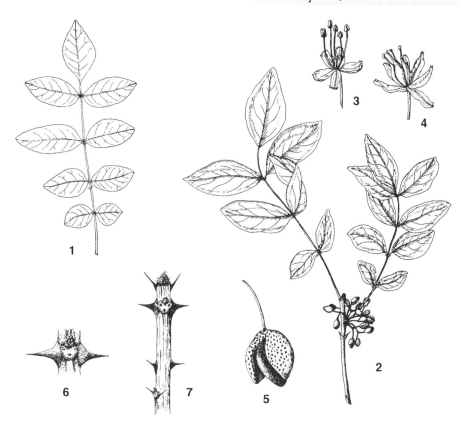

1. Leaf, × 1/3
2. Flowering female shoot, × 1/3
3. Male flower, × 4
4. Female flower, × 4
5. Fruit, follicle, × 2
6. Winter lateral bud with paired prickles and leaf scar, × 2
7. Winter twig, × 1

KEY CHARACTERS

- tall, erect, armed, aromatic shrub, sprouting from roots and forming multistemmed clones
- leaves pinnately compound, 5–11 leaflets, dotted with translucent glands; aromatic when crushed, bitter to taste
- twigs stout with persistent, large, broad-based prickles, paired at the nodes; short-shoots prominent on larger stems
- buds with rusty-red woolly or matted hairs so that bud scales are indistinct
- fruit a firm-walled or somewhat fleshy follicle, reddish-brown, strongly aromatic
- habitat seasonally wet floodplains, circumneutral to basic soils, diverse disturbed sites

Chromosome No. $2n = 68, 136$; $n = 34, 68$.

Similar Species. The Hercules-club prickly-ash, *Zanthoxylum clava-herculis* L., is a small tree occurring in the Atlantic and Gulf coastal plain states. The most similar shrub is the small prickly-ash, *Z. parvum* Shinners, which occurs in Texas. The lime prickly-ash, *Z. fagara* (L.) Sarg., is a prickly, aromatic evergreen shrub or tree, ranging from southern Florida and west along the Gulf coastal plain to Texas, Mexico, and Baja California. In Japan a similar deciduous shrub is *Zanthoxylum piperitum* Daniell & Benn. The pepper used in the Sichuanese and Yunnanese cuisine of China is derived in part from at least two species, *Z. simulans* Hance and *Z. bungeanum* Maxim.

Families of Shrubs and Vines

Important activities in plant systematics are identification and classification. Classification is the arrangement of plant entities in a logically organized scheme of relationships (Judd et al. 2008). For convenience, we have presented the plants alphabetically by genus and identified, at the top left of each description, the family within which each genus is classified. The families are listed below in alphabetical order, together with each genus represented in the species we describe. The number of species in each genus is given in parentheses following the genus name. The families of all Michigan flora are described by Voss and Reznicek (2012), and the worldwide phylogenetic arrangement of families is presented by Judd et al. (2008).

Overall, 66 genera and 36 families are represented. Of the species we treat, the number of genera in each family ranges from 1 to 10. The number of genera per family is as follows: 1 genus in 23 families, 2 genera in 9 families, 3 genera in 2 families, 9 genera in 1 family, and 10 genera in 1 family.

Adoxaceae
  Sambucus (2)
  Viburnum (6)

Anacardiaceae
  Rhus (4)
  Toxicodendron (2)

Aquifoliaceae
  Ilex (2)

Araliaceae
  Aralia (1)

Berberidaceae
  Berberis (1)

Betulaceae
  Alnus (2)
  Betula (1)
  Corylus (2)

Bignoniaceae
  Campsis (1)

Cannabaceae
  Celtis (1)

Caprifoliaceae
  Lonicera (4)
  Symphoricarpos (1)

Celastraceae
  Celastrus (1)
  Euonymus (2)

Cornaceae
  Cornus (5)

Cupressaceae
  Juniperus (2)

Diervillaceae
  Diervilla (1)

Elaeagnaceae
  Elaeagnus (1)
  Shepherdia (1)

Ericaceae
  Andromeda (1)
  Arctostaphylos (1)
  Chamaedaphne (1)
  Epigaea (1)
  Gaultheria (2)
  Gaylussacia (1)
  Kalmia (2)
  Rhododendron (1)
  Vaccinium (7)

Fagaceae
  Quercus (1)

Grossulariaceae
  Ribes (4)

Hamamelidaceae
  Hamamelis (1)

Hypericaceae
  Hypericum (2)

Lauraceae
  Lindera (1)

Menispermaceae
  Menispermum (1)

Myricaceae
  Comptonia (1)
  Myrica (1)

Oleaceae
  Ligustrum (1)
  Syringa (1)

Ranunculaceae
  Clematis (1)

Rhamnaceae
  Ceanothus (2)
  Frangula (1)
  Rhamnus (2)

Rosaceae
  Amelanchier (3)
  Aronia (1)
  Crataegus (5)
  Dasiphora (1)
  Physocarpus (1)
  Prunus (3)
  Rosa (5)
  Rubus (6)
  Sorbus (2)
  Spiraea (3)

Rubiaceae
  Cephalanthus (1)

Rutaceae
  Ptelea (1)
  Zanthoxylum (1)

Salicaceae
  Salix (12)

Sapindaceae
  Acer (2)

Smilacaceae
  Smilax (2)

Solanaceae
  Solanum (1)

Staphyleaceae
  Staphylea (1)

Taxaceae
  Taxus (1)

Thymelaeaceae
  Dirca (1)

Vitaceae
  Parthenocissus (3)
  Vitis (2)

ECOLOGY OF SHRUBS AND VINES

All woody plants are characterized by lignified woody tissue that maintains their structure by perennial cycles of growth and by longer life compared to herbs. Though similar in their woodiness, shrubs are markedly different from trees in many characteristics, especially their ecological role in landscape ecosystems around the world. Among the most notable attributes, the following are worth attention.

- Multiple stems and bushy form (i.e., weak hormonal inhibition of sprouting)
- Low stature; easily observed throughout life; begin to produce flowers and fruit when still at or below eye level
- High tolerance of and adaptation to extreme physical conditions—drought, oxygen deficiency, high and low temperatures, and severe disturbances
- Worldwide, outside wet tropics, often associated primarily with arid or dry and drought- and fire-prone ecosystems
- Characteristically shade-intolerant but once well established typically tolerant of light to moderate shade from encroaching vegetation; the shade tolerance condition is incompatible physiologically with drought tolerance
- Sexually precocious—early-in-life flowering and fruiting (i.e., a short juvenile phase of physiological development)
- Predominantly insect pollinated
- Annual flowering and fruiting or a relatively short interval between large flower and fruit crops
- Enormous asexual ability, with multiple strategies of clone formation, spread, and maintenance
- Great abundance of fruits and seeds produced under favorable conditions
- Preponderance of fleshy fruits
- Predominantly and widely bird- and mammal-dispersed seeds
- Providers par excellence of food, cover, and nesting habitat for birds and mammals
- Establishment, reproduction, competitive ability, and persistence highly favored by natural disturbances, especially fire
- Opportunistic and aggressive ability to colonize diverse ecosystems following natural and human disturbances

- In forest communities, occurring exclusively in understory and ground-cover-layer positions, whereas climbing vines ascend into the canopy
- Many wetland-adapted species in Michigan and the Great Lakes Region
- In temperate, forested landscapes significantly more indicative of physical site conditions (i.e., soil-water regime, nutrients, microclimate, and light irradiance) than trees but less than herbs
- Interspecific hybrids common in many genera
- Highly prized in horticulture around the world, easily domesticated

It is beyond the scope of this book to consider each attribute because of the many interrelated properties and processes that bind shrubs and vines inseparably to the ecosystems that support them. Nevertheless, their establishment and ecological strategies, worldwide and regionally, and certain other features in comparison with trees are worth consideration.

Woody vines of Michigan and the Great Lakes Region are similar in many of the above attributes: insect pollination, early and often annual flowering, bearing fleshy fruit that is predominantly bird and mammal dispersed, vegetative parts easily observed, highly favored by natural and human disturbances to forest canopies, and highly regarded in horticulture and gardening. They differ in significant ways.

- Their life form is linear or sticklike—lacking an extensive supporting structure compared to those of shrubs or trees. They rely on surface features or shrubs and trees for support, thereby conserving carbon and nitrogen.
- Their stems are thin, tough, and flexible.
- They are adapted to creep or trail along the surface and, when support is available, also climb tall objects, to 15–20 m or more, by means of roots, stems, and modified leaves (i.e., tendrils).
- Some species grow as low shrubs when support is lacking and climb when it is present.
- They spread rapidly, and, under favorable conditions, some species quickly colonize large areas (e.g., *Vitis* and native *Parthenocissus* species).
- They are adapted to sites favorable in soil-water, nutrients, and temperature, often in river floodplains where open canopies predominate and long growing seasons prevail.
- Many species flower and fruit far above easy observation in the high light irradiance of the midforest position or upper canopy, although they are shade-tolerant enough to establish on the forest floor.
- Generally, they lack the ability to spread asexually by roots and root-collar sprouts.

Trees, shrubs, and vines play markedly different roles in the composition, structure, and function of forests, woodlands, shrublands, prairies, tundra, and other landscape ecosystems. In examining these differences, we note that individual shrubs and vines are small and overall cover less of the Earth's surface than trees. The word *shrub* derives from the Old English *scrybb,* meaning "brushwood" (akin to "scrub"), or from the Danish *skrubbe,* meaning "less than elegant." This etymology suggests the object of work: clearing shrubs for other, more desirable purposes. Recall that the primary defining characteristic of shrubs is the occurrence of multiple stems sprouting from the base. Frequent sprouting from the root collar often leads to thickened and large root plates, which European settlers struggled to remove (see "Notes" in description of New Jersey tea).

Obviously, there is a continuum of heights of woody plants, depending not only on species but on the physical and biotic conditions of site-specific ecosystems where they grow. Some plants typically fall in the intermediate category of tall shrub–small tree, such as choke cherry, striped maple, serviceberry, nannyberry, and many willows. Others may have one trunk reaching over 5 or even 10 m—even the "lowly" buttonbush, 10.7 m, and sandbar willow, 11.7 m (Ehrle 2006). The unique vine life form may range from 1 to 20 m in height. Conversely, in extreme sites, some otherwise tall trees become shrublike with multiple stems. For example, in alpine and high-latitude ecosystems, low temperatures, deep snow, and high winds relegate individuals of Engelmann spruce (*Picea engelmannii*) and subalpine fir (*Abies lasiocarpa*) to bushy forms known as *krummholz* (crooked wood). It is these and other extreme sites of the Earth's surface to which the shrub life-form is especially well adapted.

GEOGRAPHIC AND ECOLOGICAL DISTRIBUTION OF SPECIES

Under the heading "Distribution" in the description of each species, the present percentage of occurrence is given for each of four regional landscape ecosystems and the total for the state of Michigan. The basis of this occurrence is the map of regional landscape ecosystems for the state (Fig. 19). The boundaries of the four regions were determined by integrating factors of climate, landform, soil, and vegetation (Albert, Denton, and Barnes 1986). A brief description of the ecological regions follows. In Figure 19, a thick black boundary line in lower Michigan separates Region I (38 counties) and Region II (30 counties); the northernmost counties of Region I are Muskegon, Kent, Montcalm, Isabella, Midland, and Bay. A thick black boundary line in upper Michigan separates Region III (7 counties) and Region IV (8 counties); the westernmost counties of Region III are Menominee, Delta, and Alger). Thus, for many species we see a regional ecological difference in occurrence among the warm southern part of Michigan (Region I, growing season ca. 154 days), northern lower Michigan (Region II, growing season ca. 127 days), the lake-moderated eastern upper Michigan (Region III, growing season ca. 115

Fig. 19. Map of the four Regional Landscape Ecosystems of Michigan. Thick gray lines separate natural boundaries between the four Regions. Thick dashed lines and shading indicate boundaries of counties between Regions I and II in lower Michigan and between Regions III and IV in upper Michigan. In Regions I, II, III, and IV there are 38, 30, 7, and 8 counties, respectively. A brief description of each Region is given in Appendix A.

days), and the colder western upper Michigan (Region IV, growing season ca. 107 days) (Albert, Denton, and Barnes 1986). A brief characterization of each region is presented in appendix A to help you think regionally about the occurrence of shrub species.

Although individual species of trees, shrubs, and vines have distinctive distributions, as a group, native shrubs and vines are different from trees. Seventy-four native tree species of Michigan are described in *Michigan Trees* (Barnes and Wagner 2004). Of these, 34 percent are species of southerly range and have at least 75 percent of their county occurrence in southern lower Michigan (Region I). In contrast, of the 112 native shrub and vine species we describe, only 11 percent have their dominant northernmost occurrence in Region I. Thus, as a group, the native shrubs and vines in Michigan tend to have a more northerly distribution than the trees do. One important reason is that proportionally more shrubs than trees occur naturally in relatively harsh, wet to drought-prone sites in the interior or along the coasts of lakes Michigan, Superior, and Huron. Ecosystems with these physical conditions are of far greater occurrence and diversity in northern lower and upper Michigan (Regions II, III, and IV) than in southern lower Michigan.

Figure 19 also illustrates the distinctive distribution of many species. Examples of several different patterns of occurrence are illustrated in Figure 20. Some species, such as red-osier and meadowsweet, are distributed nearly everywhere in the state (Fig. 20*A*). Other patterns include species that are only found in southern lower Michigan (Fig. 20*C*), species reported only from southern and northern lower Michigan (Fig. 20*D*), and species that typically occur on sites of counties adjacent to the Great Lakes (Fig. 20*E*). Only two species occur exclusively in upper Michigan (Fig. 20*E*).

Some of our species extend well beyond the Great Lakes Region and into Eurasia, and such distributions have their origins much farther back in time than when Michigan most recently reappeared from under the glaciers. Some, including such common species as *Arctostaphylos uva-ursi, Dasiphora fruticosa, Myrica gale, Rhododendron groenlandicum, Rosa acicularis,* and *Sambucus racemosa,* are found irregularly across all the temperate and boreal latitudes of the globe, showing what is termed a circumpolar distribution. Closely related and morphologically similar species may also occur widely separated in different continents such as eastern North America and eastern Asia, or in different biomes, such as tropical rain forests. In order to explain such broad distribution patterns, it is important to consider the climates and environments in which our shrub and vine species and genera evolved.

HISTORICAL BIOGEOGRAPHY OF SHRUBS AND VINES

Apart from our Jurassic-age "living fossil" conifers such as *Taxus* (Florin 1958) that originated during the reign of dinosaurs, most of our shrub and vine genera originated during the Tertiary period (65 to 2.7 million years ago,

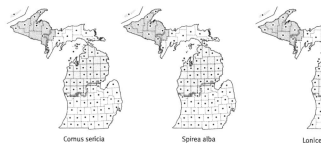

A. Widespread throughout the state
(Regions I to IV)

Cornus sericia

Spirea alba

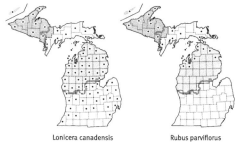

B. Occurence increasing northward
(Regions I to IV or Regions II to IV)

Lonicera canadensis

Rubus parviflorus

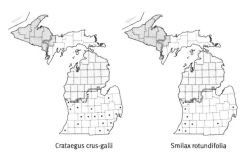

C. Southern Lower Michigan only (Region I)

Crataegus crus-galli

Smilax rotundifolia

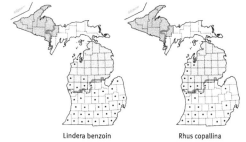

D. Lower Michigan only (Region I and II)

Lindera benzoin

Rhus copallina

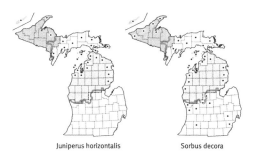

E. Often coastal occurence near the Great
Lakes (especially Regions II and III)

Juniperus horizontalis

Sorbus decora

F. Upper Michigan only (Region IV)

Vaccinium ovalifolium

Fig. 20. Selected shrub species exhibiting different patterns of occurrence in the four regional landscape ecosystems of Michigan (Albert et. al. 1986). Thick gray lines delineate natural boundaries between the four Regions. Dashed lines mark the boundary of counties between Regions I and II in lower Michigan and between Regions III and IV in upper Michigan. In Regions I, II, III, and IV there are 38, 30, 7, and 8 counties, respectively. A brief description of each Region is in Appendix A.

or Ma). During this time, northern latitudes were warmer than today and experienced much less winter-summer seasonality. We distinguish two major climatic and floristic phases during the early development of the modern flora: (1) semitropical, frost-free climates of the early Tertiary, and (2) mild-temperate climates of the mid- to late Tertiary. The gradual and global cooling and drying of the late Tertiary, capped by the Quaternary Ice Ages, greatly reduced the diversity and connectivity of northern temperate floras.

BOREOTROPICAL FLORA (65 TO 34 MA)

The first half of the Tertiary (the Paleocene and Eocene epochs, 65 to 33.9 Ma) was one of the warmest periods in Earth history over the past half billion years (Graham 1993). Greenhouse conditions were likely caused by high levels of carbon dioxide released from the Earth by intense volcanism. Our region experienced a semitropical and humid climate, without significant freezing, and harbored a warm-adapted vegetation, known as the boreotropical flora (Wolfe 1985) to distinguish it from the true (equatorial) tropical flora.

Like modern tropical forests, boreotropical forests harbored monkeys, crocodiles, and other warm-adapted animals, as well as palms, gingers, figs and their specialized insect associates (Graham 1993). Lianas would have flourished (Tiffney 1985; Graham 1993). Boreotropical forests reached latitude 50° north (north of upper Michigan), and because the northern continents were more closely connected than today, semi-tropical vegetation was continuously distributed between North America and Eurasia. Evidencing their historical connectivity, fossil floras of Asia and North America from this time share 50 percent of their species (Graham 1993). As climates progressively cooled through the remainder of the Tertiary, most boreotropical taxa became restricted the equatorial region. Possible modern relicts of this period include species from primarily tropical families, including pawpaw from the custard-apple family (Annonaceae), prickly-ash and wafer-ash from the citrus family (Rutaceae), trumpet-creeper (Bignoniaceae), moonseed (Menispermaceae), and bristly sarsaparilla (Araliaceae). Like their tropical relatives, these shrubs and vines are dependent on animals for pollination; they contain rich secondary chemistry and, in some cases, extrafloral nectaries to protect their leaves against the intense herbivore pressure typically found in species-rich tropical forests.

TEMPERATE CLIMATES

During the mid-Tertiary (beginning of the Oligocene, ca. 32 Ma), with decreasing concentrations of atmospheric carbon dioxide, the Earth transitioned from a greenhouse to an icehouse climate. Glacial ice began to accumulate at the poles, and our region became cooler, drier, and more seasonal.

Cold-adapted, deciduous vines and shrubs largely replaced the boreotropical elements. By middle Miocene times (ca. 15 Ma), a remarkable floristic continuity and similarity still existed from eastern Asia to North America, via the Bering Land Bridge near modern-day Alaska, and possibly from Europe to North America via the North Atlantic routes of migration (Tiffney 1985). This was about to change.

ICE AGE EXTINCTIONS

The increasing cold and seasonality of the late Tertiary and successive (up to twenty) glaciations of the Pleistocene caused regional extinctions of many temperate plant genera (Latham and Ricklefs 1993). Europe harbored few climatic refuges and suffered dramatic losses, including the regional extinctions of *Aralia, Hamamelis, Kalmia, Lindera, Staphylea,* and *Zanthoxylum* (Latham and Ricklefs 1993). Increasing aridity in western North America prior to the Pleistocene led to the extinction of *Carpinus, Kalmia, Lindera,* and other genera. In contrast, temperate genera in eastern North America and eastern Asia experienced reduced extinction because suitable habitats were available to them even at the peak of continental ice cover. This partly explains why so many similar species occur in Japan and China despite being much more distant than western North America or Europe.

ONSET OF THE HOLOCENE

The continental ice sheet that formed during the most recent glacial episode, the Wisconsin, reached its southern limit near the Ohio River about 23,500 calendar years ago (Larson and Kincare 2009) and obliterated the Michigan flora. Thereafter, the climate gradually warmed, with a sudden warming event about 13,000 years ago marking the beginning of the Holocene epoch (Hupy and Yansa 2009). Climatic warming led to the return of species from southern refuges. While the timing of the migration of wind-pollinated tree species has been extensively documented (Davis 1981, 1983; Delcourt and Delcourt 1987; Jackson et al. 2000; Webb, Cushing, and Wright 1983; Webb, Shuman, and Williams 2003), the postglacial migrations of shrubs and vines are poorly known, in part because they do not leave abundant pollen in the fossil record. To an airborne observer, the migration of plant species would have revealed an extremely complex and nonrandom pattern of regional and fine-scale occurrence due not only to climatic warming but also to concomitant physical changes in landforms, soils formed therein, and water forms of lakes, rivers, and streams.

A broad and useful characterization of major changes in species and communities in Michigan is given by Hupy and Yansa (2009). The general phases (based on calendar years) of plant colonization given here provide a temporal basis for understanding natural and human disturbances leading to the geographic and ecological occurrence of today's shrubs and vines. The dates are for the state as a whole; southern lower Michigan, being the warmest regional ecosystem, would show the earliest plant and human occurrence.

17,000 to 13,000 years ago: Cold-climate plant colonization
Cold climate; willows and arctic-adapted plants first arrive in southern Michigan

13,000 to 11,000 years ago: Spruce-sedge parkland and boreal forest
With warming, 13,000–11,000 years ago in the southern Great Lakes Region, including southern lower Michigan, fire-dependent trees, shrubs, and herbs (jack and red pines, oaks, and associated shrubs, vines, and herbs) arrive (Delcourt and Delcourt 1987).

11,000 to 6,800 years ago: Early Holocene species diversification
Michigan ice free, warming climate, diverse temperate conifers and deciduous species replace spruces and move northward.

6,800 to 3,200 years ago: Mid-Holocene warming
Mid-Holocene warming, peak warmth and aridity (Hypsithermal period); extensive areas of prairies, oak savannas, oak-hickory forests in flux from Native American burning in southern Great Lakes region, southern Lower Michigan; wild squash cultivated; mixed oak and pine forests in Lower Michigan; mesic hardwoods increasing.

3,200 to 150 years ago: Late Holocene vegetation dynamics

Marked increase in precipitation and associated decline in temperature; expansion of mesic beech-sugar maple forests in southern Lower Michigan; Native American land use activities extensive to circa 1600–1700 when European diseases decimate Indian populations; fire and primitive agriculture common but not revealed in fossil pollen record; European settlement disturbances noticeable in the pollen record.

NATIVE AMERICANS AND LANDSCAPE MANAGEMENT BY FIRE

Before the arrival of humans, the communities of landscape ecosystems were structured by diverse natural disturbances (e.g., wind and ice storms, floods, and wildfires) associated with geologic landforms, soil development, hydrologic settings, and regional and local climates. Thereafter, indigenous peoples markedly affected woody plant occurrence and community composition for thousands of years. Their forms of landscape management were followed by even greater and more diverse forms of disturbance and change by European settlers for a few hundred years (for the Americas as a whole, see Mann 2005; for the eastern United States, see Delcourt and Delcourt 2004; and Abrams and Nowacki 2008; for New England, see Cronon 1983; for the Great Lakes Region, see Riley 2013).

In overview, the first plants entered southern lower Michigan about 17,000 years ago in a tundralike environment. Besides sedges and other arctic herbs, the pioneer plants would have included dwarf shrubs of the genus *Dryas*, willows, and alders as tundralike vegetation gave way to sedge-spruce woodlands. Shrubs and vines increased in diversity and colonized northward with the warming climate and developing landforms and soils. Grasses, sedges, and shrubs reached their grand period of occurrence, abundance, and diversity in the mid- to late Holocene. This would include the warm and dry mid-Holocene, 6,000 to 3,200 years ago in southern lower Michigan, and the following cooler-wetter time, 3,200 to 400 years ago. During these periods Native Americans used fire extensively for many purposes (Lewis 1993). The most extensive land use at this time would have been in the southern lower Michigan (Region I, Fig. 19) and nearby southern Ontario (Riley 2013). With longer growing seasons and a warmer climate than other Michigan regions, it was first to become ice free and then support initial fire-dependent forests of hard pines (jack, red) and oaks with associated shrubs and grasses by ca. 10,000 years ago (Delcourt and Delcourt 1987). The landforms of outwash and lake plains, ice-contact terrain, and coarse-textured moraine provided flat to gently rolling topography, dry and dry-mesic fire-prone ecosystems, fertile soils from calcareous parent material, and abundant lakes, rivers, and streams. Thus, these favorable physical conditions eventually facilitated development of small and large family or tribal settlements and cultures with extensive use of fire and primitive agriculture where appropriate.

The first Michigan inhabitants, Paleoindians of ca. 12,000 to 10,000 years ago in the spruce-sedge parklands, were small bands of hunter-fisher-gatherers (Lovis 2009). For thousands of years thereafter, and before arrival of Europeans, Indians spread throughout Michigan and the Great Lakes Region (Cornell 2009; Lovis 2009; Riley 2013). They experienced the migration of tree, shrub, and vine species and ever-changing local community composition in tune with Michigan's markedly diverse regional and local ecosystems. These early hunters also contributed to the extinctions of thirty-four megafaunal genera, including in Michigan major herbivores and landscape modifiers such as mastodons and mammoths, musk-oxen, and bear-sized giant beavers (Koch and Barnosky 2006). The disappearance of these mammals in the fossil record coincides with an increase in charcoal and a shift to fire-adapted vegetation in the early Holocene (Gill et al. 2009).

Inhabiting the Earth's physical features of Michigan and the Great Lakes Region, Native Americans shaped a cultural landscape, influencing the distribution and abundance of plants and animals and managing succession to foster a diversity of harvestable plants and animals. Fire was their primary tool. Lightning was a natural ignition source. But it would be hard to deny that native burning was the most powerful tool for manipulating plants, animals, and their communities (Kimmerer and Lake 2001; Abrams and Nowacki 2008). More than seventy uses of fire are documented in Lewis 1993. These include clearing travel corridors, fireproofing settlements, killing trees and preparing fields for planting, hunting, creating and maintaining prairies and forest openings to attract game, and generally creating a patchy landscape with many stages of succession to enhance diversity and the yield of shrub and tree fruits and seeds.

Shrubs and vines thrived in the open, fire-prone environments of savannas, prairies, and forest edges. They spread extensively by seed and asexual means following periodic surface fires. The exception was grassland (i.e., prairie ecosystems) created or maintained by annual burning by Indians. Bearing predominantly fleshy fruits, shrubs and vines were favored by mammals and birds as well as Indians, whose care of the landscape increased with increasing populations over thousands of years. So important was indigenous burning that Abrams and Nowacki (2008) marshal convincing evidence for the widespread use of fire in managing vegetation directly or indirectly in many parts of eastern North America.

European diseases struck Indian villages with horrible ferocity (Mann 2006). Following the sudden decline in their populations and land care, European settlement in the eighteenth and nineteenth centuries markedly reduced burning, and succession rapidly favored trees and closed-canopy forests. Specific examples of their cultural landscape and effects on shrub and vine occurrence in Michigan and the Great Lakes Region have been obscured due to the enormous changes introduced by European settlement (agriculture, grazing, draining, logging, postlogging fires, landscape fragmentation by many means, and fire exclusion), ever increasing over three hundred years (Dickman and Leefers 2003; Lewis 2009; Dickman 2009). In Michigan what is to be seen today of landscapes and communities created by Native Americans even five hundred or a thousand years ago? The indirect legacy of Indian fire management is likely best indicated by the occurrence, circa 1800, of grasslands, mixed-oak savannas, and barrens in southern lower Michigan based on the General Land Office Survey, interpreted and mapped by Comer, Albert, and Austin (1998; see map on the website) and Albert and Comer (2008).

NATIVE, INTRODUCED, AND INVASIVE SPECIES

Of the 138 species we describe in detail, 127 (92 percent) were native in Michigan and the Great Lakes Region before the arrival of European explorers or settlers; 11 have been introduced since then from other continents. Another introduction is oriental bittersweet (*Celastrus orbiculatus*), which we compare to the native American bittersweet but do not describe in detail. Nine of these introductions have become naturalized in our landscape ecosystems and are regarded as invasives. The term *invasive* is defined as a species introduced from a different area, most often a different continent, which becomes established, increases in density, and expands rapidly across a new habitat (Myers and Bazely 2002). Two specific features are added to the definition by Cronk and Fuller (1995): their occurrence in natural or seminatural ecosystems (i.e., not the human-made artificial ecosystems of resource farming such as agricultural fields and plantations), and the fact that they produce a significant change in ecosystem composition, structure, and processes. Overall, most introduced species in North America do not become invasive (Myers and Bazely 2003).

Three introduced species that have not become invasive are Boston ivy (introduced in 1862), bridal-wreath (introduced prior to 1866), and lilac (cultivated since 1563) (Rehder 1940). The nine invasive species we describe are common buckthorn (cultivated for centuries), common privet (cultivated since ancient times), Tartarian honeysuckle (1752), autumn-olive (1830), Amur honeysuckle (1855 or 1860), oriental bittersweet (1860), Japanese barberry (1864), multiflora rose (introduced before 1868), and glossy buckthorn (long cultivated). When observed today in field sites, without knowing their origin, one might well assume that they are native plants. Three important factors characterizing invasive species are small mean seed mass, minimum juvenile period, and short mean interval between large seed crops (Rejmanck and Richardson 1996). Fleshy fruits dispersed by birds and other animals are also a significant factor for our invasive shrub and vine species.

As Spurr (1964) noted, from the ecocentric viewpoint of the Earth or a landscape ecosystem, there is no meaning to the concepts of native and introduced species. Regardless of how a plant is introduced, if it successfully colonizes, establishes, reproduces, and spreads in an ecosystem it is a native member of that ecosystem; it has become naturalized. From a human or biocentric viewpoint, introduced species that become invasive may cause intolerable or devastating effects—reducing biodiversity, replacing native species, or markedly changing ecosystem processes. There are many areas, including parks, preserves, and natural areas, where managers and landowners work to control or eliminate invasives to maintain or restore the pre-European-settlement composition of landscape ecosystems (Myers and Bazely 2003).

In North America, we are most familiar with introduced fungal and insect species that have caused and are causing catastrophic losses to tree species—chestnut blight, white pine blister rust, Dutch elm disease, gypsy moth, fir and hemlock adelgids, emerald ash borer, and beech bark disease, among others. Around the world, many introduced tree, shrub, and vine species also have had devastating effects (Cronk and Fuller 1995). Our nine shrub and vine invasive species are far less destructive. Bearing fleshy fruits that are widely dispersed by birds, they are plants of the understory and ground cover layers, especially in seminatural (disturbed) oak, oak-hickory, and oak-pine forests where fire has been excluded, as well as open areas along forest edges, roads, and power lines. Their presence represents a change in forest composition. In addition, we reason that by their persistence and habit, native plant survival and richness are reduced due to low light irradiance below their crowns. Furthermore, invasives remove soil-water and nutrients otherwise available to native species. Specific evidence of the detrimental effects of Amur honeysuckle include reducing herb and woody plant abundance and richness below their crowns (Collier, Vankat, and Hughes 2002) and a reduction in the survival and fecundity of herbs (Gould and Gorchov 2000).

SEXUAL REGENERATION

Establishment is the end point of the process of regeneration. In sexual regeneration the interconnected processes include sexual reproduction, seed dispersal, seed germination, and establishment. The basic steps of the sexual regeneration process are as follows.

1		2		3		4		5
Reproduction	→	Fruit and seed dispersal	→	Residence in the active or dormant seed bank	→	Seed germination	→	Plant establishment

Sexual reproduction (i.e., the exchange of genes, most often among genetically different individuals) includes the processes of pollination, fertilization, seed production, and ripening of seeds for dispersal. Seeds are often dispersed by birds and mammals in summer or early fall before they are ready to germinate. However, an after-ripening process and further changes during winter enable seeds residing in the dormant seed bank of the forest floor or other surface to complete the maturation process and germinate in the spring. Seeds dispersed by wind in the spring, such as those of species of *Salix*, are physiologically ready for germination once they reach a favorable site. Germination is the process whereby seeds imbibe water and, internally prepared by hormonal actions, extend a radicle through their seed coats and into the soil. Nearly simultaneously the young stem extends into the air above ground. Once rooted in soil and air, plant life, as we know it, begins. Finally, in the course of two to five years, a seedling may become "established" (i.e., set up permanent residence). However, millions of seedlings never make it through this complex process during the inseparable connection of juvenile plant and the life-giving Earth.

Once established, the shrub individual is in the juvenile phase of development, but, unlike trees, shrubs and vines typically pass quickly into the adult phase. However, as noted above, the lower part of the stems of some shrubs remains permanently in the juvenile phase, as illustrated by the dense stem armature of prickly gooseberry and red raspberry (Fig. 11). Once shrubs and vines are in the adult phrase, they are distinguished by their characteristic leaf morphology and flowering-fruiting parts, which are presented in the diagrams for each species.

ASEXUAL REPRODUCTION

The asexual or vegetative regeneration process is fundamentally different. All shrubs and vines have the ability to reproduce asexually, often in several ways, such as sprouting, nodal and tip rooting, layering, fragmentation, and seed apomixis. In vegetative reproduction, meristems that develop into genetically identical replicas (the ramets) of the original sexually produced individual (the ortet) form inside vegetative buds that are located on the stems, rhizomes, roots, or juncture of stem and root (i.e., the root collar or root crown) of the parent individual. These meristems may be activated in two ways. Most characteristically, they develop or "sprout" following some disturbance, often fire. The disturbance kills or wounds aboveground stems, thereby changing the hormonal balance that has kept the meristems dormant inside existing buds. Fire, stem cutting, or injury to the crown may reduce or eliminate the flow of the hormone auxin to the roots. Although auxin inhibits sprouting from the roots or root collar, other factors may act to stimulate resprouting. These include stresses due to lack of light (i.e., shade), soil-water, or oxygen. Also high early spring temperatures may activate vegetative buds on roots or rhizomes that are near the ground surface. Auxin is produced in the crowns of woody plants and translocated to the roots, but by late spring its abundance and inhibitory effect are weak.

The original seedling may produce genetically identical stems and form multistemmed clones either in compact clumps, characteristic of many shrub species, or by spreading spatially by sprouting from roots or rhizomes to form colonies or thickets. Such sprouting takes place at the root collar or from rhizomes and roots. In addition, individuals may root at nodes of stems growing on the ground surface or at the ends of arching branches and give rise to new ramets. Once the connecting root or rhizome dies, the ramet becomes independent and forms a root system of its own, which may further spread the clone. Clones that have a clump habit (see, e.g., Fig. 7) or are spreading in space by roots or rhizomes are most common. The clump habit is typical in *Ilex, Lonicera, Lindera, Hamamelis, Amelanchier, Physocarpus,* and *Sambucus.* The clonal spreading habit by roots and rhizomes is characteristic of species of *Cornus, Vaccinium, Rubus, Rhus, Gaylussacia, Arctostaphylos, Parthenocissus, Toxicodendron, Zanthoxylum,* and many others.

By vegetative regeneration, shrubs and vines occupy ecological space and extend their network of branches into light gaps where they can more abundantly reproduce sexually. Also asexual reproduction enables a species to maintain populations in extreme or unfavorable geographic regions or microsites within an area where seedling regeneration is rare or impossible. Infertile or sterile hybrids are also perpetuated spatially and temporally by asexual reproduction. Clone-forming individuals with well-developed root plates, such as New Jersey tea and prairie redroot, or an extensive underground network of rhizomes and roots are exceedingly difficult to remove physically or chemically. Injury or incomplete removal often serves to stimulate further asexual sprouting or clonal spread. Sexual and asexual reproduction are complementary and enhance the persistence of shrub clones for decades or centuries.

In clonal reproduction by seed (i.e., seed apomixis or agamospory – parthenogenesis in animals), unfertilized but viable seeds develop in the plant's ovary and, following normal dispersal and establishment, give rise to seedlings of the same genetic constitution as the parent. Such seedlings may occur near the parent (dispersal by gravity or mammals) or far from it (dispersal by birds or water). Thus, such apomictic clones have no resemblance in habit to the typical clump or belowground spreading, multiramet, thicket-forming clone. Such seedling ramets of the parent individual may develop in species of *Rubus*, *Amelanchier*, and *Crataegus*.

Although shrub stems are typically short-lived, one to ten years, new stems take their place by nearly continuous asexual reproduction. However, a given clone, by repeated asexual regeneration, may live for decades or centuries. For example, a clone of *Vaccinium myrsinites* Lam. about 1 km across was estimated to be at least a thousand years old by Darrow and Camp (1945).

SPECIFIC SITE FACTORS

The establishment of the seedling is the single most critical part of the plant's life. Light irradiance is often the most important factor, promoting flowering or the production of cones, especially for shrubs that carry out their life cycles in the forest understory or other ecosystems (savannas, shrublands, or prairies) where strong competition with existing vegetation occurs. The ability of a shrub individual to establish, grow, and reproduce in shade (e.g., a forest understory or forest edge) depends on its tolerance of or requirement for shade. Such an ability is in part genetic and in part a modification by the environment and other factors, such as compensation by soil-water or nutrients.

Shade (i.e., reduced light intensity) is one major factor, and a given species' ability to tolerate shade is characterized generally by one of five arbitrary shade-tolerance classes (Barnes et al. 1998, 399): *very tolerant, tolerant, midtolerant, intolerant,* and *very intolerant*. Shade tolerance is a very complex plant attribute and is influenced by many interrelated factors of physical site conditions and plant physiology (Walters and Reich 1996; Valladares and Niinemets 2008). In forest ecosystems, one naturally thinks of shade as the most likely cause of poor shrub growth, lack of flowering, or death, but is it? The possibility of oversimplification is remedied in forest ecology by the term *understory tolerance*. We recognize that several even more site-specific limiting factors, such as soil-water and nutrients, determine or act together with shade to affect plant establishment, growth, and survival (Barnes et al. 1998, 395–406). Nevertheless, this subjective tolerance rating is useful in understanding why plants grow where they do in nature—especially where they are able to establish, maintain themselves, and reproduce. The effect of light on sexual reproduction is extremely important. Worldwide, shrubs predominate in dry-mesic to dry, drought- and fire-prone sites, which are characterized by high light irradiance. Although most shrubs and vines are able to maintain themselves by asexual reproduction in the shade, few are adapted to reproduce sexually in moderate to deeply shaded conditions. In horticultural practice, the light climate where a plant will prosper to fulfill its expected flowering or fruiting potential is of the utmost importance.

Many species that establish in high light irradiance following disturbance may gradually be overtopped by tree associates. One often finds an individual of a shade-intolerant shrub species surviving in a shaded understory (e.g., fragrant sumac and downy arrow-wood). Why this unexpected situation? It requires understanding the conditions present when the individual became established. With well-developed root systems and other site factors compensating for low light irradiance, individuals may be able to tolerate light to moderate shade and survive vegetatively. However, they are unlikely to reproduce sexually, and with increasing shade they die or struggle to maintain themselves by asexual reproduction.

In the forest understory, soil-water is often even more critical than light, especially for germinants and young seedlings, because of the rapid drying of surface soil horizons. Intense competition occurs by midsummer when well-established, intensive, fine root systems of trees permeate the surface horizons. On very dry sites, root systems become extensive to provide water, and leaf characters are modified to conserve water. In ecosystems that are continuously wet during the growing season (e.g., some floodplain ecosystems, bogs, and fens), soil-water for foliage and all living cells is in short supply. Belowground oxygen deficiency curtails fine root activity because it is required for respiration, which goes on night and day. Thus, plants at home in wetlands conserve water by foliage adaptations, anaerobic respiration, providing oxygen to roots via lenticels, and developing adventitious roots—all ways to maintain water balance in the plant.

Nutrients are required by all species, especially nitrogen. Nutrient ions are absorbed and transported along with water to all living cells. Therefore, soil-water and nutrients are closely related in the sense that without soil water there are no nutrients. A surrogate for determining nutrient availability (i.e., whether the soil site is relatively infertile or fertile), pH, is the relative proportion of H+ ions (not a nutrient ion) to OH- ions available in the soil solution. It is measured along the logarithmic pH scale from 0 to 14. In an acidic soil, the proportion of H+ ions to OH- ions is high, and pH is low (e.g. 3.5 to 6.5). A neutral soil has approximately equal numbers of H+ and OH- ions and a pH of 7.0. The term *circumneutral* (pH 6.8 to 7.2) is applied in recognition of natural variation around the neutral threshold of 7.0. Basic soils (those relatively low in H+ ions), exhibit a pH range from 7.3 to circa 8.5. Acidic soils tend to be nutrient poor (infertile), and decomposition of litter (all types of organic matter on or in the soil—leaves, roots, twigs, fruits, etc.) on the surface and in the soil is slow. Nutrient ions in the litter are thus retained and not readily released into the soil for plant use. In contrast, basic soils tend to be nutrient rich (fertile). A circumneutral or basic soil reaction indicates relatively rapid decomposition of litter by microorganisms and the release and availability of nutrient ions for plant uptake.

Temperate zone woody plants establish, grow, and reproduce over a relatively wide range of soil pH, although fertile soils favor certain species compared to acidic, infertile soils. Much more than trees, many shrub species are very good indicators of either fertile or infertile sites. In general, species of the family Ericaceae are strongly adapted to grow and compete in soils with deep acidic layers, both in uplands (e.g., low sweet blueberry, trailing arbutus, sweetfern, black huckleberry) and wetlands (bog-rosemary, Labrador-tea, bog-laurel). Shrubs in many other families are characteristic of or indicate sites that are in the basic range (bog birch, shrubby cinquefoil, maple-leaved viburnum). Selected shrub species of wet and wet-mesic ecosystems that are characteristic of basic and acidic soils are listed in Table 1. Most of the shrubs we describe are strongly competitive and favored in circumneutral (pH 6.8 to 7.2) to moderately basic soils. Note that soils have vertically oriented horizons, and at a given place horizons typically differ, often markedly (e.g., pH 7.5 to 4.5), in their soil reactions. Almost all woody plants have roots penetrating both acidic and basic horizons and survive and thrive in a relatively broad range of nutrient availability as indicated by the soil pH reaction. Nevertheless, some exceptional species are characterized by narrower limits of tolerance for their optimal development. Notably, these are the rhododendrons, in the range of extremely to strongly acidic (pH 4.0 to 5.0), of which our native is Labrador-tea (*Rhododendron groenlandicum*).

Soil-water status is rated on an arbitrary but widely used five-part scale: *dry, dry-mesic, mesic, wet-mesic,* and *wet*. Usually, there is a strong connection between soil-water availability and drainage class. Dry soils tend to be excessively or somewhat excessively drained, dry-mesic sites somewhat excessively to well drained, mesic sites moderately well drained, wet-mesic sites somewhat poorly drained, and wet sites poorly drained or undrained. Pine barrens, where such shrubs as low sweet blueberry, bearberry, sweetfern, and sand cherry grow, are typically dry and excessively or somewhat excessively drained depending on the sand fraction (very coarse, coarse, medium, fine, or very fine) and the amount of gravel in the rooting zone.

Disturbance by fire, wind, ice storm, and flooding were common occurrences in pre-European-settlement ecosystems occupied by shrubs, and communities experienced periodic or occasionally catastrophic changes in species composition. Shrub species of that time that were most successful in

Table 1. Selected Shrub and Woody Vine Species of Wet and Wet-Mesic Ecosystems That Occur on Circumneutral to Basic Soils versus Acidic Soils and Substrates

SPECIES OCCURRING PRIMARILY IN CIRCUMNEUTRAL OR BASIC AOILS	SPECIES OCCURRING PRIMARILY IN CIRCUMNEUTRAL OR ACIDIC SOILS
Alnus incana	*Andromeda glaucophylla*
Betula pumila	*Aronia prunifolia*
Celastrus scandens	*Chamaedaphne calyculata*
Cephalanthus occidentalis	*Ilex mucronata*
Clematis virginiana	*Kalmia angustifolia*
Dasiphora fruticosa	*Kalmia polifolia*
Euonymus obovatus	*Rhododendron groenlandicum*
Hypericum kalmianum	*Vaccinium angustifolium*
Hypericum prolificum	*Vaccinium corymbosum*
Lonicera dioica	*Vaccinium macrocarpon*
Menispermum canadense	*Vaccinium myrtilloides*
Myrica gale	*Vaccinium ovalifolium*
Parthenocissus quinquefolia	*Vaccinium oxycoccos*
Physocarpus opulifolius	
Rhamnus alnifolia	
Ribes americanum	
Ribes lacustre	
Ribes triste	
Rosa palustris	
Salix candida	
Salix eriocephala	
Salix myricoides	
Salix serissima	
Sambucus canadensis	
Shepherdia canadensis	
Smilax hispida	
Solanum dulcamara	
Spiraea alba	
Toxicodendron vernix	
Vitis riparia	

Note: Included are species described in the body of the book. Most of these plants can grow under a wide variety of site conditions. However, through complex interactions with the physical environment and associated biota, we often perceive certain species growing in a distinctive, recurrent site where they maintain themselves and reproduce by sexual and asexual means. Their occurrence is due in part to seed source availability, dispersal method, and competition and mutualistic relationships with other plants and animals at a given site for a certain time.

colonizing disturbance-prone ecosystems are those that are most successful today. Small-seeded, wind-dispersed shrubs, such as willows, benefited most but were limited primarily to their riverine and wetland sites. Bird-dispersed shrubs spread widely (e.g., *Crataegus, Rosa, and Rubus*) and where forest canopies had been removed, shrub fields expanded greatly.

Given appropriate light conditions, all shrub species grow best on moist, fertile sites that are not too cold, too hot, too wet, or too dry—the mesic condition where soil-water is available throughout the growing season. We perceive that prior to the arrival of Europeans and native peoples in North America, all woody plant species tended to be strongly restricted by natural processes to sites where they were adapted to grow and reproduce. For certain shrubs this meant a severe restriction to extreme sites—often dry, drought- and fire-prone, or wet habitats. Cranberries were restricted to open bogs and sweet-fern to cold, dry, fire-prone, nutrient-poor sites. Other species had characteristics enabling them to be opportunistic and compete, establish, grow, and reproduce in a much wider spectrum of habitats, especially following natural disturbances. Overall, however, site is an excellent characteristic for identifying shrubs and vines because many of them are indicator species—plants with affinities for particular site conditions.

THE ROLE OF SHRUBS AND VINES IN LANDSCAPE ECOSYSTEMS

Around the world, shrubs are the dominant life form of semiarid and arid regions where full sunlight prevails for establishment, growth, and reproduction. Also they dominate or are abundant in extreme northerly or alpine ecosystems. They are favored in diversity and abundance by any disturbance that causes or maintains open, nonforested conditions: fire, windstorm, flooding, avalanche, insect or disease epidemic, or livestock grazing. In contrast, vines (i.e., lianas) are most abundant and diverse, worldwide, in tropical forests, and elsewhere they tend to occur under favorable conditions of temperature, soil-water, and nutrients.

Vast areas of shrublands cover the Earth in the dry parts of South America, Africa, Mediterranean Europe, Australia, India, China, and the steppes of Eurasia. Shrub ecology and management in these areas have been described by McKell (1989). Shrubs are also the dominant life-form, or are strongly represented, in many regional ecosystems of North America, including the many chaparral ecosystems of California and the southwestern United States (Keeley 2000); the warm deserts of the Southwest and adjacent Mexico (MacMahon 2000); the woodlands of California (Barbour and Minnich 2000); the valleys and lower mountain slopes of the intermontane West (West and Young 2000); the tall-shrub tundra, low-shrub tundra, and low-shrub dwarf tundra of the Arctic (Bliss 2000); shrublands throughout the boreal forest (Elliott-Fisk 2000); and the southeastern and Gulf coastal plains from New Jersey to eastern Texas (Christensen 2000).

An important natural disturbance associated with many of these subcontinental and regional ecosystems is fire. Wildfire favors grasses and shrubs in particular because they are admirably adapted to it by asexual and sexual reproduction. Annual or very frequent fire occurrence favors and maintains prairies, whereas longer fire intervals favor shrubs and trees (Wells 1970; Sims and Risser 2000). Many tree species, as well as their shrub associates, are dependent on fire for their persistence at intervals from three to over two hundred years. Livestock grazing also favors shrubs, and wherever overgrazing has gone unchecked shrubs tend to replace grasses. Considerable restoration activity today is directed to using prescribed fire to eliminate shrubs and trees from former prairies. The importance of fire in restoring prairies and savannas in Michigan is described in O'Connor, Kost, and Cohen 2009.

In contrast to the shrub-dominated arid or boreal/tundra areas of the world, Michigan, as well as the Great Lakes Region as a whole, is markedly different. However, the basic principles apply: shrubs typically occur (1) where site conditions are too extreme for trees; (2) where forests naturally occur and some shrubs and vines are adapted to forest understories where disturbance by fire or windstorm provides canopy gaps for their regeneration; and (3) where disturbance-maintained habitats of many kinds, especially forest edges that are home to birds and mammals, are common. Vines typically occur under site conditions favorable for their establishment and in situations 2 and 3 above. We are fortunate to have the opportunity to observe and study shrubs and vines in a great diversity of regional and fine-scale landscape ecosystems ranging from very dry to wet. These range from fully sunlit prairies to the shaded understories of southern and northern forests, dry and wet, where fire may or may not have structured the landscape. Shrub associates also include a great diversity of trees, vines, and herbs.

Today, in comparison to the forests of pre-European-settlement times (Comer, Albert, and Austin 1998; Albert and Comer 2008), shrub occurrence is quite different because of massive human disturbances (ca. 1800–1940). These include extensive logging of conifer and hardwood forests, catastrophic fires in massive accumulations of logging debris, and recurrent lighter surface fires that followed logging, especially in northern lower Michigan (Dickman and Leefers 2003). In addition, large areas of lake plain wetlands in southeastern lower Michigan were drained for agriculture. Shrubs are well adapted and were well positioned to take advantage of these massive disturbances. They colonize well-lighted, cutover forests; the forest edges of increasingly fragmented communities; and a great variety of open sites provided by expanding agriculture and urbanization.

In our region, a major difference between shrubs and trees is their respective position in the vertical structure of forest communities. In forests trees occupy the dominant and subdominant overstory canopy layers. Due to their low stature, shrubs and some vines occupy the community layers known as the understory (individuals that at a breast height of 1.35 m have stem diameter at breast height [dbh] of 1.5 to 9.0 cm) and the ground cover (individuals with greater than 1.5 cm dbh). Some vines (e.g., Virginia creepers, grapes, and poison-ivy) climb tree trunks and, given substantial light irradiance, occasionally reach the subdominant or dominant layer. In landscape ecosystems not naturally dominated by trees (savannas, barrens, prairies, taiga, tundra, bogs, fens, an marshes), shrubs and vines occupy a dominant position (to ca. 5 m) in full or nearly full sunlight or they share dominance with herbaceous species.

Trees are notable, even famous, not only for their dominance but for their large size (old growth forest), longevity, and history. However, you are unlikely to find books entitled *Shrub Giants of the Pacific Northwest, Mythic Shrubs, Shrubs of Renown, Ancient Vines,* or *The Wisdom of Shrubs.* Giant oaks and elms witness the signing of important treaties and documents, which never takes place beside a burning-bush or in a patch of prickly-ash. Nevertheless, shrubs and vines take all this in stride because they have intrinsic qualities that make them important structural and functional parts of wetlands, savannas, and forest ecosystems, especially those where fire was or is the dominant regenerating disturbance. They are most diverse and abundant in extremely dry, fire-prone, and wet ecosystems where trees are naturally rare or absent.

SHADE AND UNDERSTORY TOLERANCE

Most shrub and vine species of Michigan and the Great Lakes Region are very shade-intolerant, intolerant, or midtolerant—based on requirements for successful establishment and reproduction. Shrubs are most abundant, diverse, and sexually reproductive in open sites with 75 to 100 percent full sunlight: woodlands, barrens, bogs, fens, swales, sand dunes, alvars, and edges of wetlands of all kinds. Of the 138 species we describe, 72 percent are very intolerant or intolerant. Only a few shrubs are able to survive, maintain themselves, and reproduce in the deeply shaded understory of mesic forest ecosystems; only 5 percent of species we treat are tolerant or very tolerant. These ecosystems, typically beech–sugar maple and hemlock–northern hardwoods forest types, are the domain of the spring ephemerals, which photosynthesize, flower, and fruit in early spring before trees leaf out. Shrubs characteristic of forests dominated by tree species that cast deep to moderate shade on mesic sites (e.g., sugar maple, American beech, American basswood, red maple, yellow birch, hop-hornbeam, and eastern hemlock) are shade tolerant to midtolerant and have the ability to quickly make use of gaps caused by windstorms—the primary disturbance force in such ecosystems. Each of these shrubs has its special microsite and strategy for reproduction and persistence. Each is able to flower and reproduce sexually in relatively low light conditions or maintain itself asexually until a light gap occurs. These shrubs include Canada yew, striped maple, leatherwood, and American fly honeysuckle; more fertile sites favor alternate-leaved dogwood and maple-leaved viburnum and in cool, moist sites the mountain maple. Some woody plants exhibit greater shade tolerance in the juvenile phase of physiological development (i.e., seedlings) than in the adult phase (e.g., choke cherry). This attribute makes survival "sense." For a shrub in the forest understory, this strategy is successful only as long as a gap appears in the canopy that provides sufficient light for its development into and persistence in the adult phase.

In contrast to the few shrubs that can persist under low light intensity, most shrubs require high light irradiance and have evolved and exist in open or lightly shaded sites. They are most common in number and abundance in the wetlands of fens, bogs, southern shrub-carr, and forested ecosystems where fire on dry sites or periodic flooding along rivers is the prevailing disturbance.

In the pre-European-settlement forests and before wildfire was excluded, shrubs exerted a profound influence on the composition, structure, and function of oak and oak-hickory forests compared to their relatively minor effect in mesic beech–sugar maple and hemlock–northern hardwoods forests. Below the tree canopy layers, the canopies of tall and low shrubs intercepted a high fraction of the light irradiance that penetrated tree layers and would reach the forest floor. Intensities on the forest floor, where tree establishment takes place, were typically too low for seedlings of the overstory layer to survive. As long as fire played a major role, shrubs played a significant role in the species establishing and being recruited to the understory and overstory. The shrubs by their very abundance often indirectly killed germinants and young seedling of oaks and hickories by shade and by outcompeting them for soil-water with well-developed root systems. These oak and oak-hickory forests, as well as oak-pine forests, were periodically burned by light or moderate surface fires fueled by the carpet of leaf litter and other organic matter. As large overstory trees aged, summer windstorms caused canopy gaps, and midtolerant and intolerant trees could survive in patches and in time were recruited into the understory and overstory. The recurrent fires killed shrub and oak and hickory stems alike, but both shrubs and young trees resprouted vigorously. Growing faster in height, the trees, especially oaks, soon overtopped shrubs in gaps, and maintained an oak or oak-hickory overstory with an abundant shrub ground cover and understory. Plant ecologist John Curtis (1959) characterized this role of shrubs of a former time in xeric forests of Wisconsin. His definitive statement would apply to similar forests throughout the Great Lakes region.

> The most conspicuous feature of the groundlayer of the southern xeric forests is the great abundance of shrubs. The shrub layer is frequently so thick that it impedes ready passage, especially when it contains large elements of thorny species like the blackberries (*Rubus*), gooseberries (*Ribes*), and prickly ash (*Zanthoxylum*). This prevalence of shrubs is in great contrast to their virtual lack in the best mesic forests as is shown by the sum of presence for shrub species on the prevalent species lists, where shrubs total 507 in the dry forests, 622 in the dry-mesic forests and only 46 in the mesic forests. A total of 48 species of shrubs and woody vines were found in all the xeric stands, both dry and dry-mesic.

The situation described by Curtis for the first half of the twentieth century would be less commonly observed today. The disturbances that favored shrubs—logging, surface fires, farm abandonment—often have been replaced with forest fragmentation, exclusion of fire, maturing second- and third-growth forests, and the invasion of mesophytic trees such as red maple (Abrams 1998, 2003). Actually, already in the 1950s and before, red maple was invading oak forests where fire and other disturbances were lacking (Larsen 1953). In one southern Wisconsin oak forest, Larsen reported that red maples dominated patches of the understory and by dense shade were excluding the typical shrub species of dogwoods and species of *Rubus*. Curtis (1959) commented, "The red maple colonies are continuing to spread, destroying the oak community as they grow, much as a cancer destroys its host."

Curtis correctly anticipated the inevitable signs of the widespread invasion of oak and other upland forests in which fire was excluded. Thus, it is no surprise that in today's fire-excluded upland forests mesophytic trees often dominate the subdominant and understory layers, shading out both shrubs and young trees (Abrams 1992). Forest understories throughout many parts of the Midwest, Great Lakes Region, New England, and mid-Atlantic states have become shaded and less favorable for shade-intolerant shrubs and forest trees such as upland oaks and pines (Abrams 2003; Nowacki and Abrams 2008). Moderate shade-tolerating introduced shrubs—common buckthorn, Maack's and Tartarian honeysuckles, Japanese barberry, autumn-olive—in oak forests outcompete native shrubs. A modern habitat for shrubs and vines has emerged due to extensive road systems in public and private forests, roadsides, seasonally wet roadside ditches, highway medians, old fields, and edges of cutover woodlands. These habitats provide relatively high light intensity and relatively competition-free (i.e., open) sites for establishment and rapid growth, flowering, and abundant fruit production.

Shrubs and vines of the understory and ground cover must cope with multiple physical factors of those layers—competition with tree roots, herbivory by animals, fungi, and disturbances characteristic to a particular ecosystem. Survival and persistence of shrubs, vines, and seedling trees in the understory are difficult but not surprising given the evolution of these plants as integral parts of specific ecosystems and the advantages of reproduction and persistence by both sexual and asexual reproduction.

In many conifer and conifer-hardwood swamps of presettlement times, the crowns of species such as northern white-cedar, hemlock, white and black spruces, balsam fir, and white pine cast deep shade and only the most shade-tolerant shrubs could survive. However, wind disturbance is common in swamps, and uprooting shallow-rooted trees provided gaps and a decaying nurse log for establishment of a variety of shrubs and herbs that would otherwise be excluded by shade. When large gaps occurred, species that required significant light irradiance, and normally might only grow along wetland and stream edges, were favored: speckled alder, willow species, gooseberry and currant species, red-osier, winterberry, and mountain holly. Settlement of the uplands brought winter logging to the swamps and often drainage. Thus, human-caused disturbances, together with natural tree uprooting by windstorms, increased the probability of shrubs colonizing swamp wetlands. Diverse open wetlands such as bogs, muskegs, fens, wet meadows, and shrub thickets were home to myriad shrub species, which then via disturbance had access to forested swamps.

RESPONSE OF SHRUBS AND VINES TO CLIMATE CHANGE

Thus far we have examined the distribution pattern of individual shrub and vine species in regional landscape ecosystems based on their occurrence in counties of Michigan. From now to 2100, what will be their response to our changing climate and the concomitant effects of ecosystem change that are already in progress? Although uncertainty characterizes ecosystem change, climate-based modeling projections indicate that shrub and vine species will migrate primarily north and east and some species with ranges south of Michigan will advance into the state (McKenney et al. 2007). Although the climate models that suggest these changes are under continual improvement, this general trend appears robust (McKenney et al. 2011, 2014).

The increasing abundance and importance of shrubs in landscape ecosystems are due to their many genetic adaptations and plastic responses to changing site conditions. We anticipate marked expansion of their distribution in the future due to a suite of natural history traits, of which the most significant are as follows.

- Occurrence throughout a great range of landform, soil-site, and light-irradiance conditions and natural and human disturbances of site-specific kinds
- A combination of early and abundant sexual reproduction with multiple means of asexual reproduction
- Widespread dispersal by birds and other animals
- Shade intolerance—72 percent of the species we describe are intolerant or very intolerant of shade
- The ability to respond aggressively to climatic warming, recurrent violent episodes of weather, and concomitant changes in the physical environment and biota
- Once established, the ability to tolerate and persist in ecosystems with light to moderate shade from competing plants

Given these and the full range of their natural-history traits, shrubs and vines have aggressively colonized diverse landscapes from late glacial times (Elliott-Fisk 2000) to the present. With the warming northern climate and expected increase in violent disturbances, we expect the migration of shrubs and vines to markedly exceed that of trees, especially along the boundaries of the southern regional ecosystems of Michigan, Wisconsin, Minnesota (Albert 1995), and Ontario (Crins et al. 2009). Monitoring the pattern change in shrub and vine occurrence using a landscape ecosystem approach (Rowe 1992; Barnes et al. 1998; Barnes 2009a) is urgently needed to determine the ecological scope and rate of this process.

In addition to climate-based ecosystem change and new patterns of natural disturbance, the decline in species of the Great Lakes Region is increasingly severe because of the spread of many diseases and insect pests, which through human movement and commerce have markedly changed sites and species composition in the twentieth century. These include the larch casebearer, chestnut blight, white pine blister rust, Dutch elm disease, oak wilt, butternut canker, hemlock adelgid, balsam woolly aphid, beech-bark disease, emerald ash borer, and Japanese long-horned beetle, among others. In addition, the introduction of many alien ornamental woody plants (e.g., *Elaeagnus, Frangula, Lonicera, Rhamnus, and Ribes*) and concomitant landscape fragmentation and fire exclusion have led to their naturalization and aggressive spread in cutover prairie, woodland, and forest ecosystems. The result is their replacement of native plants.

All species of shrubs and vines that we describe show northern to northeastern expansion of their range based on climatic modeling (i.e., temperature and precipitation variables; McKenney et al. 2007, 2014; http://planthardiness.gc.ca) using several general circulation models and greenhouse gas emission scenarios. We identify three groups in particular, according to their current distribution and projected range shifts for the period 2071–2100.

1. Native Michigan species with more southerly core ranges moving significantly north or northeast but still occurring in northern Michigan by 2071–2100: *Ceanothus americanus, Ptelea trifoliata, Rhus copallina, Rubus flagellaris, and Toxicodendron vernix*, among many others.

2. Native species present in Michigan but migrating northward and absent in 2071–2100: *Alnus incana, Alnus viridis, Hypericum kalmianum, Ilex mucronata, Kalmia polifolia, Myrica gale, Shepherdia canadensis,* and *Sorbus decora.*

3. Native species of southern distribution (i.e. absent in Michigan) projected to be well suited by 2071–2100: *Aralia spinosa* L., devil's-walking-stick; *Aristolochia macrophylla* Lam., Dutchman's pipe; *Chionanthus virginicus* L., fringe-tree; *Cotinus obovatus* Raf., smoke tree; *Hydrangea arborescens* L., American hydrangea; *Symphoricarpos orbiculatus* Moench, coralberry; and *Wisteria fructescens* (L.) Poiret, Atlantic wisteria.

Using the website of the Canadian Forest Service, http//planthardiness. gc.ca, readers can access maps showing the present distribution of hundreds of tree, shrub, vine, and herb species, for which sufficient data are available, and then access the projected distribution for (1) a specific thirty-year time period from 2011 to 2100, and (2) the climate-change scenario of one's choice.

APPENDIX A

The four regional landscape ecosystems are presented in Figure 19. The percentage occurrence of shrub and vine species in each of these regions is given in the "Distribution" section of each plant description. Distinctive patterns for selected species are illustrated in Figure 20. In Appendix A the names and physiographic characteristic of regions, districts and subdistricts are presented in Tables 2a and 2b. These are followed by a brief geological and ecological characterization of each regional landscape ecosystem. The data for ten climatic factors for each region are available in Table 3 (368–69) in *Michigan Trees* (Barnes and Wagner 2004). Detailed descriptions of the regional ecosystems are available in Albert, Denton, and Barnes 1986 and Albert 1995.

REGION I: SOUTHERN LOWER MICHIGAN

In the Lower Peninsula, southern Region I is distinguished from the northern Region II by warmer temperatures throughout the year and a longer and less variable growing season. Also, inland areas in the south (e.g., the Kalamazoo and Ionia Districts) have lower heat sums prior to the last spring freeze than do inland areas to the north (e.g., the Newaygo and Highplains Districts). Consequently, there is less danger of plant injury from freezing spring temperatures. The 82°F isotherm is the approximate boundary between the warm south and the colder northern half. Glacial landforms include lake plain, outwash plain, ice-contact terrain, and end and ground moraines. Broad lake plains extended inland along Lakes Huron and Michigan for 130 km (50 mi) along the Huron shoreline at Saginaw Bay. Sand dunes occur in a 1.6 to 8 km (1 to 5 mi) band along the Lake Michigan shoreline. The relatively low terrain of ground and end moraines forms the interior of the region, interspersed with narrow outwash-plain channels.

Prior to European settlement, oak savanna was the most prevalent vegetation type, followed by oak-hickory and beech–sugar maple forests. Today, however, due to fire exclusion (via land fragmentation and fire suppression), the oak savannas have become oak forests—where forests still exist in this area of extensive farming and urban and suburban residential and industrial development. Expansive wetlands of swamp forests and marshes once occurred, but those of the lake plain were drained for agriculture. Today many tree, shrub, and vine species are found primarily in this relatively warm southern region. Activity of Native Americans was likely greater in this region, especially the use of fire and primitive agriculture, than in any other regional landscape in Michigan.

REGION II: NORTHERN LOWER MICHIGAN

In northern lower Michigan, Region II, the more northerly location, greater exposure to the effects of lakes Michigan and Huron, and larger and higher physiographic features control the climate, which is considerably cooler and more variable from place to place than that of Region I. Compared to Region I, there is a greater chance of late spring freezes and in some areas a chance of frost throughout the growing season. Physiographically, large moraines, with the highest elevations in Lower Michigan (up to 26 m, 1,725 ft, near Cadillac), are predominant features (Schaetzl and Barnes 2009). A sandy high plain dominates the interior of the region. In the center of this plain (Subdistrict 8.2, Fig. 19), the growing season reaches a maximum of only 115 days, whereas it is 163 days in southeastern Michigan (District 1). Cold-tolerant species in the high plains include the trees northern pin oak, black cherry, and jack pine. Shrubs are redroot, sand cherry, bearberry, and sweetfern, among others. The upland pre-European-settlement forest consisted of hemlock–northern hardwoods forests on the moraines and jack pine barrens on the very fire-prone sites. Swamp, fen, bog, and river floodplain communities are nearly as common today as in presettlement times. Massive logging of the pines and hemlock from about 1850 to 1900 and the catastrophic fires that followed drastically changed the vegetation over large areas. Shade-intolerant shrubs, such as blueberries, black huckleberry, bearberry, sand cherry, wintergreen, trailing arbutus, prairie redroot, and sweetfern, flourished in fire-created open sites, as well as the bigtooth and trembling aspens that replaced pines in many ecosystems.

REGION III: EASTERN UPPER MICHIGAN

In upper Michigan, geological, physiographic, and climatic conditions provide the basis for distinguishing the eastern Region III from the western Region IV. Their boundary is based primarily on the geological division of Cambrian and Precambrian bedrock to the west and the younger Ordovician and Silurian bedrock to the east. The eastern region is characterized by low elevation, relatively young bedrock, and flat glacial lake plains, which cover the largest part of the region. The surface of the lake plain consists of poorly drained sand and clay soils, exposed limestone and dolomitic bedrock, or thin soils over bedrock. Due to its northern latitude and close proximity to lakes Michigan, Superior, and Huron, the region has a cool, lake-moderated climate. There is considerable lake-effect snowfall. The growing season is mostly between 110 and 130 days, and, being strongly lake moderated, it is similar to that of Region II, northern lower Michigan. However, the eastern Upper Peninsula lacks comparable high plains and exhibits a considerable expanse of lowland swamps characterized by tamarack, black spruce, and the associated shrubs Labrador-tea, large and small cranberries, showy mountain-ash, creeping-snowberry, winterberry, and *Ribes* spp., among others.

Prior to European settlement, hemlock–northern hardwoods forests, with sugar maple and beech as common dominants, were concentrated on end and ground moraines and drumlin fields. Jack pines grew on extensive sandy outwash plains, along with red and white pines where fires were less severe. Today, following logging and postlogging fires, trembling aspen is common in these cutover lands, especially where soil-water is readily available.

REGION IV: WESTERN UPPER MICHIGAN

High elevation, northern latitude, and the absence of a major body of water to the south and southwest are important determinants of the climate in this most continental of regions. It is distinguished by its ancient and resistant bedrock of Precambrian and Cambrian origin. Bedrock outcrops are common, especially in the Michigamme Highlands (District 17, Marquette and Baraga Co.), the Porcupine Mountains, and the mountains of the late Precambrian age of the Keweenaw Peninsula. Exposed bedrock knobs and sandy end- and ground-moraine ridges are characteristic, but a large clay lake plain and outwash channels and plains also occur. Many rocks are rich in iron, which gives a reddish color to the sandy soils. Although the growing season is relatively warm, winter temperatures are extremely cold. The lack of lake moderation leads to greater variation in temperature within the growing season. It may be as short as 60 days. Snowfall and rainfall are heavy adjacent to Lake Superior as a result of moisture-laden air being forced to rise rapidly over the bedrock uplands at the northern edge of the region. The Keweenaw Peninsula (District 21) has a growing season of 134 days, significantly more than in the eastern Upper Michigan (Mackinac and Luce Districts), although its growing season heat sum is the lowest of any district in the state. Prior to European settlement, hemlock–northern hardwoods forests, dominated by sugar maple, hemlock, yellow birch, basswood, beech (eastern part only), and some white pine, prevailed on the heavier-textured soils. The thin soils and bedrock knobs supported red pine, white pine, and red oak. Jack pine and northern pin oak occurred on the dry, fire-prone flat outwash plains. Over much of the region, hemlock–northern hardwoods forests are still common. However, following logging of the pines and hemlock, trembling aspen and paper birch stands flourished and are still prominent today.

Table 2a. Names and Physiographic Characteristics of Regions, Districts, and Subdistricts of the Regional Landscape Ecosystems of Lower Michigan

Number	District	Subdistrict	Site Condition
Region I: Southern Lower Michigan			
1.1	Washtenaw	Detroit	Heat island
1.2		Maumee	Lake plain
1.3		Ann Arbor	Fine- and medium-textured moraine
1.4	.	Jackson	Coarse-textured interlobate and end moraine, outwash plain, and ice-contact topography
2.1	Kalamazoo	Battle Creek	Outwash plains and ground moraine
2.2		Cassopolis	Coarse-textured end moraine and ice-contact terrain
3.1	Allegan	Berrien Springs	End and ground moraine
3.2	.	Benton Harbor	Lake plain
3.3		Jamestown	Fine-textured end and ground moraine
4.1	Ionia	Lansing	Medium-textured ground moraine
4.2		Greenville	Coarse-textured end and ground moraine
5.1	Huron	Sandusky	Lake plain
5.2		Lum	Medium and coarse-textured end moraine ridges and outwash plains
6.1	Saginaw		Lake plain and reworked till plain
Region II: Northern Lower Michigan			
7.1	Arenac	Standish	Lake plain
7.2		Higgins Lake	Fine-textured end and ground moraines
8.1	Highplains	Cadillac	Coarse-textured end moraine
8.2		Grayling	Outwash plains
8.3		Vanderbilt	Steep end and ground moraine ridges
9	Newaygo	.	Outwash plains
10	Manistee		End moraine and sand lake plain
11.1	Leelanau	Williamsburg	Coarse-textured end moraine ridges
11.2		Traverse City	Coarse-textured drumlin fields on ground moraine
12.1	Presque Isle	Onaway	Drumlin fields on coarse-textured ground moraine
12.2		Stutsmanville	Steep sand ridges
12.3		Cheboygan	Lake plain

Table 2b. Names and Physiographic Characteristics of Regions, Districts, and Subdistricts of the Regional Landscape Ecosystems of Upper Michigan

Number	District	Subdistrict	Site Condition
Region III: Eastern Upper Michigan			
13.1	Mackinac	St. Ignace	Limestone bedrock and sand lake plain
13.2		Rudyard	Clay Lake plain
13.3		Escanaba	Limestone bedrock and sand lake plain
14.1	Luce	Seney	Poorly drained sand lake plain
14.2		Grand Marais	Sandy end moraine, shoreline, and outwash plains
15.1	Dickinson	Hermansville	Drumlins and ground moraine
15.2		Gwinn	Poorly drained sandy outwash
15.3		Deerton	Sandstone bedrock and high, sandy ridges
Region IV: Western Upper Michigan			
16.0	Norway		Granitic bedrock and end moraine
17.0	Michigamme		Granitic bedrock
18.1	Iron	Iron River	Drumlinized ground moraine
18.2		Crystal Falls	Kettle-kame topography, outwash plain, and sandy ground moraine
19.1	Bergland	Bessemer	Large, high, coarse-textured ridges and metamorphic bedrock knobs
19.2		Ewen	Dissected clay lake plain
19.3		Baraga	Broad ridges of coarse-textured rocky till
20.1	Ontonagon	Rockland	Narrow, steep, bedrock ridge
20.2		White Pine	Clay lake plain
21.1	Keweenaw	Gay swamps	Coarse-textured broad ridges and
21.2		Calumet	High igneous and sedimentary bed rock ridges and knobs
21.3		Isle Royale	Island of igneous bedrock ridges and swamps

GLOSSARY

Abaxial   Away from the axis; e.g., the "lower" or dorsal surface e.g., of a leaf.

Abortion   Imperfect development or non-development of an organ.

Abscise   To cut off.

Achene   A dry, indehiscent fruit, usually one-seeded, with seed attached to fruit wall at one point only, derived from a superior ovary.

Acorn   The fruit of the oak, consisting of a nut with its base enclosed in a cup of imbricated scales.

Acuminate   Gradually tapering to a long point.

Acute   Terminating with a sharp angle; tapering to a point, but not long-pointed.

Adaxial   Toward the axis; e.g., the "upper" or ventral surface e.g., of a leaf.

Aggregate fruit   A group of separate fruits developed from one flower.

Alternate   Arrangement of leaves, branches, buds, etc., scattered singly along the stem; occurring one at each node; not opposite.

Alvar   Flat limestone rock ("pavement") with thin (if any) soil and usually graminoid vegetation with trees few and often dwarfed.

Angiosperm   A flowering plant.

Anther   The part of a stamen which bears the pollen.

Apetalous   Without petals.

Apex   The top, as the tip of a bud or the end of a leaf, which is opposite the petiole.

Apiculate   Ending in a short-pointed tip.

Apomixis   As used here, reproduction by seed without fertilization—a form of asexual reproduction.

Appressed   Lying close and flat against; (of buds) laying flat against the stem.

Arcuate veins   Curved and arranged nearly parallel to the blade margin, as in leaves of *Cornus* and *Ceanothus*.

Aril   A fleshy outgrowth that surrounds the seed, commonly brightly colored, as in *Taxus*.

Armed   Bearing thorns, spines, or prickles.

Aromatic Fragrant; having an agreeable odor.

Ascending Arising somewhat obliquely, or curving upward.

Astringent Shrinking and driving the blood from the tissues; contracting.

Awl-shaped Tapering from the base to a slender or rigid point.

Axil The upper one of the angles formed by the juncture of a leaf with a stem.

Axillary Situated in an axil.

Axis The central line of support, as a stem.

Bark The outer covering of a trunk or branch.

Berry A juicy or fleshy fruit in which the seeds are embedded in the pulp.

Biennial Living for two years.

Bisexual Having male and female sex organs in the same individual.

Blade The expanded portion of a leaf.

Bloom A powdery or waxy substance that is easily rubbed off.

Bog An acidic peatland dominated by *Sphagnum* mosses and shrubs in the family Ericaceae, often found around lakes and ponds with vegetation in ± concentric zones of increasing maturity from open water to surrounding swamp forest or upland. Its water typically derives mostly from precipitation not groundwater flow. See also fen.

Boreal Of or pertaining to Boreas (god of the north wind), hence northern.

Bract A modified leaf often subtending a flower or belonging to an inflorescence.

Branch A secondary division of a trunk.

Branchlet A small branch.

Bristle A flexible, pointed outgrowth from the epidermis or cortex of any organ.

Bur A rough, prickly seedcase or cuplike structure surrounding one or more nuts.

Calcareous Said of soil containing free lime ($CaCO3$) and having a basic pH reaction.

Calyx The outer part of a perianth, usually green in color.

Cambium A layer of cells between the wood and bark, capable of producing phloem and xylem tissue, by which the stem grows in diameter.

Campanulate Bell-shaped.

Capitate Like a pin-head (as certain stigmas on the style).

Capsule A dry fruit of two or more carpels which splits at maturity to release the seeds.

Carpel   A floral organ that contains ovules in angiosperms either borne separately or as a unit of a compound pistil.

Catkin   A spike or elongate axis bearing apetalous, unisexual flowers; falling as a unit after flowering or fruiting.

Caudate   Furnished with a tail, or with a slender tip or appendage.

Chambered   Said of pith when divided into small compartments separated by transverse partitions.

Ciliate   Fringed with hairs on the margin.

Clay   A kind of soil containing more than 40 percent clay particles, less than 45 percent sand particles, and less than 40 percent silt particles; a soil particle less than 0.002 mm in diameter.

Cleft   Cut about halfway to the middle.

Clone   The aggregate of stems originating vegetatively from one sexually produced individual. They may develop spontaneously in nature as in aspens and sumacs or be produced humans and spread, often for ornamental use, in horticultural practice.

Cluster   A group of two or more organs (flowers, fruits, etc.) on a plant at a node or end of a stem.

Collateral   Said of extra buds which occur on either side of an axillary bud.

Compound leaf   Single leaf divided into separate leaflets.

Cone   The reproductive structure of gymnosperms, consisting of an axis to which are attached many woody, overlapping scales which bear seeds.

Conifer   A gymnosperm of the order Coniferales; a cone-bearing tree.

Connate   United to another part of the same kind.

Continuous   Said of pith that is solid, not interrupted by cavities.

Cordate   Heart-shaped.

Coriaceous   Leatherlike in texture.

Corky   Made of cork cells; corklike.

Corolla   The inner part of the perianth, composed of separate or connate petals, often brightly colored.

Cortex   The outer tissue that lies between the epidermis and the vascular strands.

Corymb   A flower cluster in which the axis is shortened and the pedicels of the lower flowers lengthened, forming a flat or more or less round-topped inflorescence, the marginal flowers blooming first.

Creeping   Running at or near the surface of the ground and rooting.

Crenate   With very rounded teeth; scalloped.

Crenulate   Finely crenate.

Cross section   Said of a section cut at right angles to the long axis.

Cultivar   A variety of a plant produced and maintained by cultivation; also the corresponding taxon.

Cuneate   Wedge-shaped.

Cuspidate   Tipped with a firm, sharp point.

Cutting   A piece of the stem, root, or leaf which, if placed in contact with a rooting medium, will form new roots and buds, reproducing the parent plant.

Cyme   A broad and flattish inflorescence consisting of a determinate central axis bearing a number of pedicelled flowers, the central flowers blooming first.

Deciduous   Not persistent; falling away, as the leaves of a tree in autumn.

Decumbent   Stems or branches reclining, but the ends ascending.

Dehiscent   Opening by valves or slits.

Dentate   Toothed, with the teeth usually pointed and directed outward.

Denticulate   Finely dentate, the teeth small and shallow.

Depressed   Somewhat flattened from above.

Dichotomous   Branching regularly in pairs.

Dioecious   Having the sexes on separate plants; i.e., all flowers on a single plant either staminate or pistillate. See also monoecious.

Disk   A rounded, flat plate.

Dispersal   The spread of objects (e.g., pollen, fruits, seeds) from a fixed or constant source.

Dissected   Cut or divided into numerous segments.

Distal   At or toward the apex, i.e., toward the opposite end from that at which a structure is attached (e.g., upper half or tip of leaf blade).

Distribution   The geographical extent and limits of a species.

Divergent   Said of buds, fruits, etc., which spread apart and point away from the twig or other organ.

Dormant   A term applied to whole plants or parts which are in a resting stage.

Dorsal   Pertaining to the back or outer surface of an organ.

Downy   Covered with fine hairs.

Drumlin   A smooth, oval hill formed as glacial ice moved over an area and molded underlying till to form elongate, streamlined hills. Drumlins are relatively common near the shores of the Great Lakes.

Drupe   A fleshy or pulpy fruit in which the inner portion of the ovary wall is hard or stony.

Ecosystem   Any single perceptible ecosystem is a topographic unit, a volume of land and air plus organic contents extended areally over a particular part of the earth's surface for a certain time. See also landscape ecosystem.

Ellipsoid   An elliptical solid.

Elliptical   Oval or oblong with regularly rounded ends.

Emarginate   Notched at the apex.

Entire   Blade margin smooth, lacking teeth or other protrusions.

Epidermis   The outer cell layer or covering of plants.

Epigeal   Living or growing or germinating above the soil surface. Epigeal germination refers to seed germination above the ground, and cotyledon leaves develop above the ground.

Escape   Any plant formerly cultivated that grows wild in nature.

Esker   A long, narrow ridge chiefly composed of stratified glacial drift deposited by subglacial rivers.

Evergreen   Having foliage which does not fall at the end of the growing season, green leaves in winter.

Excavated   Hollowed out; hollow.

Excurrent   With axis forming an undivided major trunk, as in conifers.

Exfoliate   To peel away, as of the outer layers of bark.

Exserted   Projecting beyond a covering.

Fen   A peatland nourished by calcareous groundwater flow through its near-surface peat layers and dominated by sedges and non-Ericaceous shrubs, with little if any *Sphagnum* and more alkaline compared to the typical acid bog. Gradations between bog and fen may occur, and some large peatlands are mosaics.

Fertile   Capable of bearing fruit; normally reproductive. Of a site, rich or well supplied with nutrients.

Fertilization   The union of a sperm (contained in pollen) and an egg (contained in ovule).

Filament   The part of a stamen which bears the anther.

Flaky   With loose scales easily rubbed off, as with bark.

Flora   The plant species of a particular region or area.

Flower   An axis bearing stamens or pistils or both (calyx and corolla usually accompany these).

Fluted   Grooved longitudinally, with alternating ridges and depressions.

Fluvial   Of, found in, or produced by a river, as in fluvial processes or landforms.

Follicle   A dry, dehiscent fruit which opens along one side.

Forked   Divided into nearly equal branches.

Form   The general appearance of a plant; habit.

Frost   Ice formed when moisture condensing at ground level freezes. The occurrence of frost does not necessarily correspond to the occurrence of freezing temperature as measured in standard meteorological shelters. Frost may occur when air at the ground surface drops below freezing.

Fruit   The matured ovary of flowering plants.

Fusiform   Thick, but tapering toward each end.

Glabrate   Almost without hairs; with occasional hairs.

Glabrescent   Becoming glabrous.

Glabrous   Without hairs.

Gland   A small protuberance consisting of one or more secreting cells.

Glandular   Bearing glands.

Glaucous   Covered or whitened with bloom which may be rubbed off.

Globose   Spherical or nearly so.

Globular   Nearly globose.

Glutinous   Sticky, gluey.

Habit   The general appearance of the plant, best seen from a distance; form.

Habitat   The place and the sum total of associated physical ecosystem factors, as modified by the biota, where a plant grows; site.

Hardwood   A term used in forestry to indicate a broad-leaved, deciduous tree. Not necessarily defined by "hardness" of the wood itself.

Heartwood   The dead, central portion of the trunk or large branch.

Heat sum   Accumulated heat over a certain threshold: the product of temperature above a certain base or threshold (0 or 5ºC) and the time duration of that temperature

Herb   A plant with no persistent live stem on or above ground; a nonwoody plant.

Hip   An aggregate of achenes surrounded by an urn-shaped receptacle.

Hirsute   Covered with rather coarse or stiff, usually regularly long, hairs.

Hispid   Pubescent with bristly, rigid hairs.

Hoary   Grayish white with a fine, close pubescence.

Husk   The rough outer covering of a fruit or seed.

Hybrid   A cross between two species, usually yielding an intermediate form.

Hypogeal   Growing or germinating below the soil surface. Hypogeal germination refers to seed germination below the ground and cotyledons remain below the ground.

Ice-Contact terrain   Irregular, abruptly-hilly topography formed when a continental ice sheet (glacier) stagnates and ice disintegrates in place. Outwash materials of sands, gravel, and stones are deposited by water in "contact" with ice, i.e., adjacent to an ice block (kame) or in a stream within the ice (esker). Typical landforms are kettles (ice-block depressions, often lakes), kames, and eskers. Also termed kettle-kame topography.

Imbricate   Overlapping, like the shingles on a roof.

Indehiscent   Not opening by valves or slits; remaining persistently closed.

Indigenous   Native and original to a region.

Inequilateral   Unequal-sided; oblique at the base; asymmetrical.

Inflorescence   The flowering part of a plant, and especially its arrangement; a flower cluster.

Infructescence   A fruit cluster having the same arrangement as the inflorescence.

Internode   The portion of a stem or rachis between two nodes.

Intolerant   Incapable of enduring under adverse conditions, as shade, drought, lack of nutrients or oxygen.

Introduced   Brought in from another region.

Invasive   Said of an introduced species that negatively affects the development or occurrence of a native species.

Involucre   A circle of bracts subtending a flower or cluster of flowers.

Kame   A steep-sided hill or complex of hills composed of water-laid parent material of sand and gravel originally deposited in stagnant glacial ice, as sediment in a pond on the glacier's surface. See also Ice-Contact terrain.

Keeled   A ridge ± centrally located on the long axis of a structure, such as a sepal or an achene.

Kettle   A steep-sided basin created when a block of glacial ice is surrounded or buried by deposits of sand and gravel; may or may not be filled with water (e.g., a kettle lake). See also Ice-Contract terrain.

Lanceolate   Lance-shaped, broadest above the base and tapering to the apex, but several times longer than wide.

Landscape ecosystem   Any single perceptible ecosystem is a topographic unit, a volume of land and air plus organic contents extended areally over a particular part of the earth's surface for a certain time. Also called geographic ecosystem.

Landform   A terrain feature which has definable shape (geomorphology) and parent material (e.g., eskers, drumlins, moraines).

Lateral   Situated on the side.

Leaf   A photosynthetic and transpiring organ; an expanded, usually green, organ borne on the stem of a plant.

Leaflet   One of the small blades of a compound leaf.

Leaf scar   The scar left on a twig by the falling of a leaf.

Legume   A simple, dry, podlike fruit composed of a solitary carpel that usually splits open along both sutures (as in the Fabaceae).

Lenticel   A corky, porous spot on bark, appearing as dots or warts, which admit air to the interior of a twig or branch.

Linear   Long and narrow, with parallel edges (as pine needles).

Loam   The textural class name for soil having a moderate amount of sand, silt, and clay. Loam soils contain 7 to 27 percent of clay, 28 to 50 percent of silt, and less than 52 percent of sand.

Lobe   Part of a blade separated by sinuses, i.e., a division of a blade that is broadly attached, not contracted or stalked at its base.

Lobed   Provided with a lobe or lobes.

Lustrous   Glossy; shiny.

Marl   A deposit of calcium carbonate resulting from the activity of photosynthetic plants in altering the carbonate/bicarbonate balance in a lake or pond.

Marsh   A wetland dominated by coarse, non-woody vegetation. See also swamp.

Meadow   A treeless area less wet than a marsh and dominated by smaller grasses or sedges. Wet meadows are often dominated by sedges and may grade into fens or bogs; if dominated by grasses, they may grade into prairies in southern Michigan. When upland, meadows are usually successional or formed by clearing and often include many introduced species.

Membranous   Thin, rather soft, and somewhat translucent.

Meristem   Embryonic cells that form new tissues.

Mesic   Said of a site that is well supplied with soil water throughout the growing season, not overly dry or wet.

Midrib   The prominent central vein of many leaves (often best seen from the lower side).

Monoecious   Unisexual, with separate male and female flowers on the same individual.

Moraine   An accumulation of drift deposited chiefly by direct glacial action. The initial constructional form of a moraine is independent of the floor beneath it. It may have a ridgelike form (end moraine) or be of low relief (ground moraine).

Mucro   A short, sharp, slender point.

Mucronate   Tipped with a small, abrupt point.

Multiple fruit   A cluster of ripened ovaries of separate flowers inserted on a common

Receptacle   The haploid or gametophytic number of chromosomes; ordinary cells of a seed plant have this number of pairs of chromosomes. Many plants have more than the basic two sets or complements of chromosomes (diploid) and the number of these is indicated with the suffix -ploid; triploid (3n) = 3 sets; tetraploid (4n) = 4 sets; pentaploid (5n) = 5 sets; octoploid (8n) = 8 sets; etc.

Naked   Lacking a covering, as buds without scales or leaves without pubescence.

Naturalized   Said of introduced plants which are reproducing and establishing themselves in a new environment.

Nectary   An organ that produces nectar.

Node   The region of the stem where one or more leaves arise.

Nut   A dry, hard, indehiscent (usually 1-seeded) fruit, often larger than normally termed an achene (or nutlet).

Nutlet   An achene or similar tiny 1-seeded indehiscent fruit; also used for the stony carpels embedded in the pome of *Crataegus*.

Oblanceolate   Lanceolate, with the broadest part toward the apex.

Oblique   Slanting, or with unequal sides.

Oblong   Longer than broad, with sides approximately parallel.

Obovate   Ovate, with the broadest part toward the apex.

Obovoid   An ovate solid with the broadest part toward the apex.

Obtuse   Blunt or rounded at the apex.

Opposite   Arrangement of leaves, branches, buds, etc., on opposite sides of a stem at a node.

Orbicular   Circular in outline or nearly so.

Outwash   Materials, mostly sands and gravels, deposited by meltwater streams flowing away from a glacier.

Oval   Broadly elliptical.

Ovary   The part of a pistil that contains the ovules.

Ovate   Shaped like the longitudinal section of a hen's egg, with the broad end basal.

Ovoid   Solid ovate or solid oval; egg-shaped.

Ovule   The part of a flower which after fertilization becomes the seed.

Palmate   Radiately lobed or divided; hand-shaped.

Panicle   A loose inflorescence with two or more orders of branching and pedicellate flowers.

Paniculate   Arranged in panicles or resembling a panicle.

Papillose   Covered with papillae, i.e., with short, rounded, blunt projections.

Pedicel   The stalk of a single flower in a compound inflorescence.

Pedicellate   Borne on a pedicel.

Peduncle   A primary flower stalk, supporting either a cluster or a solitary flower.

Pendent, Pendulous   Hanging or drooping

Perfect   Said of a flower with both stamens and pistil.

Perianth   The calyx and corolla of a flower considered together.

Persistent   Staying on the plant, as leaves remaining through the winter.

Petal   One of the divisions of a corolla.

Petiole   The stem or stalk of a leaf.

Petiolule   The stem or stalk of a leaflet.

pH   The reaction of the soil, whether basic neutral, or acidic, based on the concentration of hydrogen ions in solution; literally the proportion of hydrogen ion concentration (pH value is the logarithm of the reciprocal of the hydrogen ion concentration).

Phototrophic   Responding toward light.

Physiography   The physical geography of an area.

Pilose   Hairy with long, soft hairs.

Pinnate   Arranged in two rows, one on each side of a common axis, as veins in a leaf or leaflets in a compound leaf; featherlike. In an odd-pinnate leaf, there is a terminal leaflet; in an even-pinnate one, there is no terminal leaflet. Twice-pinnate; with the primary divisions again pinnate.

Pistil   The seed-bearing organ of a flower, normally consisting of ovary, style, and stigma.

Pistillate   Provided with a pistil, but usually without stamens.

Pith   The softer central part of a twig or stem.

Pollen   The grains (microspores, continuing sperm) produced in the anther.

Polygamous   With flowers, sometimes perfect, sometimes unisexual, the sexes borne on the same (polygamomonoecious) or on different (polygamodioecious) individuals.

Pome   A fleshy fruit with a papery core, derived from an inferior compound ovary and receptacle, as an apple.

Prickle Armature that is a sharp-outgrowth of the epidermis or cortex of any organ.

Prostrate Lying flat on the ground.

Proximal At or toward the base, i.e., toward the end at which a structure is attached e.g., lower half or base of leaf blade.

Puberulent Minutely pubescent.

Pubescent A covering of short, soft hairs; downy.

Punctate Dotted with translucent or colored dots or pits.

Raceme Usually a narrow inflorescence with only one order of branches of flowers on pedicels of about equal length at maturity, arranged on a common, elongated axis, the lowermost flowers blooming first.

Rachis The central axis of a spike or raceme of flowers or of a compound leaf.

Radicle The first root formed in a seed by the embryo.

Ramet Name given to a shoot or sucker of a clone, arising from a root or rhizome.

Receptacle The more or less expanded portion of an axis which bears the organs of a flower.

Recurved Curved downward or backward.

Reflexed Bent sharply backward.

Remote Scattered; not close together.

Resin ducts Long, narrow channels between the elements of the wood, filled with resin.

Reticulate Arranged as in a network.

Revolute Rolled backward from the edge.

Rhizome A horizontal underground stem, usually rooting at the nodes.

Rib A prominent vein of a leaf or the extended edge of a fruit husk, fruit, or seed.

Root collar Region of the juncture of trunk base and the major roots.

Rugose Wrinkled.

Rust A fungal disease usually with orange-colored spores masses.

Samara An indehiscent winged fruit.

Sand A soil textural class, the largest kind of soil particle, between 0.05 and 2.0 mm in diameter; a kind of soil containing at least 85 percent sand particles.

Sapwood The outer portion of a trunk or large branch of a tree between the heartwood and the bark, containing the living elements of the wood.

Scabrous Rough to the touch when rubbed in at least one direction.

Scales   Small modified leaves, usually thin and scarious, seen in buds; the flakes into which the outer bark often divides.

Scurfy   Covered with small branlike scales.

Seed   The ripened ovule.

Segment   One of the parts of a structure that is cleft or divided.

Sepal   One of the divisions of a calyx.

Serrate   Toothed, the teeth sharp and pointing forward.

Serrulate   Finely serrate, the teeth small and shallow.

Sessile   Attached directly, with a stalk.

Sheath   A thin enveloping part, as of a leaf; any body enwrapping a stem.

Shoot   The repeating unit of plant construction, comprising stem, leaves, and buds.

Shrub   A woody perennial plant usually branched several times at or near the base giving a busy appearance, usually less than 5 meters tall.

Silt   A soil textural class; soil particle between 0.05 and 0.002 mm in diameter.

Simple   Of one piece; not compound.

Sinuate   Strongly wavy.

Sinuous   In form like the path of a snake.

Sinus   The cleft or space between two lobes of a blade.

Site   The place and the sum total of associated physical ecosystem factors, as modified by the biota, where a plant grows; habitat.

Smooth   Lacking hairs or projections.

Soil   A dynamic natural body on the surface of the earth in which plants grow; composed of mineral and organic materials and living organisms.

Soil texture   The relative proportions of the various soil particles (sand, silt, and clay) in a soil.

Spatulate   Wide and rounded at the apex but gradually narrowed downward.

Spike   A simple inflorescence of sessile flowers arranged on a common, elongated axis.

Spine   A sharp-pointed stiff outgrowth from a stem derived from a leaf or part of a leaf.

Spray   The aggregate of smaller branches and branchlets.

Spur shoot   A short, stubby branch with leaf scars greatly crowded as a result of little or no growth in length.

Stalk   A lengthened plant part on which an organ grows or is supported (e.g., a peduncle or rachis to which the fruit is attached).

Stamen   The pollen-bearing organ of a flower, normally consisting of filament and anther.

Staminate   Provided with stamens, but usually without pistils.

Stellate   Star-shaped.

Stem   The organ of the plant that makes up the aerial portions (sometimes also the rhizomes) and bears the leaves and buds.

Sterile   Nonproductive, opposite of fertile. Of a site, poor or not well-supplied with nutrients.

Stigma   The part of a pistil that receives the pollen.

Stipules   Leaflike appendages on either side of a leaf at the base of the petiole, always in pairs.

Stipule-scar   The scar left by the fall of a stipule.

Stolon   An elongate, propagative stem with long internodes, rooting at the tip.

Striate   Marked with fine longitudinal stripes or ridges.

Style   The part of a pistil connecting ovary with stigma.

Sub-   A prefix applied to many botanical terms to indicate "somewhat" or "slightly," as subcordate.

Subtend   To lie under.

Succulent   Fleshy and juicy.

Sucker   A shoot arising from the base or a subterranean part of a plant.

Superposed   Placed above, as one bud above another at a node.

Suture   A junction or line of dehiscence.

Swamp   A forested wetland area whose soil, typically organic, is permanently or periodically saturated (not flooded) with water.

Taxon   Any taxonomically recognized unit, regardless of rank; e.g., genus, species, variety, cultivar.

Tendril   A long, slender, coiling structure serving as the organ of attachment in some climbing plants.

Terete   Circular in cross section.

Terminal   Situated at the end of a shoot or branch.

Thorn   A stiff, woody, sharp-pointed projection which represents a modified stem.

Tissue   A mass of cells which has a distinctive character and function, e.g., cork or wood.

Tolerant   Capable of enduring unfavorable conditions, as shade, drought, salt, etc.

Tomentose   Densely pubescent with soft, matted curled hairs; woolly.

Toothed   With teeth or short projections.

Tree   A woody perennial plant with a single trunk (typically unbranched near the base), usually exceeding 5 meters high.

Truncate   Ending abruptly, as if cut off at the end.

Trunk   The main stem of a tree.

Twig   The woody portion of a branch.

Umbel   A simple inflorescence of flowers on pedicels which radiate from the same point.

Unarmed   Without thorns, spines, or prickles.

Undulate   With a wavy margin or surface.

Valvate   Said of buds in which the scales merely meet at the edges without overlapping.

Variety   A subdivision of a species; this rank is below the rank of subspecies

Vascular bundle   A strandlike portion of the conducting system of a plant.

Veins   Threads of vascular tissue in a leaf, petal, or other flat organ.

Vestigial   The aborted remnant of a structure or function.

Villous   Covered with long, soft hairs, not matted together.

Viscid   Sticky, glutinous.

Whorl   An arrangement of three or more similar leaves or branches arising at a single node.

Wing   Any membranous or thin expansion bordering or surrounding an organ.

Wood   The hard internal tissue of a perennial stem.

Woolly   Covered with long and matted or tangled hairs; tomentose.

Xeric   Characterized by deficiency of soil water, or aridity. Said of a site that is dry or droughty.

LITERATURE CITED

Abrams, M. D. 1992. Fire and the development of oak forests. *BioScience* 42:346–53.

Abrams, M. D. 1998. The red maple paradox. *BioScience* 48:355–64.

Abrams, M. D. 2003. Where has all the white oak gone? *BioScience* 53:927–39.

Abrams, M. D., and G. J. Nowacki. 2008. Native Americans as active and passive promoters of mast and fruit trees in the eastern USA. *The Holocene* 18:1123–37.

Adams, R. P. 2008. *Junipers of the World: The Genus Juniperus*. 2nd ed. Trafford Publishing, Vancouver. 402 pp.

Albert, D. A. 1995. Regional landscape ecosystems of Michigan, Minnesota, and Wisconsin: A working map and classification. USDA, For. Serv., North Central For. Exp. Sta., Gen. Tech. Report NC-178. 250 pp., map.

Albert, D. A. 2003. Between land and lake: Michigan's Great Lakes coastal wetlands. Michigan Natural Features Inventory, Michigan State University Extension, East Lansing, Extension Bulletin E-2902. 96 pp.

Albert, D. A., and P. J. Comer. 2008. *Atlas of Early Michigan's Forests, Grasslands, and Wetlands: An Interpretation of the 1816–1856 General Land Office Surveys*. Michigan State Univ. Press, East Lansing. 107 pp.

Albert, D. A., S. R. Denton, and B. V. Barnes. 1986. *Regional Landscape Ecosystems of Michigan*. School of Natural Resources, University of Michigan, Ann Arbor. 32 pp.

Argus, G. W. 2010. *Salix*. pp. 23–162. *In* Flora of North America Editorial Committee (ed.), *Flora of North America*. Vol. 7. Oxford Univ. Press, New York.

Arnold, M. L. 1992. Natural hybridization as an evolutionary process. *Ann. Rev. Ecol. Syst.* 2:237–61.

Bailey, V. L. 1962. Revision of the Genus *Ptelea* (Rutaceae). *Brittonia* 14:1–45.

Barbour, M. G., and R. A. Minnich. 2000. Californian upland forests and woodlands. pp. 161–202. *In* M. G. Barbour and W. D. Billings (eds.), *North American Terrestrial Vegetation*. 2nd ed. Cambridge Univ. Press, Cambridge.

Barnes, B. V. 1991. Deciduous forests of North America. pp. 219–344. *In* E. Röhrig and B. Ulrich (eds.), *Ecosystems of the World*. Vol. 7: *Temperate Deciduous Forests*. Elsevier, New York.

Barnes, B. V. 2009a. Tree response to ecosystem change at the landscape level in eastern North America. *Forstarchiv* 80:76–89.

Barnes, B. V. 2009b. Vegetation history and change. pp. 36–49. *In* K. J. Nadelhoffer, A. J. Hogg Jr., and B. A. Hazlet (eds.), *The Changing Environment of North-*

ern Michigan: A Century of Science and Nature at the University of Michigan Biological Station. Univ. Michigan Press, Ann Arbor.

Barnes, B. V., and B. P. Dancik. 1985. Characteristics and origin of a new birch species, *Betula murrayana*, from southeastern Michigan. *Can. J. Bot.* 63:223–26.

Barnes, B. V., and W. H. Wagner. 2004. *Michigan Trees*. 2nd ed. Univ. Michigan Press, Ann Arbor. 447 pp.

Barnes, B. V., D. R. Zak, S. Denton, and S. H. Spurr. 1998. *Forest Ecology*. 4th ed. John Wiley & Sons, New York. 774 pp.

Billington, C. 1977. *Shrubs of Michigan* (2nd Edition). Cranbrook Institute of Science. 468 pp.

Bliss, L. C. 2000. Arctic tundra and polar desert biome. pp. 1–40. *In* M. G. Barbour and W. D. Billings (eds.), *North American Terrestrial Vegetation*. 2nd ed. Cambridge Univ. Press, Cambridge.

Braun, E. L. 1950. *Deciduous Forests of Eastern North America*. McGraw-Hill, New York. 596 pp.

Braun, E. L. 1989. *The Woody Plants of Ohio*. Ohio State Univ. Press, Columbus. 362 pp.

Brown, R. G., and M. L. Brown (1972) *Woody Plants of Maryland*. University of Maryland.

Catling, P. M., S. M. McKay-Kuja and G. Mitrow. 1999, Rank and typification in North American dwarf cherries and a key to the taxa. *Taxon* 48:483–88.

Chant, S. R. 1986. Whitebeams, Mountain Ash (genus *Sorbus*). pp. 184–86. *In* B. Hora (consultant ed.), *The Oxford Encyclopedia of Trees of the World*. Crescent Books, New York. 288 pp.

Christensen, N. L. 2000. *Vegetation of the Southeastern Coastal Plain*. pp. 397–448. *In* M. G. Barbour and W. D. Billings (eds.), *North American Terrestrial Vegetation*. 2nd ed. Cambridge Univ. Press, Cambridge, UK.

Coats, A. M. 1970. *The Plant Hunters, Being a History of the Horticultural Pioneers, Their Quests, and Their Discoveries*. McGraw-Hill, New York. 400 pp.

Coats, A. M. 1992. *Garden Shrubs and Their Histories*. Simon and Schuster, New York. 225 pp.

Collier, M., J. Vankat, and M. Hughes. 2002. Diminished plant richness and abundance below *Lonicera maackii*, an invasive shrub. *Am. Nat.* 155:311–25.

Comer, P. J., and D. A. Albert, with M. B. Austin (cartographer). 1998. Vegetation of Michigan circa 1800: An interpretation of the General Land Office Surveys. Michigan Natural Features Inventory, Michigan Dept. Nat. Res., Wildlife Div. 2 maps. Website for county ("static") maps, www.msue.msu.edu.mnfi; website for Geographic Information System (GIS) analysis, www.state.mi.us/webapp/cgi/mgdl/.

Core, E. L., and N. P. Ammons. 1958. *Woody Plants in Winter*. Boxwood Press, Pacific Grove, CA. 218 pp.

Cornell, G. L. 2009. Native Americans. pp. 91–105. *In* R. Schaetzl, J. Darden, and D. Brandt (eds.), *Michigan Geography and Geology*. Custom Publishing, New York.

Crins. W. J., P. A. Gray, P. W. C. Uhlig, and M. Wester. 2009. *The Ecosystems of Ontario*. Part 1: *Ecozones and Ecoregions*. Ontario Ministry Natural Resources, Peterborough Ontario, Inventory, Monitoring, and Assessment, SIB TER IMA TR-01. 71 pp.

Cronk, Q. C. B., and J. L. Fuller. 1995. *Plant Invaders: The Threat to Natural Ecosystems*. Chapman and Hall, London. 241 pp.

Cronon, W. 1983. *Changes in the Land*. Hill and Wang, New York. 257 pp.

Cullina, W. 2002. *Native Trees, Shrubs, and Vines: A Guide to Using, Growing, and Propagating North American Woody Plants*. Houghton Mifflin, Boston. 354 pp.

Curtis, J. 1959. *The Vegetation of Wisconsin*. Univ. Wisconsin Press, Madison. 657 pp.

Dancik, B. P., and B. V. Barnes. 1972. Natural variation and hybridization of yellow birch and bog birch in southeastern Michigan. *Silvae Genetica* 21:1–9.

Darrow, G. M., and W. H. Camp. 1945. *Vaccinium* hybrids and the development of new horticultural material. *Bull. Torrey Bot. Club* 72:1–21.

Davis, M. B. 1981. Quaternary history and the stability of deciduous forests. pp. 132–53. *In* C. D. West, H. H. Shugart, and D. B. Botkin (eds.), *Forest Succession: Concepts and Applications*. Springer-Verlag, New York.

Davis, M. B. 1983. Quaternary history of deciduous forest of eastern North America and Europe. *Annals Missouri Bot. Gard*. 70:550–63.

Deam, C. C. 1924. *Shrubs of Indiana*. Department of Conservation, State of Indiana, Indianapolis. 351 pp.

DeGraaf, R. M., and G. M. Witman. 1979. *Trees, Shrubs, and Vines for Attracting Birds: A Manual for the Northeast*. Univ. Massachusetts Press, Amherst. 194 pp.

Delcourt, H. R., and P. A. Delcourt. 2000. Eastern deciduous forests. pp. 123–59. *In* M. G. Barbour and W. D. Billings (eds.), *North American Terrestrial Vegetation*. 2nd ed. Cambridge Univ. Press, Cambridge.

Delcourt, P. A., and H. R. Delcourt. 1987. *Long-Term Forest Dynamics of the Temperate Zone*. Springer-Verlag, New York. 439 pp.

Delcourt, P.A., and H. R. Delcourt. 2004. *Prehistoric Native Americans and Ecological Change*. Cambridge Univ. Press, Cambridge. 203 pp.

Del Tredici, P. 1977. The buried seeds of *Comptonia peregrina*, the Sweet Fern. *Bull. Torrey Bot. Club* 104:270–75.

Denton, S. R., and B. V. Barnes. 1987. Spatial distribution of ecologically applicable climatic statistics in Michigan. *Can. J. For. Res*. 17:598–612.

Dickman, D. I. 2009. Forests and forestry from a historical perspective. pp. 614–30. *In* R. Schaetzl, J. Darden, and D. Brandt (eds.), *Michigan Geography and Geology*. Custom Publishing, New York.

Dickman, D. I., and L. A. Leefers. 2003. *The Forests of Michigan*. Univ. Michigan Press, Ann Arbor. 297 pp.

Dirr, M. A. 1990. *Manual of Woody Landscape Plants: Their Identification, Ornamental Characteristics, Culture, Propagation, and Uses*. Stipes Publishing, Champaign, IL. 1,007 pp.

Eastman, J. 1992. *Forest and Thicket: Trees, Shrubs, and Wildflowers of Eastern North America.* Stackpole Books. Harrisburg, PA. 212 pp.

Ehrle, E. B. 2006. The Big Trees and Shrubs of Michigan. *Mich. Bot.* 45:65–152.

Elias, T. S., and H. Gelband. 1975. Nectar: Its production and functions in trumpet creeper. *Science* 189:289–91.

Elliott-Fisk D. L. 2000. The taiga and boreal forest. pp. 41–73. *In* M. G. Barbour and W. D. Billings (eds.), *North American Terrestrial Vegetation.* 2nd ed. Cambridge Univ. Press, Cambridge.

English-Loeb, G., A. P. Norton, D. Gadoury, R. Seem, and W. Wilcox. 2005. Tritrophic interactions among grapevines, a fungal pathogen, and a mycophagous mite. *Ecol. Appl.* 15:1679–88.

Fairchild, D. 1947. *The World Was My Garden: Travels of a Plant Explorer.* Charles Scribner's Sons, New York. 494 pp.

Farrar, J. L. 1995. *Trees in Canada.* Fitzhenry and Whiteside, Markham, ON. 502 pp. Published in the United States as *Trees of the Northern United States and Canada.* Iowa State Univ. Press, Ames. 502 pp.

Flint, H. L. 1997. *Landscape Plants for Eastern North America.* 2nd ed. John Wiley & Sons, New York. 842 pp.

Flora of China. 2014. http://www.efloras.org. Accessed January 1, 2014.

Florin, R. 1958. On Jurassic taxads and conifers from north-western Europe and eastern Greenland. *Acta Horti Bergiani* 17:257–402.

Fralish, J. S., and S. B. Franklin. 2002. *Taxonomy and Ecology of Woody Plants in North American Forests.* John Wiley & Sons, New York. 612 pp.

Franklin, J. F., and C. B. Halpern. 2000. Pacific Northwest forests. pp. 123–59. *In* M. G. Barbour and W. D. Billings (eds.), *North American Terrestrial Vegetation.* 2nd ed. Cambridge Univ. Press, Cambridge.

Gilbert, E. F. 1966. Structure and development of sumac clones. *Amer. Midl. Nat.* 75:432–45.

Gill, J. D., and W. M. Healy (comps.). 1974. Shrubs and vines for northeastern wildlife. USDA For. Serv. Gen. Tech. Report NE-9, Northeastern For. Exp. Sta., Upper Darby, PA. 180 pp.

Gill, J. L., J. W. Williams, T. Jackson, K. B. Lininger, and G. S. Robinson. 2009. Pleistocene megafaunal collapse, novel plant communities, and enhanced fire regimes in North America. *Science* 326:1100–1103.

Gleason, H. A., and A. Cronquist. 1991. *Manual of Vascular Plants of Northeastern United States and Adjacent Canada.* 2nd ed. New York Botanical Gardens, Bronx. 910 pp.

Gould, A., and D. Gorchov. 2000. Effects of the exotic invasive shrub *Lonicera maackii* on the survival and fecundity of three species of native shrubs. *Amer. Midl. Nat.* 144:36–50.

Graham, A. 1993. History of the vegetation: Cretaceous (Maastrichtian)-Tertiary. pp. 57–70. *In* Flora of North America Editorial Committee (ed.), *Flora of North America.* Vol. 1. Oxford Univ. Press, New York.

Grimm, W. C. 1966. *How to recognize shrubs.* Castle Books, NY. 318 pp.

Gysel, L.W., and W. A. Lemmien. 1964. An eight-year record of fruit production. *J. Wildl. Management* 28:175–77.

Hardin, J. W. 1968. *Diervilla* (Caprifoliaceae) of the southeastern U.S. *Castanea* 33:31–36.

Hardin, J. W. 1973. The enigmatic chokeberries (*Aronia*, Rosaceae). *Bull. Torrey Bot. Club* 100:178–84.

Hardin, J. W., and L. L. Phillips. 1985. Hybridization in eastern North American *Rhus* (Anacardiaceae). *Assoc. Southeastern Biologists Bull.* 32:99–106.

Hayes, D. W. 1960. *Key to Important Woody Plants of Eastern Oregon and Washington.* USDA, For. Serv. Agr. Handbook No. 148. Washington, DC. 227 pp.

Herms, D. A., and W. J. Mattson. 1992. The dilemma of plants: To grow or defend. *Quart. Rev. Biol.* 67:283–335.

Hibbs, D. E. 1979. The age structure of a striped maple population. *Can. J. For. Res.* 9:504–8.

Hibbs, D. E., and B. C. Fischer. 1979. Sexual and vegetative reproduction of striped maple (*Acer pensylvanicum* L.). *Bull. Torrey Bot. Club* 106:222–27.

Hoch, W. A., E. L. Zeldin, and B. H. McCown. 2001. Physiological significance of anthocyanins during autumnal leaf senescence. *Tree Physiology* 21:1–8.

Holmgren, N. H. 1998. *The Illustrated Companion to Gleason and Cronquist's Manual: Illustrations of the Vascular Plants of Northeastern United States and Adjacent Canada.* New York Botanical Garden, Bronx. 937 pp.

Hook, D. D. 1984. Waterlogging tolerance of lowland tree species of the South. *South J. Appl. For.* 8:136–49.

Hora, B. (ed.). 1980. *The Oxford Encyclopedia of Trees of the World.* Crescent Books, New York. 288 pp.

Hupy, C. M., and C. H. Yansa. 2009. The last 17,000 years of vegetation history. pp. 91–105. *In* R. Schaetzl, J. Darden, and D. Brandt (eds.), *Michigan Geography and Geology.* Custom Publishing, New York.

The International Plant Names Index. 2014. http://www.ipni.org. Accessed January 2014.

Jackson, R. C. 1976. Evolution and systematic significance of polyploidy. *Ann. Rev. Ecol. Syst.* 7:209–34.

Jackson, S. T., J. T. Overpeck, I. T. Webb, J. W. Williams, B. C. S. Hanson, R. S. Webb, and K. H. Anderson. 2000. Vegetation and environment in Eastern North America during the last glacial maximum. *Quaternary Science Reviews* 19:289–508.

Judd, W. S., C. S. Campbell, E. A. Kellogg, P. F. Stevens, and M. J. Donoghue. 2008. *Plant Systematics: A Phylogenetic Approach.* 3rd ed. Sinauer Associates, Sunderland, MA. 611 pp.

Keeley, J. E. 2000. Chaparral. pp. 203–53. *In* M. G. Barbour and W. D. Billings (eds.), *North American Terrestrial Vegetation.* 2nd ed. Cambridge Univ. Press, Cambridge.

Kershaw, L., A. MacKinnon, and J. Pojar. 1998. *Plants of the Rocky Mountains.* Lone Pine Publishing, Edmonton, AB. 384 pp.

Kimmerer, R. W., and F. K. Lake. 2001. The role of indigenous burning in land management. *J. Forestry* 99:36–41.

Kingsbury, J. M. 1964. *Poisonous Plants of the U.S. and Canada.* Prentice-Hall, Upper Saddle River, NJ. 626 pp.

Koch, P. L., and A. D. Barnosky. 2006. Late Quaternary extinctions: State of the debate. *Ann. Rev. Ecol. Evol. Syst.* 37:215–50.

Kost, M. A., D. A. Albert, J. G. Cohen, B. S. Slaughter, R. K. Schillo, C. R. Weber, and K. A. Chapman. 2007. Natural communities of Michigan: Classification and description. Michigan Natural Features Inventory Report No. 2007–21. Michigan State University Extension, East Lansing. 314 pp.

Kost, M. A., J. G. Cohen, B. S. Slaughter, and D. A. Albert. 2010. *A Field Guide to the Natural Communities of Michigan.* Michigan Natural Features Inventory, Lansing. 189 pp.

Lanner. R. M. 2002. *Conifers of California.* Cachuma Press, Los Olivos, CA. 274 pp.

Larsen, J. A. 1953. A study of an invasion by red maple of an oak woods in southern Wisconsin. *Amer. Midl. Nat.* 49:908–14.

Larson, G. J., and K. Kincare. 2009. Late Quaternary history of the eastern midcontinent region, USA.. pp. 69–90. *In* R. Schaetzl, J. Darden, and D. Brandt (eds.), *Michigan Geography and Geology.* Custom Publishing, New York.

La Rue. C. D. 1948. The lilacs of Mackinac Island. *Amer. Midl. Nat.* 39:505–8.

Latham, R. E., and R. E. Ricklefs. 1993. Continental comparisons of temperate-zone tree species diversity. pp. 294–314. *In* R. E. Ricklefs and D. Schluter (eds.), *Species Diversity in Ecological Communities.* Univ. Chicago Press, Chicago.

Leicht-Young, S. A., N. B. Pavlovic, R. Grundell, and K. J. Frohnapple. 2007. Distinguishing native (*Celastrus scandens* L.) and invasive (*C. orbiculatus* Thunb.) bittersweet using morphological characters. *J. Torrey Bot. Soc.* 134:441–50.

Levin, D. A. 1976. The chemical defenses of plants to pathogens and herbivores. *Ann. Rev. Ecol. Syst.* 7:121–59.

Lewis, H. T. 1993. Patterns of Indian burning in California: Ecology and ethnohistory. pp. 55–116. *In* T. C. Blackburn and K. Anderson (eds.), *Before the Wilderness: Environmental Management by Native Californians.* Malki-Ballena Press, Banning, CA.

Lewis, K. E. 2009. The settlement experience. pp. 412–29. *In* R. Schaetzl, J. Darden, and D. Brandt (eds.), *Michigan Geography and Geology.* . Custom Publishing, New York.

Lovis, W. A. 2009. Between the glaciers and Europeans: People from 12,000 to 400 years ago. pp. 389–401. *In* R. Schaetzl, J. Darden, and D. Brandt (eds.), *Michigan Geography and Geology.* Custom Publishing, New York.

Macleod, R. D. 1952. *Key to the Names of British Plants.* Pitman & Sons, London. 104 pp.

MacMahon, J. A. 2000. Warm deserts. pp. 285–322. *In* M. G. Barbour and W. D. Billings (eds.), *North American Terrestrial Vegetation.* 2nd ed. Cambridge Univ. Press, Cambridge.

Mann, C. C. 2005. *1491*. Knopf: New York. 541 pp.

Marles, Robin J., C. Clavelle, L. Monteleone, N. Tays, and D. Burns. 2000. *Aboriginal Plant Use in Canada's Northwest Boreal Forest*. Univ. British Columbia Press, Vancouver. 368 pp.

McKell, C. M. (ed.). 1989. *The Biology and Utilization of Shrubs*. Academic Press, New York. 656 pp.

McKenney, D. W., J. H. Pedlar, K. Lawrence, K. Campbell, and M. F. Hutchinson. 2007. Potential impacts of climate change on the distribution of North American trees. *BioScience* 57:939–48.

McKenney, D. W., J. H. Pedlar, K. Lawrence, P. Papadopol, K. Campbell, and M. F. Hutchinson. 2014. Change and evolution in the plant hardiness zones of Canada. *Bioscience* 64(4): 341–50.

McKenney, D. W., J. H. Pedlar, R. B. Rood, and D. Price. 2011. Revisiting projected shifts in the climate envelopes of North American trees using updated general circulation models. *Global Change Biology* 17:2720–30.

Meeker, J. E., J. E. Elias, and J. A. Heim. 1993. *Plants Used by the Great Lakes Ojibwa*. Great Lakes Indian Fish and Wildlife Commission, Odanah, WI. 440 pp.

Meyer, F. G. 1997. Hamamelidaceae. pp. 358–67. *In* Flora of North American Editorial Committee (ed.), *Flora of North America*. Vol. 3. Oxford Univ. Press, Oxford.

Milius, S. 2002. Why turn red? *Science News* 162(17): 264–65.

Moore, M. 1979. *Medicinal Plants of the Mountain West*. Museum of New Mexico Press: Santa Fe, NM. 200 pp.

Mosquin, T. 1966. Reproductive specialization as a factor in the evolution of the Canadian flora. pp. 43–65. *In* R. L. Taylor and R. A. Ludwig (eds.), *The Evolution of Canada's Flora*. Univ. Toronto Press, Toronto.

Musselman, L. J. 1968. Asexual reproduction in the burning bush, *Euonymus atropurpureus*. *Mich. Bot.* 7:60–61.

Myers, J. H., and D. Bazely. 2003. *Ecology and Control of Introduced Plants*. Cambridge Univ. Press. Cambridge. 313 pp.

Newsholme, C. 2002. *Willows: The Genus Salix*. Timber Press, Portland, OR. 224 pp.

Nowacki, G. J., and M. D. Abrams. 2008. The demise of fire and "mesophication" of forests in the eastern United States. *BioScience* 58:1–16.

O'Connor, R. P., M. A. Kost, and J. G. Cohen. 2009. *Prairies and Savannas in Michigan: Rediscovering Our Natural Heritage*. Michigan State Univ. Press, East Lansing. 139 pp.

O'Dowd, D. J., and M. F. Willson. 1991. Associations between mites and leaf domatia. *Trends in Ecology & Evolution* 6:179–82.

Orshan, G. 1989. Shrubs as a growth form. pp. 249–65. *In* C. M. McKell (ed.), *The Biology and Utilization of Shrubs*. Academic Press, New York.

Palmer. E. J. 1925. Synopsis of North American Crataegi. *J. Arnold Arb.* 6:5–128.

Palmer, E. J. 1946. *Crataegus* in the northeastern and central United States and adjacent Canada. *Brittonia* 5:471–90.

Parish, R., R. Coupé, and D. Lloyd (eds.). 1996 *Plants of Southern Interior British Columbia and the Inland Northwest*. Lone Pine Publishing, Vancouver. 463 pp.

Peattie, D. C. 1991. *A Natural History of Trees of Eastern and Central North America*. Houghton Mifflin, Boston. 606 pp.

Peet, R. K. 2000. Forests and meadows of the Rocky Mountains. pp. 75–121. *In* M. G. Barbour and W. D. Billings (eds.), *North American Terrestrial Vegetation*. 2nd ed. Cambridge Univ. Press, Cambridge.

Petit, R. J., and A. Hampe. 2006. Some evolutionary consequences of being a tree. *Ann. Rev. Ecol., Evol. Syst.* 37:187–214.

Phillips, H. W. 2003. *Plants of the Lewis and Clark Expedition*. Mountain Press Publishing, Missoula, MT. 277 pp.

Phipps, J. B. 1983. Biogeographic, taxonomic, and cladistics relationships between East Asiatic and North American *Crataegus*. pp. 667–700. *In* Biogeographical Relationships between Temperate Eastern Asia and Temperate Eastern North America. *Annals Missouri Bot. Gard.* 70(3). St. Louis, MO.

Phipps, J. B., with Robert J. O'Kennon and Ron W. Lance. 2003. *Hawthorns and Medlars*. Timber Press, Portland OR. 139 pp.

Plants Databse. 2014. http://plants.usda.gov. Accessed January 1, 2014.

The Plant List. 2013. http.//www.theplantlist.org. Accessed January 1, 2013.

Pollan, M. 2002. *The Botany of Desire*. Random House. New York. 271 pp.

Rehder, A. 1940. *Manual of Cultivated Trees and Shrubs*. 2nd ed. Macmillan, New York. 996 pp.

Rejmanek, M., and D. M. Richardson. 1996. What attributes make some plant species invasive? *Ecology* 77:1655–61.

Riley, J. L. 2013. *The Once and Future Great Lakes Country: An Ecological History*. McGill-Queens Univ. Press, Montreal. 488 pp.

Rohrer, J. R. 2000. The sand cherry in Wisconsin and neighboring states. *Mich. Bot.* 39:59–69.

Rosendahl, C. O. 1963. *Trees and Shrubs of the Upper Midwest*. Univ. Minnesota Press, Minneapolis. 411 pp.

Rowe, J. S. 1972. *Forest Regions of Canada*. Can. For. Serv., Dept. Env. Publ. No. 1300, Ottawa. 172 pp. + map.

Rowe, J. S. 1992. The ecosystem approach to forestland management. *Forestry Chronicle* 68:222–24.

Ryskamp, M. P., and D. P. Warners. 2012. Relic or recruit? Newly discovered *Aronia arbutifolia* (Ell.) Pers. (red chokeberry) in Kent County raises questions regarding past and future distributions. *Mich. Bot.* 51:100–109.

Satake, Y., H. Hara, S. Watari, and T. Tominari. 1993. *Wild Flowers of Japan: Woody Plants*. Heibonsha, Tokyo.300 pp. In Japanese.

Schaetzl, R. J., and B. V. Barnes. 2009. Landforms, physical landscapes, and glacial history. pp. 25–25. *In* K. J. Nadelhoffer, A. J. Hogg Jr., and B. A. Hazlet (eds.), *The Changing Environment of Northern Michigan: A Century of Science and Nature at the University of Michigan Biological Station*. Univ. Michigan Press, Ann Arbor.

Scholtens, M. C., and K. Williams (eds.). 2008. *Grapevine to Pine Point: The University of Michigan Biological Station.* Lulu Press: Raleigh, NC. 246 pp.

Schultheis, L. M., and M. J. Donoghue. 2004. Molecular phylogeny and biogeography of *Ribes* (Grossulariaceae), with an emphasis on gooseberries (subg. Grossularia). *Systematic Botany* 29:77–96.

Sims, P. L., and P. G. Risser. 2000. Grasslands. pp. 323–56. *In* M. G. Barbour and W. D. Billings (eds.), *North American Terrestrial Vegetation.* 2nd ed. Cambridge Univ. Press. Cambridge.

Smith, H. H. 1928. Ethnobotany of the Meskwaki Indians. *Bull. Public Museum, City of Milwaukee* 4:175–326.

Smith, W. R. 2008. *Trees and Shrubs of Minnesota: The Complete Guide to Species Identification.* Univ. Minnesota Press, Minneapolis. 703 pp.

Soltis, D. E., V. A. Albert, J. Leebens-Mack, C. D. Bell, A. H. Patterson, C. Zheng, D. Sankorff, C. W. dePamphilis, P. Kerr Wall, and P. S. Soltis. 2009. Polyploidy and angiosperm diversification. *Amer. J. Botany* 96:336–48.

Soper, J. H., and M. L. Heimburger. 1994. *Shrubs of Ontario.* Royal Ontario Museum, Toronto. 495 pp.

Spurr. S. H. 1964. *Forest Ecology,* Ronald Press, New York. 352 pp.

Stebbins, G. L. 1972. Evolution and diversity of arid-land shrubs. pp. 111–20. *In* C. M. McKell, J. P. Blaisdell, and J. R. Goodin (eds.), *Wildland Shrubs: Their Biology and Utilization.* USDA For. Serv. Gen. Tech. Report INT-1, Intermountain For. Rge. Expt. Sta., Ogden, UT.

Stebbins, G. L. 1985. Polyploidy, hybridization, and the invasion of new habitats. *Ann. Missouri Bot. Gard.* 72:824–32.

Stuart, J. D., and J. O. Sawyer. 2001. *Trees and Shrubs of California.* California Natural History Guides. Univ. California Press, Berkeley. 467 pp.

Sullivan, J. R. 1983. Comparative reproductive biology of *Acer pensylvanicum* and *A. spicatum* (Aceraceae). *Amer. J. Bot.* 70:916–24.

Tiffney, B. H. 1985. The Eocene North Atlantic land bridge: Its importance in Tertiary and modern phytogeography of the northern hemisphere. *J. Arnold Arb.* 66:243–73.

US Department of Agriculture, Forest Service. 2008. *The Woody Plant Seed Manual: A Handbook on Seeds of Trees and Shrubs.* F. T. Bonner, and R P. Karrfalt (eds.). USDA For. Serv. Agr. Handbook No. 727. Washington, DC. 1,223 pp.

Valladares, F., and Ü. Niinemets. 2008. Shade tolerance, a key plant feature of complex nature and consequences. *Ann. Rev. Ecol. Evol. Syst.* 39:237–57.

Vidaković, M. 1991. *Conifers: Morphology and Variation.* Grafički zavod Hrvatske, Zagreb, Croatia. 755 pp.

Vines, R. A. 1960. *Trees, Shrubs, and Woody Vines of the Southwest.* Univ. Texas Press, Austin. 1,104 pp.

Voss, E. G. 1985. *Michigan Flora.* Pt. 2: *Dicots (Saururaceae-Cornaceae).* Cranbrook Inst. Sci. Bull. 59. Cranbrook Institute of Science, Bloomfield Hills, MI, and University of Michigan Herbarium, Ann Arbor. 120 pp.

Voss, E. G. 1996. *Michigan Flora.* Pt. 3: *Dicots (Pyrolaceae—Compositae).* Cranbrook Inst. Sci. Bull. 59. Cranbrook Institute of Science, Bloomfield Hills, MI, and University of Michigan Herbarium, Ann Arbor. 622 pp. Voss, E. G., and A. A. Reznicek. 2012. *Field Manual of Michigan Flora.* Univ. Michigan Press, Ann Arbor. 990 pp.

Wagner, W. H., Jr. 1968. Hybridization, taxonomy, and evolution. pp 113–38. *In* V. H. Heywood (ed.), *Modern Methods in Plant Taxonomy.* Academic Press, New York.

Wagner, W. H.., Jr. 1974. Dwarf hackberry (Ulmaceae: *Celtis tenuifolia*) in the Great Lakes Region. *Mich. Bot.* 13:73–99.

Wagner, W. H., Jr. 1975. The spoken "×" in hybrid binomials. *Taxon* 24:296.

Walters, M. B., and P. B. Reich. 1996. Are shade tolerance, survival, and growth linked? Low light and nitrogen effects on hardwood seedlings. *Ecology* 77:841–53.

Weatherbee, E. E. 2006. *Guide to Great Lakes Coastal Plants.* Univ. Michigan Press, Ann Arbor. 180 pp.

Webb, I. T., E. J. Cushing, and H. E. Wright. 1983. *Holocene Changes in the Vegetation of the Midwest. in Late Quaternary Environments of the United States.* pp. 142–65. Univ. Minnesota Press. Minneapolis.422 pp.

Webb, I. T., B. Shuman, and J. W. Williams. 2003. Climatically forced vegetation dynamics in eastern North America during the Late Quaternary period. pp. 459–78. *In the Quaternary Period in the United States.* Elsevier, Amsterdam.

Wells, P. V. 1970. Postglacial vegetation history of the Great Plains. *Science* 167:1574–82.

West, N. E., and J. A. Young. 2000. *Intermountain Valleys and Lower Mountain Slopes.* pp. 255–84. *In* M. G. Barbour and W. D. Billings (eds.), *North American Terrestrial Vegetation.* 2nd ed. Cambridge Univ. Press, Cambridge.

Wilson, B. F. 1984. *The Growing Tree.* Univ. Massachusetts Press, Amherst. 138 pp.

Winkworth, R. C., and M. J. Donoghue. 2005. *Viburnum* phylogeny based on combined molecular data: Implications for taxonomy and biogeography. *Amer. J. Botany* 92:653–66.

Wolfe, J. A. 1985. Some aspects of plant geography of the Northern Hemisphere during the late Cretaceous and Tertiary. *Annals Missouri Bot. Gard.* 62:264–79.

Xiang, Q. Y. J., D. T. Thomas, W. H. Zhang, S. R. Manchester, and Z. Murrell. 2006. Species level phylogeny of the genus *Cornus* (Cornaceae) based on molecular and morphological evidence: Implications for taxonomy and Tertiary intercontinental migration. *Taxon* 55:9–30.

Zimmerman, M. H., and C. L. Brown. 1971. *Trees: Structure and Function.* Springer-Verlag, New York. 336 pp.

INDEX

Note: Page numbers in italic indicate a detailed description. An asterisk (*) indicates the species is illustrated. A dagger (†) indicates a description of the genus or group.

Printed and bound by CPI Group (UK) Ltd, Croydon, CR0 4YY

13/04/2025

14656541-0001